WALTER G. VINCENTI

What Engineers Know and How They Know It

Analytical Studies from Aeronautical History

THE JOHNS HOPKINS UNIVERSITY PRESS

BALTIMORE AND LONDON

© 1990 The Johns Hopkins University Press
All rights reserved. Published 1990
Printed in the United States of America on acid-free paper

Johns Hopkins Paperbacks edition, 1993
04 03 02 01 00 99 98 97   6 5 4 3

The Johns Hopkins University Press
2715 North Charles Street, Baltimore, Maryland 21218-4319
The Johns Hopkins Press Ltd., London

Library of Congress Cataloging-in-Publication Data

Vincenti, Walter G. (Walter Guido), 1917–
What engineers know and how they know it : analytical studies from aeronautical
history / Walter G. Vincenti.
p.   cm.—(Johns Hopkins studies in the history of technology : new ser., no. 11)
Includes biographical references.
ISBN 0-8018-3974-2 (alk. paper) ISBN 0-8018-4588-2 (pbk.)
1. Aeronautics—History.   I. Title.   II. Series.
TL515.V44   1990   629.13′09—dc20   89-49003

A catalog record for this book is available from the British Library.

Contents

Preface and Acknowledgments

This book concentrates, as the title indicates, on the intellectual content of engineering. I hope that historians, philosophers, sociologists, economists, fellow engineers, and general readers, indeed, anyone who wants to understand better the nature and sources of engineering knowledge, will find it useful. Since the historical studies come from my own field of aeronautics, they form at the same time a contribution to the history of aviation. They may therefore have value for people interested in the development of this important branch of twentieth-century technology. Those in the latter category may prefer to omit the epistemological analysis and concentrate on the narrative sections of chapters 2 through 6.

Obviously, no endeavor of this kind takes place in an intellectual vacuum. My work here has been especially influenced by the thinking (in more or less the order in which I became aware of it) of Edwin Layton, Edward Constant, Donald Campbell, Rachel Laudan, and John Staudenmaier. I am much indebted to all of them. I hope I do not misrepresent or misuse their ideas.

Many people gave direct aid and encouragement. I attempt to express my gratitude in the notes, particularly an initial note to each chapter. I trust I have included everyone. Three people, however, deserve more general mention. Edwin Layton has been, in the words of Ring Lardner in his play *Elmer the Great,* "my best friend and severest critic" throughout the project. Edward Constant has played a similar role for chapters 2, 3, 7, and 8, as has Rachel Laudan for chapters 1, 7, and 8. I am deeply grateful for their criticisms and suggestions. Faults that remain are strictly my own.

Others have helped in various ways. Otto Mayr, by inviting me to speak at a historical session of the American Society of Mechanical Engineers, provided impetus for my first venture into formal historical writing (chapter 5). Thomas Hughes, past general editor of these Johns Hopkins Studies, suggested I think about putting that and my subsequent articles into a book. Robert Post, editor-in-chief of *Technology and Culture,* provided unfailing interest and encouragement. James Adams, present chairman of Stanford University's Program in Values, Technology, Science, and Society (VTSS), lent sympathetic advice, as did other faculty in the program, including those mentioned in the dedication. Other Stanford colleagues, most notably my engineering deans, William Kays and the late Joseph Pettit, and my departmental chairmen, Nicholas Hoff, Arthur Bryson, and Robert Cannon, proved uniformly supportive. Much-needed funds for secretarial help came from the VTSS program and from Stanford's Committee on the Status of Emeriti, the latter on the basis of a grant from the Ford Foundation; the work itself was very ably done by Marmee Eddy and Darlene Lamun, as well as by Virginia Mann, who has accommodated my old-fashioned, low-tech writing habits for a considerable time. The editors at the Johns Hopkins University Press—Henry Tom, Barbara Lamb, Mary Hill, and Kimberly Johnson—carried out their tasks in exemplary fashion. Most of the thought and labor to compile the index was capably supplied by Thomas McFadden. I appreciate all that these people have done.

Finally, I want to thank Joyce Vincenti, my wife of forty-three years, for her love and support. Thank heaven for that railroad passenger agent in Chicago.

**What Engineers Know
and How They Know It**

CHAPTER 1

Introduction: Engineering
As Knowledge

Engineering knowledge, though pursued at great effort and expense in schools of engineering, receives little attention from scholars in other disciplines. Most such people, when they pay heed to engineering at all, tend to think of it as applied science.[1] Modern engineers are seen as taking over their knowledge from scientists and, by some occasionally dramatic but probably intellectually uninteresting process, using this knowledge to fashion material artifacts. From this point of view, studying the epistemology of science should automatically subsume the knowledge content of engineering. Engineers know from experience that this view is untrue, and in recent decades historians of technology have produced narrative and analytical evidence in the same direction. Since engineers tend not to be introspective, however, and philosophers and historians (with certain exceptions) have been limited in their technical expertise, the character of engineering knowledge as an epistemological species is only now being examined in detail. This book is a contribution to that effort.

My involvement in the study of engineering knowledge stems in part from a question put to me by my Stanford economics colleague Nathan Rosenberg over lunch in the early 1970s: "What is it you engineers really *do*?" What engineers do, however, depends on what they know, and my career as a research engineer and teacher has been spent producing and organizing knowledge that scientists for the most part do not address. My attempts to deal with Rosenberg's question led me therefore—without at first realizing just what I was doing—to examine the cognitive dimension of engineering. Given a long-standing interest in history, it was also instinctive for me to approach the problem historically. To my pleasant surprise, I found myself in step with the work being produced by historians of technology.

In the view developed by these historians, technology appears, not as derivative from science, but as an autonomous body of knowledge, identifiably different from the scientific knowledge with which it interacts. The idea of "Technology as Knowledge"—title of an influential paper by Edwin Layton, one of the view's early champions—credits technology with its own "significant component of thought." This form of thought, though different in its specifics, resembles scientific thought in being creative and constructive; it is not simply routine and deductive as assumed in the applied-science model. In this newer view, technology, though it may *apply* science, is not the same as or entirely *applied* science.

This view of technology—and hence engineering—as other than science accords with statements sometimes made by engineers, such as the following by a British engineer to the Royal Aeronautical Society in 1922: "Aeroplanes are not designed by science, but by art in spite of some pretence and humbug to the contrary. I do not mean to suggest for one moment that engineering can do without science, on the contrary, it stands on scientific foundations, but there is a big gap between scientific research and the engineering product which has to be bridged by the art of the engineer." The creative, constructive knowledge of the engineer is the knowledge needed to implement that art. Technological knowledge in this view appears enormously richer and more interesting than it does as applied science.[2]

The newer view comes from the work of historians over several decades. The historiographic development has been examined in an extended study by John Staudenmaier and a shorter review by George Wise. Both come to the conclusion, as expressed by Wise, that "treating science and technology as separate spheres of knowledge, both man-made, appears to fit the historical record better than treating science as revealed knowledge and technology as a collection of artifacts once constructed by trial and error but now constructed by applying science."[3] The evidence to be presented here supports this conclusion. The reality of the distinction is emphasized for me by the fact that the school of engineering at my own university, as at all such institutions, finds it necessary to maintain its own library, separate from those of the departments of physics and chemistry. This separation is more than a convenience. Engineers, though they require many of the same books, journals, and documents as physicists and chemists, need others not kept in the science libraries. Despite the historical and institutional evidence for its autonomy, however, the features that distinguish technological knowledge have not been laid out in detail.

The view of technology as an autonomous form of knowledge is bound closely with the debate over the relation between technology and science, which has been a long-standing concern for historians of technology. Staudenmaier sees the view as having emerged out of that debate and become a major theme in itself, with the science-technology relation reduced to a subtheme. Wise regards the science-technology relation as still an organizing issue for research, with the view of technology as a special kind of knowledge defining the technological side of the relation. However that may be, viewing technological knowledge as autonomous leaves the relation between technology and science still open to specification. Technological knowledge then takes its place as a component on one side of what can be called an "interactive model" of the relationship. In this model, which has been summarized concisely by Barry Barnes, technology and science are autonomous forms of culture that interact mutually in some complicated and still-to-be-spelled-out fashion. The nature of technological knowledge constrains but does not define the relationship.[4]

Things look very different if the knowledge content of technology is seen as coming entirely from science. Such a view immediately defines the science-technology relation—technology is hierarchically subordinate to science, serving only to deduce the implications of scientific discoveries and give them practical application. This relation is summarized in the discredited statement that "technology is applied science." Such a hierarchical model leaves nothing basic to be discussed about the nature of the relationship. A model with such rigidity is bound to have difficulty fitting the complex historical record.[5]

But focus on technology as knowledge has ramifications beyond the science-technology question. Hugh Aitken, for example, makes it basic to the historical method adopted in his book *The Continuous Wave*. To recount the early technical and institutional history of radio, Aitken regards "history of technology as one branch of intellectual history or the history of ideas." From this approach, he explains the origins of inventions crucial to radio by examining "the flows of information that converged at the point and at the time when the new combinations came into existence." As the work of Aitken and other historians makes clear, however, the ideas we deal with are not disembodied—they are, as Layton points out, the ideas of people (and communities of people). Emphasis on knowledge thus brings history of technology into symbiotic relation, not only with intellectual history and philosophy, but with social history and sociology as well. Such emphasis is critical, in particular, for understanding technological change, a fundamental concern in one way or another for all these disciplines. As remarked by Rachel Laudan, "shifts in the knowledge of the practitioners play a crucial role in technological development." People who aspire to understand such development—economists and policy makers, for instance—might do well to focus accordingly when they delve (in Rosenberg's graphic phrase) "inside the black box" of technology. If these ramifications are valid, as I believe they are, laying out the features of engineering knowledge very much needs doing.[6]

In addressing this task, I will structure the inquiry around the goal of design. For engineers, in contrast to scientists, knowledge is not an end in itself or the central objective of their profession. Rather, it is, as illustrated by the quotation from the British engineer, a means to a utilitarian end—actually, several ends. Engineering can, in fact, be defined in terms of these ends, as in the following quotation from another British engineer, G. F. C. Rogers:

> Engineering refers to the practice of organizing the design and construction [and, I would add, operation] of any artifice which transforms the physical world around us to meet some recognized need.[7]

Here I take "organize" to be meant in the sense of "bring into being" or "get together" or "arrange." The first end, "design," has to do with the plans from which the artifice is built, as in the many drawings (or computer displays) of an airplane and its components. "Construction" (which I shall call "production") denotes the process by which these plans are translated into the concrete artifice, as in manufacture of the actual airplane. "Operation" deals with the employment of the artifice in meeting the recognized need, the related example here being the maintenance and flight operations of the airplanes of an airline. Definitions of engineering sometimes mention other ends such as "development," and "applications" or "sales"; these can usually be subsumed under one of the foregoing three, which will be sufficient for present purposes.

Of the three, design is frequently taken as central. Layton, in treating technology as knowledge, takes it as such (with minor mention of other ends). He adds in a later paper, however, that recent attempts among engineers to "reestablish design as the central theme of engineering" are "not without ideological overtones." Other scholars contend that rhetorical emphasis on design by engineers is primarily an attempt to gain status, that "engineers have seized on design as a way to liken their activity to that of scientists, to assert that they too are engaged in creative activity." Whatever the truth of the situation, I will restrict my focus here almost entirely to design. To attempt more would extend impractically an already lengthy study. Great numbers of engineers do, in fact, engage in design, and it is there that requirements for much engineering knowledge originate in an immediately technical sense. Though extraengineering needs—economic, military, social, or personal—may set the original problem, for many workaday engineers things come into focus at the level of concrete design. My emphasis on design, however, should not be taken to imply anything derogatory about other areas of engineering. For a complete epistemology of engineering, production and operation will require equal attention. For the time being, however, my concern will be limited mainly to engineering *design* knowledge.[8]

"Design," of course, denotes both the content of a set of plans (as in "the design for a new airplane") and the process by which those plans are produced. In the latter meaning, it typically involves tentative layout (or layouts) of the arrangement and dimensions of the artifice, checking of the candidate device by mathematical analysis or experimental test to see if it does the required job, and modification when (as commonly happens at first) it does not. Such procedure usually requires several iterations before finally dimensioned plans can be released for production. Events in the doing are also more complicated than such a brief outline suggests. Numerous difficult trade-offs may be required, calling for decisions on the basis of incomplete or uncertain knowledge. If available knowledge is inadequate, special research may have to be undertaken. The process is a complicated and fascinating one that needs more historical analysis than it has received.[9]

Design is important here, however, mainly as it conditions the knowledge required for its performance. Knowledge itself forms the primary focus; while requirements from design must be kept in mind at all times as determining that knowledge, details of how the process takes place are secondary. I have never attempted to design an airplane in my entire career as a research engineer (though I participated in planning and designing large aeronautical research facilities). The atmosphere in which I worked, however, and the knowledge I helped produce, were conditioned by the needs of airplane designers who visited our laboratory. My colleagues and I were keenly and continuously aware of the practical purposes we served. The situation in this book is somewhat similar. Though only one of the historical studies deals directly with the design process, the needs of design play the determining role throughout.

To keep matters manageable, I shall further limit attention to what can be called *normal design*. In *The Origins of the Turbojet Revolution*, Edward Constant defined "normal technology"—"what technological communities usually do"—as comprising "the improvement of the accepted tradition or its application under 'new or more stringent conditions.'" Normal design (my extension, not Constant's) is then the design involved in such normal technology. The engineer engaged in such design knows at the outset how the device in question works, what are its customary features, and that, if properly designed along such lines, it has good likelihood of accomplishing the desired task. A designer of a normal aircraft engine prior to the turbojet, for example, took it for granted that the engine should be piston driven by a gasoline-fueled, four-stroke, internal-combustion cycle. The arrangement of cylinders for a high-powered engine would also be taken as given (radial if air-cooled and in linear banks if liquid-cooled). So also would other, less obvious features (e.g., tappet as against, say, sleeve valves). The designer was familiar with engines of this sort and knew they had a long tradition of success. The design problem—often highly demanding within its limits—was one of improvement in the direction of decreased weight and fuel consumption or increased power output or both. Normal design is thus very different from *radical design*, such as that confronting the initiators of the turbojet revolution described by Constant. The protagonists of that revolution had little to take for granted in the way that designers of normal engines could. In radical design, how the device should be arranged or even how it works is largely unknown. The designer has never seen such a device before and has no presumption of success. The problem is to design something that will function well enough to warrant further development.[10]

Though less conspicuous than radical design, normal design makes up by far the bulk of day-to-day engineering enterprise. The vast design offices at firms like Boeing, General Motors, and Bechtel engage mainly in such activity. In the words of one reader of this material, "For every Kelly Johnson [a highly innovative American airplane designer who will figure in chapter 3] there are thousands of useful and productive engineers designing from combinations of off-the-shelf technologies that are then tested, adjusted, and refined until they work satisfactorily." In addition, knowledge for normal design is more circumscribed and easier to deal with. Though it may entail novelty and invention in considerable degree, it is not crucially identified with originality in the same way as knowledge for radical design. My restriction to normal design thus relates to both substance and expedience—there are sufficient matters of importance to confront at this stage without opening the Pandora's box of technical invention.[11]

I do not mean to suggest that normal and radical design, and the knowledge they require, can be sharply separated; there are obviously middle levels of novelty where the distinction is difficult to make. The difference, nevertheless, is sufficiently real to serve as a basis for analysis. I likewise do not mean to suggest that normal design is routine and deductive and essentially static. Like technology as a whole, it is creative and constructive and changes over time as designers pursue ever more ambitious goals. The changes, however, are incremental instead of essential; normal design is evolutionary rather than revolutionary. As we shall see, even within such limits the kinds of knowledge required are enormously diverse and complex. The activities that produce the knowledge, unlike the activity it is intended to support, are also sometimes far from normal and day-to-day.

One more point is fundamental. Design, apart from being normal or radical, is also multilevel and hierarchical. Interacting levels of design exist, depending on the nature of the immediate design task, the identity of some component of the device, or the engineering discipline required. For airplanes, which are typical for devices that constitute complex systems, the levels run more or less as follows from the top down:

1. Project definition—translation of some usually ill-defined military or commercial requirement into a concrete technical problem for level 2.
2. Overall design—layout of arrangement and proportions of the airplane to meet the project definition.
3. Major-component design—division of project into wing design, fuselage design, landing-gear design, electrical-system design, etc.
4. Subdivision of areas of component design from level 3 according to engineering discipline required (e.g., aerodynamic wing design, structural wing design, mechanical wing design).[12]
5. Further division of categories in level 4 into highly specific problems (e.g., aerodynamic wing design into problems of planform, airfoil section, and high-lift devices).

Such successive division resolves the airplane problem into smaller manageable subproblems, each of which can be attacked in semi-isolation. The complete design process then goes on iteratively, up and down and horizontally throughout the hierarchy. Problems at upper levels tend to be conceptual and relatively unstructured. People outside engineering think of design primarily in such terms; historians tend to focus predominantly on project definition and overall design, elsewhere as well as for airplanes. At the lower levels, where the majority of engineering effort takes place, problems are usually well defined, and activity tends to be highly structured. Whether design at a given location in the hierarchy is normal or radical is a separate matter—normal design can (and usually does) prevail throughout, though radical design can be encountered at any level. The hierarchical character of design has direct consequences for the knowledge required, though I do not explore those consequences as thoroughly as will eventually be desirable. The historical studies here deal mostly with activity at the lower levels to help redress the neglect of this large and essential area.[13]

The five case studies that make up the backbone of the work come from the history of aeronautics in the first half of this century. All but one (chapter 3), which is original here, appeared with minor differences in *Technology and Culture*, journal of the Society for the History of Technology, from 1979 to 1986. My method, unpremeditated at the outset, has been to examine in each instance the growth of knowledge over time in sufficient depth and with sufficient cognitive background to exhibit the content of the knowledge. Like Aitken, I concentrated on ideas rather than artifacts and sought to trace the flow of information. Each case was selected to emphasize a different area of knowledge and to be suitably limited and intellectually accessible for detailed examination and exposition. The resulting narrative constitutes the main body of each piece. Following the narrative, I reflect in each case on why and how the knowledge was obtained, on the structure of the knowledge, and on what all this suggests about engineering knowledge both specifically and generally. The final two chapters do the same for the entire body of material.

Although I began each case with a tentative theme and some notion of what I might find, I have tried insofar as possible to be free of preconceptions and a priori conclusions. My ideas changed greatly as the study progressed. When I began, it was with the notion of exploring the science-technology relationship; the fact that knowledge was actually my central concern did not dawn on me fully till after the first two studies. I thus reproduced the experience that Staudenmaier sees happening in the historical community. The determining role of design—especially of normal design at the lower levels of hierarchy— was similarly not clear to me till the end of the third study. My method has thus been essentially empirical and inductive, and the focus of the book is an outgrowth of the work itself. My efforts are thus in keeping with the historian's approach, which, as characterized by Otto Mayr, "is fundamentally inductive rather than deductive; it begins with microscopic research done in depth and detail on the level of individual episodes, in hopes that the empirical data thus gathered will lead to generalizations on some higher level." The conclusions I reach are compatible with those obtained from a philosophical approach by the British engineer G. F. C. Rogers quoted earlier.[14]

Understanding the intellectual content of engineering in the depth required will call for mental effort. The book is not light reading any more than engineering is an easy profession. Since the ideas are essential, I endeavor throughout to make them as accessible as possible to the serious lay reader. This requires sometimes that I simplify explanations or omit qualifications that might be important for engineers. I try to say nothing, however, to which my engineering colleagues will take serious exception. I am not always able to tell the whole truth, but I try to tell nothing but the truth. I hope this approach will meet the reader's needs. One cannot expect to comprehend engineering knowledge without coming to grips with the ideas involved.

Unfortunately, any attempt to exhibit the form and logic of ideas in historical narrative runs the risk of presenting events as tidier and less complicated than they undoubtedly were. As any engineer knows, the technological learning process always requires more effort in fact than appears necessary in hindsight. More knowledge is invariably produced than users ever require—members of the knowledge-producing community have to teach themselves to some extent through their individual doing. Errors and misconceptions inevitably arise and must be detected and surmounted; the number of these that end up in even the unpublished archival record can never constitute more than a small part of those encountered. The learning, in short, while it is going on is messy, repetitious, and uneconomical. It is easy in retrospect to read more logic and structure into the process than was most likely evident to those involved. Novelist Wallace Stegner has epitomized the difficulty nicely in terms of the Doppler effect: "The sound of anything coming at you—a train, say, or the future—has a higher pitch than the sound of the same thing going away."[15] I have tried where possible to convey something of how the intellectual process confronted the participants as it came at them. At best, however, I can hope to recapture only a small part of what my experience as a research engineer tells me must have gone on.

Engineering knowledge reflects the fact that design does not take place for its own sake and in isolation. Artifactual design is a social activity directed at a practical set of goals intended to serve human beings in some direct way. As such, it is intimately bound up with economic, military, social, personal, and environmental needs and constraints. Staudenmaier refers to these as "contextual factors that constitute the artifact's ambience" and sees technological activity as characterized (even defined) by a "tension between technical design and its ambience."[16] The present book deals mainly with the internal knowledge required by the design side of this tension. At the same time, contextual factors have a determining influence that must not be ignored. In normal design, this ambience exercises its greatest direct effect at the upper levels of hierarchy, where projects are defined and laid out. (The same would be true, by and large, for radical design.) The performance, size, and arrangement of an airplane, for example (and hence the knowledge needed to lay it out), are direct consequences of the commercial or military task it is intended to perform. At the lower levels that concern us here, the contextual influence, though still present, tends to be weaker and less direct; at these levels, knowledge derives predominantly from the internal needs of design itself. How the wing of an airplane is shaped, for instance, is a purely technical problem (though a complex and demanding one, as we shall see) once the airplane's size and performance have been specified. My stories and analysis concentrate primarily on the knowledge required to satisfy such internal requirements. At the same time, I attempt to be alert to the context within which these requirements occur, though I do not (with one exception) explore the design-ambient tension at first hand. My concern is for the what and how of engineering knowing; the why in the contextual (though not internal) sense I leave mainly to others. Exploration of the full epistemological consequences of context will need extended consideration of the upper levels of the design hierarchy. The accounts and analysis here, plus the contextual signposts along the way, may help in that direction.[17]

In succeeding chapters, then, I shall examine historical cases from my own field of aeronautical engineering to exhibit engineering knowledge for the thoughtful nonengineer, to show how that knowledge has been obtained, and to reflect on what the evidence suggests about the nature of such knowledge generally. The knowledge of concern is that used in the central activity of engineering design; specifically, in normal design at the well-structured, lower levels of the design hierarchy, which represents the bulk of day-to-day engineering. The focus is thus on the interrelated acquisition, articulation, and utilization of engineering knowledge as seen in terms of historical process. Since knowledge derives from thinking, I hope at the same time to give scholars concerned with technological affairs a deeper understanding of how engineers think and how their thinking relates to their doing.

The historical cases appear as chapters 2 through 6. Each of the five demonstrates how the needs of design impel the growth of a particular kind of knowledge. Chapter 2 ("Design and the Growth of Knowledge"), the one chapter that displays the design process as such, shows how knowledge relates to design and how the uncertainties that inevitably exist in practical knowledge affect that relationship. It also shows how engineering knowledge grows typically through a complex interplay of experiment and theory. Chapter 3 ("Establishment of Design Requirements") illustrates how objective specifications for design at lower levels of hierarchy become set, in this case in a situation involving a large subjective human element. This chapter provides the one case that explores the design-ambient tension at first hand. The story is a complicated one involving activity by an extended engineering community over a protracted time. Chapter 4 ("A Theoretical Tool for Design") describes the historical development of an analytical design tool by adaptation and reformulation of theoretical knowledge from science. In so doing, it provides an example of how the thinking of engineers necessarily differs from that of scientists. Chapter 5 ("Data for Design") exemplifies how engineers acquire the quantitative empirical data needed to carry out design. The acquisition derives in this case from a systematic experimental methodology little examined by historians but widely used by engineers when theoretical methods cannot supply requisite data. Chapter 6 ("Design and Production") goes outside design to include production and shows how the two functions are related in the knowledge they require. The knowledge in question, though of considerable quantity and from a modern science-intensive industry, contains nothing that could be described as scientific in a contemporary sense. The final two chapters then offer analysis prompted by the entire body of evidence, chapter 7 providing a categorization of engineering design knowledge and the activities that generate it and chapter 8 a model of how engineering knowledge grows in a general way. Chapters 4, 5, and 6, though they come later here, were done earlier than chapters 2 and 3; they are the ones that focused my attention onto knowledge and design. Together with chapter 2, they constitute the four that have been published elsewhere. They are reproduced here (except for a passage of several paragraphs in chapter 5, which is so noted) with only minor changes outside the introductory paragraphs. To do otherwise, besides extending the task impractically, would have run the risk of destroying the unity and integrity of the individual pieces. Fortunately, I think, they fit as they stand into a reasonably unified whole. As we shall see, engineering knowledge is so vast and diversified as to defy a strictly unified treatment in any event.

As the foregoing outline suggests, I interpret the word *knowledge* broadly. In particular, I take it to include both of what philosopher Gilbert Ryle calls "knowing how" and "knowing that," that is, knowledge of how to perform tasks as well as knowledge of fact.[18] Knowing that appears throughout the studies in the ideas and information whose growth is being traced. Knowing how shows up as both knowledge of how to do design and knowledge of how to generate the new knowledge—the ideas and information—that such doing requires. The latter sort is visible implicitly in the learning processes in all the cases and explicitly in a number of places, especially the methodological discussion of chapter 5 and the general model of chapter 8. Since I display the design process in only one case, knowledge of how to do design is less evident, though I necessarily include it in the categorization of chapter 7. It will have to figure intimately, of course, in the increased historical analysis of design that is badly needed.[19]

A problem of terminology also needs to be mentioned. I take engineering in the present discussion to be a form of human activity defined as in the statement quoted from G. F. C. Rogers near the beginning of this chapter. A key term there for present purposes is the word *organizing*, for which we could also read *devising* or *planning*. This word selects engineering out from the more general activity of "technology," which embraces *all* aspects of design, production, and operation of an artifice. Draftspersons, shop workers, and pilots, for example, though all technologists, do not organize in the engineering sense and are therefore not engineers. All engineers, that is, count as technologists, but not all technologists count as engineers.[20] The activity of engineering, by the same token, falls within that of technology, and engineering knowledge forms part of the broader realm of technological knowledge. Historians and philosophers of technology, unfortunately, seldom make such a distinction. To use their work so far in this chapter, I have had to go along with this ambiguity. In what follows I shall try to be more careful and speak only of engineering and engineering knowledge unless I intend something broader. In drawing on the existing literature, however, I will still find it necessary at times to revert to the usual practice—and I can't promise I won't do so inadvertently elsewhere. Throughout the book, however, the topic (with the exception of attention to airplane piloting in chapter 3 and shop procedures in chapter 6) is the epistemology of *engineering*.

Finally, some remarks on the scope of the study are in order: The kinds of knowledge described by the cases can hardly be considered exhaustive. Other kinds exist that could be examined to advantage. Device-specific theories that depend on special approximations (e.g., elementary beam theory, where the height and deflection of the beam are taken to be small compared with the span) are a case in point. Another is the analytical procedures, such as optimization, that figure in the design process. In-depth examination of such specific kinds of knowledge, some of which I note briefly in chapter 7, might alter the details of the discussion; I doubt if it would change the general thrust.

The examples here also come from one particular branch of engineering—aeronautical. Other branches, especially those such as civil, petroleum, and mining engineering that have to cope with intractable features special to individual environments, may involve considerations beyond those given here. I suspect these will require additions and modifications to the present generalizations, however, rather than fundamental revision.

A more serious objection historiographically is that the examples, coming as they do from aeronautics, represent almost entirely twentieth-century engineering. The epistemology in the present book is, in fact, the epistemology of engineering more or less as it exists today. Analyzing the growth of engineering knowledge from ancient times to the recent past will doubtless raise questions not included here.

A beginning in that direction has been given in an essay by Bertrand Gille.[21]

Summary of these and earlier restrictions may be useful. Briefly stated, the historical cases in the book focus on the knowledge required to do *normal design engineering* in *aeronautics* in the *twentieth century*. The restrictions look considerable at first (though the material needing examination for an exhaustive study of even this scope would be vast). Extrapolation of the conclusions, however, may be either problematic or straightforward, depending on the restriction. Almost all the elements of normal design may be expected to be necessary for radical design; the complications of novelty, however, will add the usual perplexing concerns of creative invention. Extension from aeronautics to other branches of modern engineering and from design to production and operation should be relatively easy; they may call for detailed modification and addition, but I doubt that fundamentally new ideas will be needed. The generalized categorizations in chapter 7 are intended to move tentatively in these directions, especially concerning other branches of engineering. Extending the categories backward from the twentieth century may be more troublesome as the roles of mathematical theory and transfer from science disappear and technical arts predominate; this may call either for simple deletion of some elements or fundamental rethinking of others or a combination of both. The model for knowledge growth in chapter 8, by being relatively abstract and saying something about changes with time, applies potentially to all engineering and in all periods; for the same reason, it must be viewed as relatively conjectural and subject to controversy. The validity of the earlier material, however, does not stand or fall on the basis of the model.

I do not imply by the above that I expect to say the final word in any respect—the epistemology of engineering is obviously at its beginnings. I do hope, though, to have provided a framework to aid its future development. The historical accounts may also serve as a reservoir of detailed information for others to use.

In the material that follows, specialized technical terms are defined where they are introduced. This definition can then be located by referring to that term in the index.

CHAPTER 2

Design and The Growth of Knowledge:
The Davis Wing and the Problem
of Airfoil Design, 1908–1945

Among the many important decisions confronting designers of an airplane is choice of shape for the fore-and-aft sections of the wing. During the 1930s most American designers made this choice from an extensive catalog of sections whose aerodynamic properties had been measured in wind tunnels of the National Advisory Committee for Aeronautics. In 1938, however, one major company, the Consolidated Aircraft Corporation of San Diego, chose for its B-24 bomber a somewhat mysterious section devised by a lone inventor named David R. Davis. The choice depended on some unusual test results, unexplained at the time, from the wind tunnel at the California Institute of Technology. The B-24 went on to become the most numerous and one of the most successful bombers of World War II. The Davis section, after its moment in the sun, disappeared quietly and with little effect on the evolution of wing design. This situation, curious at the time and largely forgotten today, is an interesting sidelight in the history of aeronautics. More important for present concerns, it provides a useful vehicle for studying engineering knowledge in relation to design.

When I began this study, my purpose was to see how normal, day-to-day design conditions the knowledge required for its performance. In view of the ambiguity surrounding Consolidated's choice, I hoped especially to learn something about how uncertainties in knowledge affect and are affected by the design process. Engineers frequently have to make decisions of great practical consequence in the face of incomplete and uncertain knowledge; it seemed likely that this necessity might have epistemological implications. At the same time, as an aeronautical engineer, I was curious to see if the unusual performance of the Davis section could be explained in light of subsequent understanding. In pursuit of this goal, a second and related theme emerged: how knowledge grows in relation to concrete demands from design. Design thus relates clearly to what philosopher Karl Popper sees as "the central problem of epistemology [which] has always been and still is the problem of the growth of knowledge."[1]

The design here occurred at the lowest levels of the hierarchy described in chapter 1. Aerodynamic wing design, at Consolidated as elsewhere, proceeded by a highly structured, time-honored combination of theory and experiment. Such methodological structure was possible because engineers already knew the general form of the solution, that is, what an airplane wing looked like, and how to specify the problem in concrete engineering terms—at Consolidated, the least possible drag at a particular cruising lift. Problems of this sort can be described as "well defined." The questions that remained, though still difficult, had to do with specifics of wing geometry. Uncertainties arose mainly from the explicit knowledge on which designers depended.

As explained in chapter 1, this is the only case in the book to exhibit the engineering design process. It therefore appropriately comes first. Although I shall not deal with design activity in the later cases, the reader should imagine some more or less similar process in the background, requiring and conditioning the generation of knowledge.

My two initial questions dictate organization of the material that follows. The first section explores the design process at Consolidated, with emphasis on what was and was not known. This section shows the nature of uncertainties in aerodynamic knowledge in the 1930s. It also illustrates how an engineering design community functions in the face of uncertainty at a given period in time. The second section then outlines the general history of airfoil design to determine where the Davis section belongs and to attempt to explain its performance. In so doing, we see how an engineering community acts to increase knowledge and reduce uncertainty as time proceeds. The final section then enlarges on the themes of uncertainty and growth. We observe, among other things, that growth of knowledge and reduction of uncertainty in design are actively related, even when increased performance is not at issue. We also see how the events exemplify a variation-selection model for knowledge growth, a topic that will be explored at length in chapter 8.

Consolidated Aircraft and the Davis Profile

In the summer of 1937 engineers at Consolidated (later Consolidated Vultee and now the Convair Division of General Dynamics) were engaged in extensive design study of wing optimization for long-range naval patrol aircraft. Production for the navy of the company's PBY two-engine flying boat, which in World War II would be built in larger numbers than any other water-based airplane, was nearing 100, and construction of the prototype XPB2Y-1 four-engine boat was well along. With the rapid advances taking place in airframe, engine, and propeller design, a potential market existed for still higher-performance flying boats, not only for the navy but for the growing intercontinental commercial service. At the same time, Reuben H. Fleet (figure 2-1), founder and president of Consolidated, seems to have had his eye on sales to the army air corps of long-range, land-based bombers. The Boeing B-17 was beginning to demonstrate what was possible with such airplanes. Some engineers at Consolidated already felt that the days of the flying boat would eventually be numbered and that the growth and even survival of the company required air corps as well as navy business. In the rapidly changing aeronautical world of the late 1930s, a study of wings for long-range aircraft could serve a complex of purposes.[2]

To define the shape of a wing for construction, the aircraft designer must decide on the *planform*—that is, the outline of the wing when viewed from above—and on the profile of the fore-and-aft sections, referred to as an *airfoil profile, airfoil section,* or simply *airfoil.*[3] The shape of the wing in turn determines its aerodynamic performance. Decisions on shape must therefore be made with the desired performance in view, and this requires some method for evaluating the performance of different designs. For unswept wings of the sort used in the 1930s, aerodynamic performance can be calculated fairly accurately from theoretical ideas that reduce this problem also to one of planform and section. This reduction, which is an approximation aerodynamically, affords a number of simplifications. The most notable is that it allows the aerodynamic drag to be treated as the sum of two parts, *induced drag* and *profile drag,* which have very different causes.

For a given aerodynamic lift, the magnitude of the induced drag is fixed by the planform. This drag supplies the work required to generate the energy of the continuously lengthening vortices that trail from the tip regions of any lifting wing of finite span. Induced drag is thus the price that must be paid for lift on such a wing. Theoretical calculation of induced drag does not require consideration of the viscosity (or internal friction) of the airstream. The calculation, while complicated, is thus not insuperably difficult. Practical methods for general planforms were well developed by the mid-1930s.[4]

Profile drag, by contrast, is a property of the airfoil section. It depends on the shape of the section and is assumed to be the same as would exist on a hypothetical wing having that selfsame section over an infinite span. (In such a limiting case the tips and hence the induced drag vanish.) The profile drag is a function of the viscosity of the

Fig. 2-1. Reuben H. Fleet (1887–1975), *right.* With George J. Mead, advisor on aircraft production for the National Defense Council. From Ticor Title Insurance Company.

airstream; unlike induced drag, which exists independently of viscosity, it would theoretically disappear if air were frictionless. Since calculation of practical viscous flows was beyond the reach of theory in the 1930s (difficulties exist even today), profile drag and other aerodynamic characteristics of airfoil sections had to be found by testing wings in a wind tunnel and subtracting out the calculated planform effects. Development of airfoils was thus largely an empirical activity.

Such aerodynamic ideas, plus structural considerations, formed the basis for the Consolidated study.[5] As could be calculated from theory, induced drag decreases, other things being equal, as the planform is made longer and more slender. Increased bending of the longer wing under lift, however, requires a heavier structure and, if the additional weight is not to get out of hand, a thicker airfoil. But increases in thickness tend to increase profile drag, thus counteracting and possibly nullifying the reduction in induced drag. The optimum wing for a given flight condition thus requires a complicated tradeoff between a number of conflicting requirements. In quest for their optimum, Consolidated engineers made calculations for numerous wings aimed at maximum possible flight range with as high a cruising speed as feasible. These included planforms with aspect ratio up to 12, an unusually high value for the time. (*Aspect ratio* is the engineer's measure of planform slenderness and is defined as the span of the wing divided by the average streamwise width, or *chord*.) Airfoil sections were chosen from the catalog of profiles and associated data supplied by the National Advisory Committee for Aeronautics (NACA) from its wind tunnels at Langley Field, Virginia. As the Consolidated engineers well knew, however, optimization calculations of the sort described are approximations at best. The study therefore included evaluation of several of the most promising wings in the 10-foot wind tunnel at the Guggenheim Aeronautics Laboratory of the California Institute of Technology (usually called GALCIT).[6]

It was at this point that David Davis entered the picture. Davis (figure 2-2) was an entrepreneur and self-taught inventor and designer of the type common in the pioneering days of aviation but disappearing by the 1930s. He had learned to fly in Los Angeles in the early 1910s and, in 1920, with family money, became the partner and financial supporter of Donald Douglas in founding the Davis-Douglas Aircraft Company. This became the Douglas Aircraft Company the following year when Davis (for reasons that are unclear) withdrew after helping design and flight test the first Douglas airplane, the Cloudster. Davis was brought to Consolidated and introduced to Reuben Fleet by Walter Brookins, the first civilian American taught to fly by the Wright brothers and since 1930 Davis's partner in the Davis-Brookins Aircraft Cor-

Fig. 2-2. David R. Davis (1894–1972).

poration of Los Angeles. Brookins's wife had been acquainted with the then Major Fleet, an old-time flyer himself, when she was secretary to his commanding officer at McCook Field, Dayton, Ohio, in the early 1920s. Brookins thought Fleet might be approachable for that reason.[7]

The main—and apparently sole—asset of the Davis-Brookins company was a patent, filed in 1931 and issued in 1934, for a family of airfoil shapes defined by mathematical equations of Davis's devising.[8] With these equations, Davis claimed, he had arrived at airfoils of superior performance to others then in use, performance that made them especially suitable for long-range aircraft. Fleet's initial reaction, like that of his chief engineer Isaac M. ("Mac") Laddon, was naturally skeptical. Laddon's engineers could see no physical basis for Davis's equations, and the chance of a lone and professionally untrained inventor improving on the extensive research of the NACA must have seemed unlikely. Most of all, the equations in Davis's patent contained two unspecified, assignable constants for which Consolidated engineers would need values in order to draw and examine his airfoils, and Davis refused to divulge this information. He was not about to reveal his essential secret in the absence of some commitment from Fleet and Laddon. Davis proposed instead that he build a wind-tunnel model to the same planform and spanwise thickness distribution as one of the Consolidated models but incorporating his own airfoil. He would then deliver this model, still without specifying the shape of its sections, to GALCIT for testing, together with the Consolidated models. All this was to be done at Consolidated's expense. If Davis's wing proved superior and Consolidated signed a license to use it, he would then supply them with the shape of the profile. (The Davis contribution was thus only the airfoil profile and not the entire wing, as some people have assumed. I have retained the misleading term *Davis wing* in my title and occasionally elsewhere because the episode has customarily been identified by that name.) On the basis of this proposal, Fleet and Laddon decided to go ahead. Airfoil design was still largely empirical, and there was always the outside chance that Davis might be onto something.

If Consolidated engineers doubted that Davis's equations had a valid basis in fluid mechanics, they were—as we shall see—apparently correct. Davis, however, seems to have thought otherwise. His patent of 1934 includes, without elaboration, the statement that "the equation was developed from formulas based on the mechanical action of a rotor having rotation and translation through a fluid and giving the Magnus effect."[9] ("Magnus effect" is the name given the lift experienced by a circular cylinder rotating about its axis and moving through a fluid.) Two accounts in the popular press of the early 1940s, based on interviews with the inventor, attempted to explain Davis's reasoning by (in effect) enlarging on the patent statement in terms of a translating and rotating wheel or radius arm. These explanations, however, are either physically dubious or downright nonsensical.[10]

The lack of physical basis for the equations is clear from brief hand-lettered notes, unsigned and undated but apparently formulated by Davis prior to the patent.[11] These begin with a statement about the Magnus effect like that in the patent. They then go at once, with no fluid-mechanical or other physical reasoning, to a purely geometrical procedure based on a translating-rotating circle (the Magnus-effect cylinder). Davis appears simply to have plotted the trajectory of a point on the cylinder and noticed that a loop in this curve had an airfoil-like shape. He then devised a complicated and unlikely geometrical construction to change this shape to something closer to a typical airfoil and translated this construction into equations by ordinary algebra and trigonometry. He gave no explanation of the reasoning behind his construction, which could not possibly have depended in any logical way on fluid mechanics. He also provided no theoretical rationale connecting the airfoil problem to the considerably different Magnus effect. Knowledge of that effect appears to have served simply to focus Davis's attention onto the rotating cylinder, though he may well have thought this connection gave his work a more valid theoretical basis than was apparently the case. Although he thus derived inspiration from the Magnus effect, his procedure was essentially an exercise in geometry.[12] It must be admitted, however, that Davis's equations are themselves not at all simple or obvious. The construction on which they are based is also both ingenious and complex. Although his scheme had no valid basis in fluid mechanics, it could not have been devised without a good deal of mental effort of some sort.[13]

Whatever the nature of his thinking, Davis, like others, had to resort to experiment for his airfoils' performance. Since no wind tunnel was available, he improvised by borrowing a large Packard car from his friend Douglas Shearer, chief sound engineer at the Metro-Goldwyn-Mayer Studios and brother of the movie actress Norma Shearer. He then mounted a large, flat board horizontally on top of the car, to isolate his model from the aerodynamic disturbances of the car body, and tested his airfoils cantilevered vertically above the board. The measurements—of the distribution of pressure at the surface of the airfoil—were made by photographing an array of manometers in the car as it was driven at high speed (on lonely back roads in southern California, according to one source, and with flanged wheels on an abandoned railroad track in the desert, according to another).[14] Davis's purpose was to search out the optimum airfoil from among the family obtained by altering the values assigned to the constants in his equations. After laboriously testing a number of airfoils in the years 1935 to 1937, he decided, however, that such procedure would require the rest of his life.[15] When Consolidated engineer George S. Schairer questioned him about it later (after the company had learned the shape of the profile), Davis said he therefore "sat in a chair for three days considering the matter [of the value of the assignable constants] and concluded on theoretical grounds that plus one and minus one were best."[16] He did not say what the theoretical grounds were. He then checked this airfoil out to his satisfaction on the Packard. This was the airfoil incorporated in the model he delivered to GALCIT.

The comparative measurements at GALCIT, made in late August and early September 1937, came as a shock to Professor Clark B. Millikan and his wind-tunnel staff. In Millikan's words from his report to Consolidated, "Certain of the results for the Davis wing are so striking that when they were first obtained, it was felt that some experimental error must have entered." The Davis model and the Consolidated model that served for comparison (the latter with NACA "21-series" sections) were therefore carefully remeasured to make certain their planforms agreed sufficiently (they did). The Davis model was then retested on two more occasions some weeks apart, with the three tests showing "practically perfect agreement." The Consolidated model, which according to Millikan had originally a poor surface finish, was polished to the same outstanding finish as the Davis model and also retested to see if the results of that model could be matched. (Understanding was growing, though a few years would elapse before it became firm and widespread, that surface condition can have as much effect on airfoil performance as the shape of the airfoil itself.) The Consolidated model, however, still showed nothing unusual.[17]

The most striking result for the Davis wing was in the relationship between lift and *angle of attack* (angle of inclination of the wing relative to the airstream). Engineers measure this relationship by the *lift-curve slope*, defined as the increase in lift per degree increase in angle of attack. To the consternation of the investigators, the Davis model gave an experimental slope practically equal to the value calculated from the usual theory of nonviscous (i.e., frictionless) flow. This finding caused concern because viscosity should, in principle, reduce the measured value below that given by theory. The expected relationship was recovered when Theodore von Kármán, director at GALCIT, pointed out that the generally used theory was an approximate one and that a more accurate nonviscous theory gave a comfortably higher result. A question nevertheless remained: the measured slope for the Davis model was still from 7 to 13 percent higher than for the great majority of wings tested at GALCIT and 6 percent higher than the previous best. Millikan could offer no explanation for this difference. He could only surmise that the high value for the Davis model came from some peculiar and unspecified variation of viscous effects with angle of attack (a possibility I shall return to later). This uncertainty was of more academic than practical importance, however, since a slightly higher lift-curve slope has no great use for the aircraft designer.

Of greater interest to Consolidated was the fact that the Davis model also showed a slightly lower minimum drag compared with the company's design and a significantly lower increase in drag with increasing lift. As a result, the Davis model exhibited about 10 percent less drag at the lift required for long-range cruise. This potentially useful finding was tempered, however, by a notorious difficulty with wind-tunnel testing: because of "scale effects" associated with viscosity, results from a small model at relatively low speeds (as was the case in the tests at GALCIT) cannot be extrapolated simply and reliably to the full-scale airplane at the speeds encountered in flight. On account of the unusually large aspect ratio and the limitation on span imposed by the 10-foot diameter of the tunnel, the average chord of the present models (the significant dimension for airfoil studies) was considerably less than customary at GALCIT. Millikan worried therefore whether the differences in performance might not be due more to reduced scale than to difference in airfoil shape.[18]

Reception of the results at Consolidated, at least at the level of Fleet and Laddon, was apparently less critical. (Debate must have occurred within the engineering staff, but records from the Consolidated days no longer exist at Convair.) In late September, three weeks after the tests, Fleet reported the development to Adm. Arthur B. Cook, chief of the navy's Bureau of Aeronautics. In reference to the flying-boat studies, he stated that "the lessened drag at higher angles of attack . . . was appreciable. We can anticipate an improvement . . . which, in the long range boat, saves approximately 2500 pounds of fuel." He said that Laddon was forwarding a similar statement to Comdr. Walter S. Diehl, a senior engineering officer at the bureau, along with a copy of wind-tunnel data taken from Millikan's report. A visitor from NACA headquarters in November also described Fleet as being "extremely enthusiastic" in telling him that Consolidated's new airfoil would "knock the spots off" of anything previous. His enthusiasm was said to be due in part to the fact that Davis's airfoil was defined by a mathematical formula. Fleet, like others with technical responsibility but little mathematical training, may have been inordinately impressed by mathematics per se.[19]

Others in the aeronautical community did not share Fleet's enthusiasm. Diehl quickly forwarded Laddon's letter and data to the NACA for their opinion. Unidentified members of the staff at the committee's research laboratory at Langley Field replied that "the comparison . . . attached to the letter from the Consolidated Aircraft Corporation is subject to too much uncertainty to be really reliable. . . . there are a number of airfoil sections that have characteristics superior to those of the N.A.C.A. 21 and hence, the possible appearance of a new one need not be surprising." Walter Brookins, who had for some weeks been trying to stir up interest in the Davis airfoil on the part of his old acquaintance Gen. Henry H. Arnold, assistant chief of the air corps, sent Arnold a copy of Millikan's report in late September that Arnold immediately forwarded to the Matériel Division at Wright Field for evaluation. Maj. A. W. Brock, Jr., replied much the same as did the people at Langley, basing his objections on a number of qualifications that Millikan had prudently included in his report. Wind-tunnel testing being an empirical activity with so many uncertainties, such skepticism was inevitable. Diehl, a man of many years' experience, expressed his own view some time later to a navy officer who had written to him concerning the Davis profile: "For your information, this is just another airfoil to us. We have similar wings submitted from time to time. There is no way of evaluating the claims for them, except by comparative tests." Diehl obviously meant tests by the government's own experts in the larger-scale facilities of the NACA.[20]

By the end of 1937, then, Consolidated had what they considered a superior airfoil but did not know its shape. Schairer attempted to discover the profile for the company by drawing "several hundred" airfoils from the equations of Davis's patent, using a wide range of the assignable constants; he concluded, however, that "there was no hope of guessing the values . . . which Davis had found best." Fleet told the visitor from the NACA that Consolidated had also built a model with an airfoil they believed approximated the Davis shape. (For this they depended on what the company's engineers had seen of the Davis model. According to the visitor, "Although the Consolidated people and the California Institute of Technology had the Davis airfoil for test, they failed to make a template of its contour," presumably because they were enjoined against it by Davis, who was present at the tests.) When tests at GALCIT showed nothing of promise, Consolidated was left wondering how accurate their approximation had been.[21]

The company learned the shape only after Laddon, on February 9, 1938, signed a letter of agreement with Davis-Brookins. This contract specified a sliding scale of royalties if Consolidated adopted the Davis airfoil for any of its airplanes. Davis pledged in return not to disclose the profile to any other company for a period of one year. That Laddon and his engineers still retained doubt is shown by a provision that canceled Consolidated's obligation if "tests of the same airfoils on larger scale models in the N.A.C.A. full scale wind tunnel indicate that the improvement evidenced in the smaller scale models is not existant [*sic*]."[22] Following this agreement, Davis furnished Laddon with the airfoil's ordinates on a sheet of paper. He said they were generated by using plus one and minus one for the constants in his equations but offered no reason for these values. (The *ordinates* of an airfoil are the offsets of points on the profile measured at right angles to the *chord line* connecting the leading and trailing edges.) The thickness of Davis's airfoil at its maximum was 16 percent of the airfoil chord. If desired, airfoils of other thicknesses were to be obtained by increasing or decreasing all ordinates proportionately. Figure 2-3 shows the Davis profile, adjusted to 15 percent thickness, in comparison with a section of equal thickness from the NACA 230- ("two-thirty") series, then regarded as among the best conventional profiles. That thickness changes were prescribed in this arbitrary way, with no reference to aerodynamic ideas, supports the view that Davis's thinking had little basis in fluid-mechanical theory.[23]

The decision on whether to use the Davis profile was undoubtedly difficult, despite Fleet's professed enthusiasm. Although evidence is incomplete, the probable considerations are clear. If the GALCIT comparison held up at full scale, the Davis airfoil offered a definite advantage for long-range performance, and such performance was

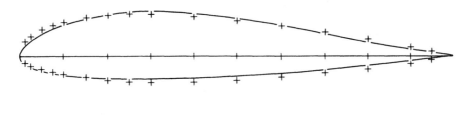

———— Davis airfoil

+ NACA 23015

Fig. 2-3. Davis airfoil compared with NACA 23015. Both airfoils have maximum thickness of 15 percent chord located at 30 percent chord. Ordinates of Davis airfoil (adjusted to 15 percent thickness) from R. P. Jackson and G. L. Smith, Jr., "Wind-Tunnel Investigation of Sealed-Gap Ailerons on XB-32 Airplane Rectangular Wing Model Equipped with Full-Span Flaps Consisting of an Inboard Fowler Flap and an Outboard Retractable Balanced Split Flap," *Memorandum Report for the Matériel Division, Army Air Corps* (unnumbered), NACA, December 13, 1941, table 1; of 23015 from E. N. Jacobs, R. M. Pinkerton, and H. Greenberg, "Tests of Related Forward-Camber Airfoils in the Variable-Density Wind Tunnel," *Report No. 610*, NACA (Washington, D.C., 1937), fig. 19.

crucial for the airplanes Consolidated had in mind. If it did not hold up, the Davis profile would likely be no worse than a conventional design. Aerodynamically there might thus be something to gain and nothing to lose by gambling. Structurally, however, there could be disadvantages. As shown in figure 2-3, the Davis shape was thinner than a conventional profile in its after portion, and this compromised the structure and mechanism of the Fowler flaps that had to move back and down from the wing for landing. The possible weight penalty from this source would be for nothing if the aerodynamic gain proved illusory.[24] An additional consideration was a desire by Fleet to have something new to add sales appeal for the navy and air corps (though the negative reaction at Wright Field, probably transmitted to Fleet by Brookins, might have dampened this idea). Continuing to explore the technical tradeoffs, Consolidated engineers arranged for tests at GALCIT, from March to May 1938, of four wings of greater thickness than before. Although the level of performance was disappointing for all wings, the three with Davis profiles proved consistently superior to the one without.[25]

In this situation of uncertainty, Fleet and Laddon decided, sometime in the first half of 1938, to gamble on the Davis airfoil for a flying boat then in preliminary design. The Model 31 (figure 2-4) was a twin-engine craft to be developed entirely at company expense ($1 million, as it turned out) and adaptable both for commercial service and as a longer-range, higher-speed replacement for the navy's PBY. Besides the novel airfoil, Consolidated engineers designed the wing to the then unusually high aspect ratio of 11.55, which gave it a long and narrow look for such a large airplane.[26] Detail design and construction of the Model 31 were authorized on July 11, 1938, and initial flight took place only ten months later on May 5, 1939. Shortly after authorization, a month of model testing of the complete configuration in numerous detailed modifications was conducted at GALCIT; associated tests of the wing alone showed the same striking results as before. The airplane lived up to expectations, but the onset of war blocked its commercial application, and a complex of nonengineering reasons kept it from military use. Only the prototype was built. As things turned out, the importance of the Model 31 lay in the design experience it provided Consolidated engineers in preparation for their land-based bomber, for which preliminary studies were in progress when go-ahead for the flying boat was given.[27]

If Fleet and Laddon had hoped to have flight experience on the Model 31 in hand for design decisions on the bomber, it was not to be. In late 1938 the air corps, conscious of the growing possibility of war in Europe, asked Consolidated if the company was interested in becom-

Fig. 2-4. Consolidated Model 31 flying boat.

ing a second source of production for the B-17. Fleet, given the advances in aeronautics generally and the knowledge from his company's studies of long-range aircraft in particular, thought his engineers could do better than the four-year-old Boeing design. Moving swiftly, in January 1939 he proposed a longer-range, higher-speed bomber incorporating the same wing and tail design as the still-unfinished Model 31. If Wright Field retained doubts about the unproven Davis wing, they did not prevent the air corps's favorable decision. Contract for the prototype B-24 was issued on March 30, 1939, and work began at a feverish pace to meet the required deadline of December 30—only nine months!—for the completed airplane. In early May, probably with some relief, Laddon wrote to General Arnold detailing the test pilot's favorable impressions from the initial flight on the Model 31, adding that "the above information is pertinent to the . . . B-24 since the same wing, flap and tail are used thereon." To check their decisions, Consolidated engineers also ran the usual tests of the airplane configuration in the GALCIT tunnel in June, with the wing showing again the performance that had become accepted as typical of the Davis profile. As if to put a seal on their knowledge (the tests in September would come too late to affect the design), they also had measurements made of two specially constructed comparative wings, with the same planform and spanwise thickness distribution employed on the two airplanes. The conventional design used NACA 230-series profiles. As before, the wing with Davis sections showed higher lift-curve slope and lower drag at cruising lift.[28]

The B-24 (figure 2-5) first flew on the afternoon of December 29, 1939, one day short of the nine-month deadline. Except for incorporating four engines instead of two, the external geometry of the wing copied that of the Model 31. In its internal structure, however, the design was new. The unusually high aspect ratio of the planform (11.55 against 7.58 for the B-17) gave the wing a distinctive appearance. Some considered it the only graceful feature of an otherwise ungainly airplane.

The B-24 proved among the most important aircraft of World War II. It served a wide variety of needs and appeared in a large number of modifications, for navy patrol as well as air corps bombing and for foreign governments as well as the United States. Its range, which was greater than for the B-17, made it particularly useful for bombing in the Pacific theater and for antisubmarine warfare. When production terminated in 1945, somewhat more than 19,000 had been built. This figure makes it the most numerous bomber in history (though the large production was due in part to decisions about manufacturing capacity made early in the war, before operating experience with the B-17 and

Fig. 2-5. Consolidated B-24 bomber.

B-24 was available). The entire output had the same Davis wing as the prototype.[29]

The Model 31 and B-24 both performed excellently for their day. Whether they were better than they would have been without the Davis airfoil, however, seems unlikely. The technical press, reporting the airplanes' generally excellent performance, assumed they were and gave Davis and his airfoil glowing praise. But the performance of an airplane is fixed by many things besides the wing, and reliable performance data are hard to measure in any event. Moreover, in contrast to the situation with the model wings in the wind tunnel, no directly comparable airplanes existed employing a conventional section. No one can say for certain how the airplanes would have performed with a different airfoil. As we shall see, however, later knowledge about the deleterious effects of surface roughness and inaccuracies on routinely manufactured wings makes it unlikely the advantages seen in the wind tunnel were realized in flight. The Davis section probably contributed little, if at all, to the B-24's outstanding range, which can be explained on the basis of the high aspect ratio and other features of the airplane. Ralph L. Bayless, reflecting on his experiences as an aerodynamicist at Consolidated in the period, now thinks a suitably chosen NACA airfoil "would have done as well, maybe better." He adds, however, that "this would not have achieved the sales approach of something new."[30]

Despite its much acclaimed use on the B-24, application of the Davis airfoil by Consolidated caused only mild excitement in the aircraft industry. The one-year prohibition in the Consolidated agreement having expired before the flight of the Model 31, Davis took advantage of the publicity surrounding that event to make public offer of "an airfoil engineering service to interested manufacturers on a royalty basis." The offer attracted no takers. The Douglas and Hughes aircraft companies did conduct studies of the Davis profile for their designs but never put it to use. At least one company—Vought-Sikorsky—inquired about the profile from the NACA but was discouraged from using it on the grounds it was inferior to new "laminar-flow" sections recently developed at Langley Field (see below). Only the Boeing Company, where George Schairer moved from Consolidated in 1939, was influenced by the events at San Diego. Although Schairer was not free to disclose the Davis ordinates to Boeing, the experience at Consolidated influenced his thinking regarding aerodynamic design of the highly successful B-29 of 1942 (and its later derivatives, the B-50 and KC-97). These aircraft all had a high aspect ratio and· an airfoil "somewhat similar" to Davis's. Aside from this indirect influence at Boeing, the Davis airfoil found no application outside Consolidated.[31]

Only one more Consolidated airplane used Davis's airfoil. The B-32, a strategic bomber designed to the same air corps specification as the B-29, was begun in mid-1940, and the prototype flew in September 1942. The configuration imitated that of the B-24, but with considerably increased size and a design top speed of 385 mph, significantly higher than the test values of about 300 mph for various modifications of the earlier machine. Results of wind-tunnel tests of the design by the NACA were interpreted as showing the Davis profile to be unsatisfactory for this higher speed. These tests, conducted in mid-1941 in the 8-foot high-speed tunnel at Langley Field, showed a relatively high drag for the wing at high speeds and other adverse effects of the compressibility of air that comes into play as flight speed increases. These results were attributed in part to the shape of the Davis profile and in part to the relatively large section thicknesses employed by Consolidated (though the fact that the successful B-29 used a "somewhat similar" profile of only slightly lesser thickness suggests that other, unrecognized effects may have been at work). Whatever the cause, the results made it doubtful the airplane would reach its projected speed. The design had already been released for production, however, and Consolidated decided it could not be changed. The B-32 in fact turned out to have numerous problems. To what extent the wing contributed to them (and whether the airplane ever attained its design performance) is not clear. The success of the B-29 made the B-32 unnecessary in any event. Only slighty over 100 were built, mostly in 1944 and 1945, and only a few saw limited duty in the Pacific by the end of the war.[32]

Independently of Consolidated, Davis also pursued study of his airfoil through tests at Davis-Brookins's expense in the low-speed wind tunnel at the University of Washington. The results—for Davis airfoils of various thicknesses incorporated in a variety of planforms, some rather extreme and unconventional—were forwarded to his company in eight reports from 1939 to 1944. Where the planforms allow comparison, the findings are much the same as at GALCIT. Also in the same period, Davis and several associates formed the Manta Aircraft Corporation to build a long-range fighter based on the Davis profile and one of his unconventional planforms. A photograph of a mock-up appeared with an article in the popular press in 1942, but nothing seems to have come of the project.[33] After its use on the B-32, the Davis airfoil disappeared quietly from the aeronautical scene. With the excitement surrounding the NACA's new laminar-flow sections and the increased preoccupation with high speeds, designers had other things to think about.[34]

Although not central to our themes, the praise in the popular press deserves a closer look. In articles in the *Saturday Evening Post* and *Popular Aviation*, Davis was presented as a "David with his slingshot" who had fought the "Goliath of aerodynamics" by questioning the accepted canons of airfoil design. His purported theoretical ideas—inadequately and confusingly elaborated in the articles—were a "heresy" and a "revolutionary conception." In successfully putting them forth, Davis had "set the academicians back on their heels" and "solved an aviation riddle which defied the most brilliant aeronautical engineering minds." This view stemmed in part from a misinterpretation of an aeronautical-engineering concept unfortunately called the *airfoil efficiency factor*, which is defined as the ratio between the measured value of the lift-curve slope and the value calculated from the approximate nonviscous theory. Since the GALCIT tests gave these values as practically equal for the Davis airfoil, the efficiency factor for Davis's brainchild came out unity. This led the popular writers to jump to the conclusion that the "efficiency" of the Davis profile was 100 percent and, when "efficiency" was reinterpreted in its usual sense of energy output divided by energy input, that "mechanically [this] meant perpetual motion." Even so, Davis had yet to receive suitable recognition "from the slow-grinding mills of science." The Davis episode thus reinforced the popularly cherished (and occasionally valid) myth of the lone and untutored inventor successfully confronting the entrenched establishment of science and engineering. Engineering knowledge—and misinterpretations of it—serves purposes other than the purely technical.[35]

Airfoil History and the Davis Profile

The events at Consolidated provide insight into an engineering design community at work in the face of typically uncertain knowledge. Examination of where the Davis profile fits into the growth of knowledge about airfoils generally is also instructive. A comprehensive and detailed history of airfoil design and its intellectual foundations has yet to be written—it would provide a fascinating study in the interplay of practical engineering and engineering science.[36] Its outline, however, is reasonably clear. I shall pursue it here to the extent necessary to assess the historical place of the Davis airfoil and in hope of answering the question about the puzzling findings at GALCIT. At the same time, we obtain an instructive view of an engineering community extending its knowledge and reducing the uncertainty under which it operates.

The forces a moving fluid exerts on the surface of an airfoil are of two kinds: *pressure* at right angles to the surface and *skin friction* tangential to the surface. Lift of an airfoil depends almost entirely on the distribution of pressure; in most normal flight conditions, this distribution is to a good approximation uninfluenced by the viscosity of the fluid. Drag, by contrast, depends primarily on the skin friction, which exists by virtue of the viscous flow in a thin *boundary layer* next to the surface of the airfoil. The magnitude of the skin friction is contingent on the nature of the flow in this layer, as I shall explain later. The distribution on the airfoil surface of both the pressure and skin friction, and thus the overall lift and drag, depend in turn on the shape of the profile and its angle of attack.

The problem confronting the airplane designer, then, is how to shape the airfoil to obtain the lift and drag characteristics needed for the airplane's mission, whether it be long-range cruise for a bomber like the B-24 or maximum speed and maneuverability for a fighter. Until well after World War II, theoretical and computational difficulties precluded a direct attack on this problem on a case-by-case basis. Instead, designers mostly chose their airfoils from catalogs (like those already mentioned) of profiles devised by research engineers or designers to achieve various categories of performance. Most of these profiles were developed by modifying previously successful forms, using rules learned from experience, such theoretical understanding and methods as were available, and a vast amount of wind-tunnel testing. The mix of methods changed with time, with fluid-dynamic theory slowly replacing rules of experience. The nonviscous theory of pressures advanced well ahead of the more difficult viscous theory of skin friction, and scale effects associated with viscosity lent continuing uncertainty to the applicability to full-scale airplanes of model results from wind tunnels. The latter problem was also complicated by the effects of model finish and of the small-scale turbulence present in wind-tunnel streams but not in the atmosphere. (Such wind-tunnel turbulence can have a distorting effect on the boundary-layer flow on a model. The atmospheric turbulence that gives an occasional rough ride to the airline passenger is of large scale and has no influence on the relatively thin boundary layer on an airplane; it has no significance for the wind-tunnel simulation problem.) One authority estimated that by 1936 efforts to solve the airfoil problem had led to design and trial in the wind tunnel or in flight of over 2,000 shapes.[37] At the time of the Davis episode, airfoil design was clearly more of an art than an engineering science.

Figure 2-6, reproduced from a textbook of 1941 by Professor Millikan, shows typical airfoils in the historical development. The early profiles (Wright, Blériot), laid out largely from knowledge obtained with gliders, were thin and highly cambered, often with a single cloth surface on one side of a supporting frame. Experience in flight and in the wind tunnels coming into increasing use in the 1910s soon showed the advantage of enclosing the frame between two surfaces. The resulting shapes, however, were still strongly influenced by structural considerations (the RAF—for Britain's Royal Aircraft Factory—6 and 15

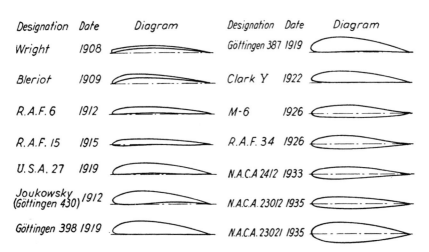

Fig. 2-6. Typical airfoils to mid-1930s. From C. B. Millikan, *Aerodynamics of the Airplane* (New York, 1941), p. 66, reproduced by permission of John Wiley & Sons.

and the USA 27). Design in this period was done mostly by eye and with little attempt at systematization, and procedures had no basis in fluid-mechanical theory. A notable exception occurred at Göttingen, with tests of a systematic family of shapes derived by Nikolai E. Joukowsky's method of conformal transformation. This sophisticated mathematical technique, based on nonviscous theory, enabled precise calculation of the surface pressure for a limited class of mathematically derived profiles. The resulting shapes (e.g., the Göttingen 430) proved to have aerodynamic disadvantages, however, and their thin after-section provided little space for a rear wing spar. Empirical modifications of these sections (the Göttingen 387 and 398 and the Clark Y) proved more satisfactory.[38]

Systematic airfoil design began in earnest following introduction in 1922 of thin-airfoil theory by Max M. Munk of the NACA. This approximate nonviscous theory (source of the approximate value for the lift-curve slope mentioned earlier) allowed an airfoil section to be analyzed in terms of an infinitely thin *mean* or *camber line* (the line halfway between the upper and lower surfaces) plus a chordwise-varying *thickness distribution* disposed equally above and below that line. The differences in pressure between the upper and lower surfaces—and hence the lift of the airfoil—depended only on the shape and angle of attack of the camber line and could be calculated to good approximation without reference to the thickness distribution. Only drag, which cannot be calculated from any nonviscous theory, depended on thickness. Munk's theory provided a new and illuminating way to think about airfoils and caused a basic shift in airfoil design. Before, airfoil designers drew or otherwise defined an airfoil's shape according to experience and judgment, with hope it would give suitable lift and drag. Now, by combining a camber line having calculated aerodynamic properties with a judiciously chosen thickness distribution, they could synthesize profiles with approximately predictable lifting characteristics. The drag and precise lift still had to be found from tests, but a new element of rationality had been introduced. Munk used this approach to design and test the M-series of twenty-seven systematically related airfoils (e.g., the M-6) at Langley Field in the mid-1920s. Researchers elsewhere quickly adopted Munk's ideas (e.g., the RAF 34).[39]

The Munk approach reached its peak in the 1930s in the work of NACA engineers at Langley Field under the leadership of Eastman N. Jacobs. Separating thickness and camber, Jacobs's researchers specified their airfoils parametrically in terms of the height and chordwise location of the uppermost point of the camber line and the magnitude of the maximum thickness. (These quantities appeared as constants in algebraic equations chosen on the basis of theory and experience to define the camber line and thickness distribution.) By varying the three parameters independently and systematically, they designed and tested extensive families of profiles (e.g., the three NACA sections of figure 2-6 and the complete family of figure 2-7). The numerical designations of the profiles, which became part of the vocabulary of aeronautical designers, provided a shorthand statement of the values of the parameters. It was from reports of wind-tunnel results for these NACA profiles that designers, including those at Consolidated, chose their airfoils at the time that Brookins and Davis called on Reuben Fleet.[40]

At this point we may ask how, if at all, did Davis's work relate to the mainstream just described? Like the Langley engineers, Davis used mathematical means to define his profile. Indeed, his simultaneous equations involving trigonometric functions, though less flexible than the algebraic polynomials used at the NACA, were less obviously capable of describing an airfoil and in this sense subtler. The assignable constants also provided a basis for parametric systemization (although a somewhat complicated one) that he could have pursued given adequate experimental facilities. Unlike the Langley people, on the other hand, Davis apparently was not aware of Munk's theoretical ideas about the separability of thickness and camber and the advantages of thinking in such terms. As the hand-lettered notes indicate, his work had no logical foundation in theoretical fluid mechanics. Contrary to how things had been done at Langley, Davis also made no appeal to experimental experience in devising his equations. His work was, in reality, a throwback to the period before Munk, when designers laid out airfoils on mainly geometrical grounds with little valid appeal to fluid-mechanical reasoning. Despite the seeming sophistication of his equations, he was outside the methodological mainstream of the 1930s both theoretically and empirically. Perhaps in part for this reason, and despite the purported success on the B-24, the Davis airfoil had no effect on the evolution of airfoil design and faded from the scene intellectually as well as practically. Ironically, however, it appears to have anticipated, at least in its performance, a second, even more basic shift in design that was beginning precisely at the time of the events at Consolidated.

To understand this shift we need to know more about the drag of airfoils and how it depends on the viscous flow in the boundary layer. Boundary layers in general are of two types of fundamentally different character, *laminar* and *turbulent*. Flow in a laminar layer takes place smoothly in parallel laminae, free of irregular eddying motion. Flow in a turbulent layer contains a large number of small eddies that move chaotically about, causing mixing transverse to the layer. In both types,

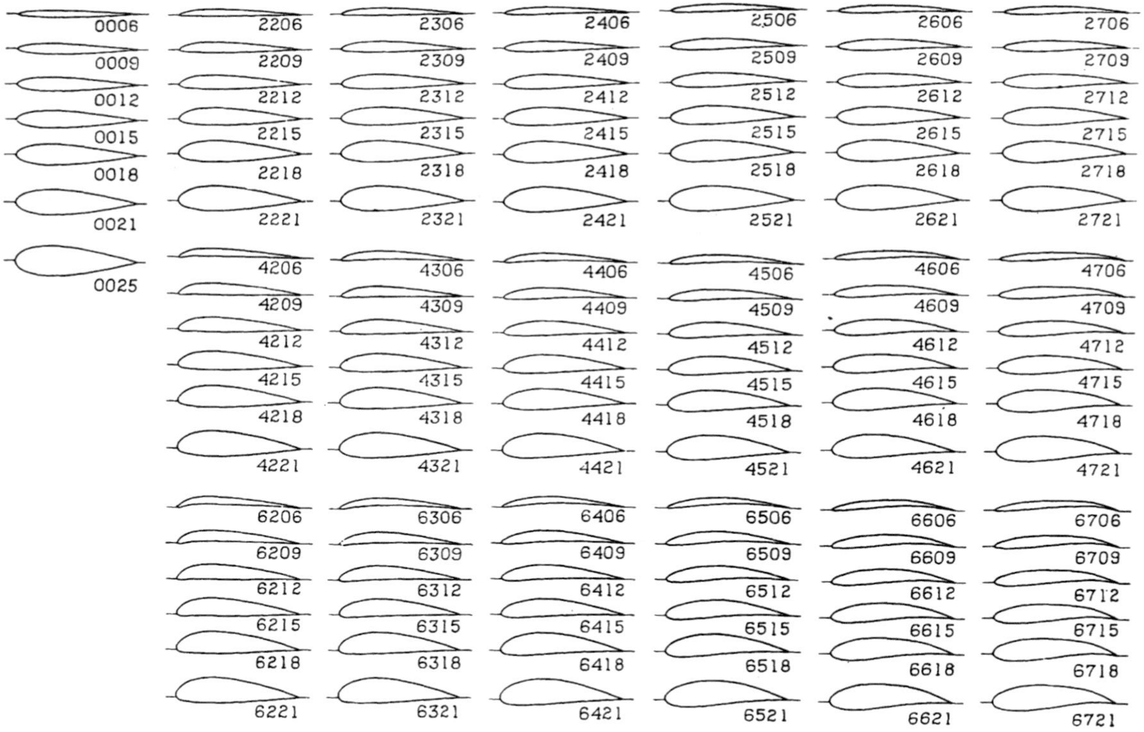

Fig. 2-7. NACA four-digit airfoil family from early 1930s. The first digit of the designation gives the maximum height of the camber line in percent chord, the second the location of this maximum height in tenths of chord aft of the leading edge, and the final two the maximum thickness in percent chord. From E. N. Jacobs, K. E. Ward, and R. M. Pinkerton, "The Characteristics of 78 Related Airfoil Sections from Tests in the Variable-Density Wind Tunnel," *Report No. 460*, NACA (Washington, D.C., 1933), fig. 3.

the local velocity of flow (time averaged for the turbulent layer) varies steeply when evaluated along a line perpendicular to the surface, becoming zero (measured relative to the surface) at the surface itself. As a result of the transverse mixing, however, the turbulent layer tends to be thicker and has higher velocities close to the surface. Most important for our concerns, the latter effect causes the turbulent layer, other things being equal, to exert higher skin friction than the laminar. Under most conditions, the boundary layer on an airfoil starts out laminar at the leading edge but turns turbulent at some point along the surface. The behavior is much like that observed in the smoke rising from a cigarette on an ashtray in a quiet room. The transition to turbulence involves a series of complex steps and depends on a number of factors in ways not completely understood even today.

By the time of the Davis episode, engineers had realized for some time that the only way to reduce airfoil drag was by reducing skin friction through prolonging the extent of laminar flow. Theoretical estimates showed clearly that, if the transition point could be moved back significantly from the 5 to 15 percent chord location common on most airfoils, appreciable gains could be made. Jacobs and his group at Langley Field had already worked toward this end by experimental studies of the boundary layer and by careful changes in airfoil shape. This work only made it clear that with such a complex problem no amount of empirical testing, however systematic, was likely to succeed except by luck. The breakthrough came in 1937 when Jacobs realized the implications of the long-standing experimental observation that transition can be delayed by a "favorable pressure gradient"; that is, by maintaining a continuously falling pressure for as long a distance as possible on the surface of the airfoil. The key therefore was to start, not with the airfoil shape, but by specifying a desired pressure distribution and then calculating the shape that provides this result.[41]

Available theoretical methods, unfortunately, could not handle such calculation. In the early 1930s, extension of the method of conformal transformation by Theodore Theodorsen, a Langley Field theoretician separate from Jacobs's group, had made it possible to go from a given shape to the corresponding pressure distribution, but the inverse problem was still out of reach. In 1937–38 Jacobs and his engineers devised increasingly precise methods to invert the calculation, based first on Munk's thin-airfoil theory and later on Theodorsen's more exact approach. Over the next five years they used these methods to design progressively better airfoils, which were tested in new and improved wind tunnels to ascertain their drag. (Wind-tunnel improvements played an essential role by providing larger effective scale, lower stream turbulence, and better simulation of infinite-span airfoil conditions by using a constant-section wing completely spanning a rectangular test section. For brevity I have not given these matters the attention they deserve.)[42] The new airfoils maintained laminar flow, over a useful range of lift, for anywhere from 30 to 70 percent of the chord, depending on the design goal. The most important visible difference from conventional airfoils was in having their maximum thickness farther aft, the shift being greater the greater the extent of laminar flow. Figure 2-8 illustrates this and other differences, relative to the same conventional airfoil used in figure 2-3, for the 747A315 "laminar-flow" airfoil.[43] (The Davis profile is also reproduced by a dashed line for later discussion.)

Although research was slowed by developmental testing for wartime aircraft, Jacobs's group, by 1945, had produced a catalog of airfoils of significantly lower measured drag than before. Already several military airplanes were using such profiles. Most important for the future, Jacobs's idea of starting from the pressure distribution changed the whole manner of thinking about airfoils. Whereas engineers had previously started with the shape, the pressure distribution now came first. Although the necessary calculations were laborious and time consuming and testing still had to be done to find the drag, airfoil design had after forty years become a logical, essentially rationalized process.[44]

The usual (though unexamined) view among engineers who remember the episode is that Davis, without understanding what he was doing, happened on a kind of laminar-flow airfoil, and this does seem to have been the case. Proof would require measuring the extent of laminar

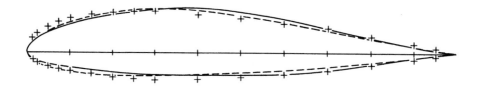

——— NACA 747A315 laminar-flow airfoil

+   NACA 23015

— — — Davis airfoil

Fig. 2-8. NACA 747A315 laminar-flow airfoil compared with NACA 23015 (see figure 2-3). Both airfoils have maximum thickness of 15 percent chord, located at 40 percent chord for 747A315 and 30 percent chord for 23015. The Davis airfoil is also reproduced from figure 2-3 to facilitate discussion. Ordinates of 747A315 from I. H. Abbott, A. E. von Doenhoff, and L. S. Stivers, Jr., "Summary of Airfoil Data," *Report No. 824*, NACA (Washington, D.C., 1945), p. 369.

flow on a Davis profile, but geometrical evidence is suggestive. Consider the laminar-flow and Davis profiles of figure 2-8 relative to the conventional NACA 23015. Though the Davis profile retains its maximum thickness at the same 30 percent chord location as the 23015, it departs in the direction of the laminar-flow profile in other respects: both profiles rise more slowly than the 23015 on the upper surface near the leading edge, have a smaller included angle at the trailing edge, and are slightly concave over the aft 40 percent of the upper surface. The departures from the conventional section are smaller for the Davis profile, but the trends are the same.

Pressure distributions (not illustrated) are more conclusive. For angles of attack near cruising lift, tests of a Davis profile at Langley Field in 1941 (with other purposes in mind) and theoretical calculations made for the present study agree in showing continuously falling pressure on the upper surface over about the first 20 percent of the chord. This result contrasts with a calculated length of only 8 percent on the 23015; it is a consequence in part of the more slowly rising surface on the Davis profile. Though not as extensive as on Jacobs's airfoils, it could be sufficient for significant laminar flow.[45]

Such increased laminar flow would account for the results at GALCIT. The reasons require modern-day knowledge of fluid mechanics and are too technical to explain completely here. Briefly, a longer run of laminar flow over the upper surface implies a thinner boundary layer over the full chord, and this result, together with the smaller trailing-edge angle, implies a reduction in adverse effects of viscosity near the trailing edge. Such reduction would be expected to lead to a higher lift-curve slope as found at GALCIT. (Laminar-flow airfoils tested at Langley Field did, in fact, exhibit generally higher lift-curve slopes than conventional profiles.)[46] The lesser friction of the longer laminar flow, plus the reduced adverse effects near the trailing edge, can also account for the smaller increase in drag with increasing lift that was the attractive feature of the Davis profile. Since this increase in drag was itself small for conventional profiles at the values of lift in question, only slight effects are called for here, and these are well within the scope of the influences cited.[47]

Published results from a B-24 model tested at Langley Field in the early 1940s reinforce the supposition of increased laminar flow on the Davis profile. In these tests the boundary layer on the upper surface was purposely disturbed by addition of a spanwise strip of roughness just behind the leading edge. This boundary-layer "trip" served to fix transition at that point and eliminate any extended laminar layer that might otherwise exist. Comparison with the undisturbed, smooth case showed a reduction in lift-curve slope of 9 percent, eliminating the high value for the Davis profile and implying that it was in fact due to laminar flow, though no note was taken of this implication at the time.[48] (No results were given for drag, but it presumably increased.) These and the earlier arguments leave little doubt that Davis had arrived at a version, though an inferior one, of the laminar-flow airfoils that were shortly to appear.[49]

Seen in retrospect, the Davis episode has many ironies. In the 1940s, the use of trips became common, both as a tool in boundary-layer study and to minimize scale effects in wind-tunnel tests. Left to its own devices, the transition point on an airfoil tends to move rearward with decreasing scale, altering the relative areas of laminar and turbulent friction and greatly complicating the extrapolation of model results back to full scale. To help with the problem, engineers adopted the trip to fix transition on the model at the location they thought it would have on the full-scale airplane. Looking back on his experience as head of the wind-tunnel crew at GALCIT, William R. Sears says, "We knew that the extrapolation to full scale, being largely empirical, had to be based on standard testing techniques . . . but we didn't yet have the idea of introducing also a standard transition location."[50] If they had, and comparison with a smooth model had at the same time not been included, no unusual behavior would likely have been noticed for the Davis profile, with very different consequences at Consolidated. Contrariwise, if a smooth model *had* been included, differences in performance could have suggested the principle of the laminar-flow airfoil also to the people at GALCIT, giving the Davis airfoil a permanent (if serendipitous) place in airfoil history.

Also ironically, the shape of the Davis profile was itself closely anticipated by the RAF 34 of 1926 (figure 2-6). Davis was aware of this resemblance during his association with Consolidated and mentioned it to Schairer. Though in use on a number of British aircraft, the RAF 34 had not been tested in a wind tunnel with sufficiently good flow to reveal its presumably superior performance under such conditions.[51]

As I have suggested, Davis evidently knew little if anything about theoretical fluid mechanics. Nowhere in the hand-lettered notes, patents, or statements to journalists did he so much as mention the boundary layer. Nonetheless, he apparently did anticipate the laminar-flow airfoil that Jacobs arrived at on theoretical grounds a few years later. Given the evidence, one can only assume this anticipation was fortuitous, that Davis was lucky where Jacobs was not. Unfortunately for his ability to exploit his luck, Davis did not understand what he had found. But then neither did Clark Millikan at GALCIT nor the trained engineers at Consolidated—even after thirty years and a vast amount of study, knowledge about airfoils in 1937 was uncertain at best.

Although laminar-flow airfoils showed up well in the wind tunnel, they were less successful in reducing drag in flight. Unfortunately, roughness and waviness disturb the laminar boundary layer and promote transition to turbulence even in a favorable pressure gradient. The surface of research models could be made smooth and fair, but both wind-tunnel tests and flight experience in the 1940s showed little significant laminar flow in the face of the constructional difficulties and inaccuracies of routinely manufactured wings. (Dust, insect debris, and other real-life circumstances had similar effect.) Tests at Langley Field, for example, revealed only small differences in drag between different sections—including a Davis profile—built to ordinary construction standards.[52] Although some airplanes are thought to have achieved drag reduction through special constructional and operational precautions, the rate of success was discouraging and led finally to their being given up for that purpose. Such experience and data support Bayless's view that an airfoil other than Davis's would have served Consolidated's aerodynamic purposes just as well.

Though "laminar-flow" airfoils fell short of their low-drag goal in practice, they did fill an important need. As we saw with the B-32, the increasing speed of military aircraft in the war raised difficult and adverse problems of compressibility. Research had suggested that these problems were related to the attainment of velocities equal to that of sound in the locally speeded-up flow about the airfoil, and it was assumed that the way to avoid or reduce such problems was to delay this equality to as high a flight speed as possible. In general, this entailed moving the maximum thickness rearward from its position on conventional airfoils, much as on the laminar-flow profiles. (This was not a feature of the Davis profile, hence its perceived unsuitability for the B-32 and, in part, its subsequent neglect.) As with laminar flow, the delay of sonic speed put constraints on the pressure distribution, and the inverse approach of Jacobs and his engineers again proved useful. Since the constraints were in the end much the same as for laminar flow (though for different physical reasons), certain of the laminar-flow profiles turned out to be useful high-speed airfoils. They were used as such on most high-performance aircraft in the late 1940s and early 1950s. As Theodore von Kármán remarked, "Eastman Jacobs, like Christopher Columbus, set out for China and discovered America."[53]

The considerable development in airfoils since 1945 requires only brief mention for our purposes. Research in the early 1950s showed that the assumption about delaying attainment of the speed of sound on the airfoil was erroneous and that it is both permissible and desirable to have sonic and supersonic velocities on the forward part of a properly shaped section. Ironically, this discovery has led back to airfoils closer to the NACA profiles of the mid-1930s, though developments are still continuing. The Davis profile may have been a better high-speed shape than it was thought at the time. In the problem of friction drag, new materials and construction methods have made smooth wings practical, and updated versions of Jacobs's laminar-flow airfoils are being used to reduce drag on low-speed aircraft, particularly sailplanes. Improved theoretical and experimental knowledge of the boundary layer has also contributed to the design of special kinds of airfoils for special purposes. Today most airfoils are tailor-made to meet specific design requirements. An indispensable tool in this ability is the modern electronic computer, which has greatly expanded the designer's capacity to make rapid and accurate airfoil calculations. In the 1940s a person using a mechanical desk calculator took three to four days to develop the ordinates of an airfoil designed to a prescribed pressure distribution; today an electronic computer does the job in a few minutes. Theoretical methods, however, are not yet perfect, especially regarding transition and the turbulent boundary layer. Wind-tunnel testing is still necessary to check and refine a design, and here scale effects still cause difficulties. The uncertainties that remain, however, are enormously less than those faced by the early designers who laid out their airfoils by eye. Airfoil design and the knowledge behind it have matured.[54]

Analysis and Discussion

Given the historical evidence, what can we say by way of analysis? Let us begin with the uncertainties that confronted the Consolidated engineers—and designers generally—in their choice of airfoils in the late 1930s. For summary we can divide these uncertainties into two categories: experimental uncertainties in wind-tunnel data and theoretical uncertainties in methods of calculation and analysis. On the experimental side, the imprecision inherent in any measurement of forces, wind speed, and the like caused little concern; by the mid-1930s techniques in this regard were reasonably accurate and, in cases of doubt, random (though not systematic) errors could be checked by repeated testing. Millikan, in fact, did this at GALCIT. Any question about the accuracy of the models could be eliminated, as Millikan also did, by careful geometrical measurement. Uncertainties from the fact that wing tests were carried out mostly with small models in a wind tunnel, however, were not so easily disposed of. Attempts to estimate the influence of scale effects and wind-tunnel turbulence were commonplace, but they involved dubious assumptions and few people felt any real confidence in them (and we now know that such confidence as did exist was unjustified). And, although engineers realized that the surface condition of models was important, they had little idea how to assess its effect on wind-tunnel data—and no suspicion at all of the implications that manufacturing irregularities would eventually have for applicability of the data in practice.

As to theory, a great deal of theoretical and experimental research had validated the separation of planform and section. Methods for calculating induced drag and other planform effects had also proved to be reliable. Such methods as had been suggested for calculating profile drag, however, were so uncertain as to be practically nonexistent. To get on with their job, designers had no choice but to rely on experiment. This move got them out of the frying pan of theoretical uncertainty but left them in the fire of the experimental uncertainties mentioned above.

Although the impression is necessarily subjective, the uncertainties in airfoil design in the second half of the 1930s loom as large as the certainties. What designers did *not* know appears as consequential in its own way as what they *did* know. In some matters, as with scale effects and tunnel turbulence, they were conscious of uncertainty, but in such attempts as they made to deal with it, they did not know as much as they thought they did. And with regard to the importance of manufacturing irregularities, they didn't even know what they didn't know.[55] Knowledge of airfoil design at the time of the Davis airfoil was more uneven than designers knew or perhaps cared to realize.

Besides these universal uncertainties, Consolidated engineers faced others peculiarly their own. Lack of explanation for the unusual results at GALCIT and Davis's failure to provide a theoretical basis for the shape of his airfoil were both unsettling. Engineers have always acted from empirical evidence when necessary, but modern ones feel uneasy without a theoretical explanation. The possible weight penalty from the Davis profile was also an unknown quantity until a detailed structural and mechanical design could be carried out. In deciding to go with the Davis profile in the face of these uncertainties, Fleet and Laddon depended critically on the GALCIT results—after all, the Davis profile *did* measure out best in the best experimental circumstances they could hope to obtain in the time available. When uncertainty creates significant doubt, engineers usually rely on tests as the ultimate authority. The technical uncertainties also gave nonengineering considerations more play at Consolidated than they might otherwise have had. As we saw, Fleet was impressed by Davis's equations, and he had the added motive of wanting something new to help sell his planes to the military. If the engineering picture had been less uncertain, these "irrational" factors presumably would have had less scope.

Having made the decision to try the Davis profile on the Model 31, Fleet and Laddon kept their options open as long as possible for the contemplated bomber and used the time to accumulate wind-tunnel evidence. As things turned out, however, they were forced to proceed on the B-24 before the Model 31 had been proved in flight. Again they conducted wind-tunnel tests as quickly as the models could be built, in time possibly to alter the detailed design, being carried out concurrently, if anything adverse were uncovered. The situation here is typical. It illustrates how uncertainties of knowledge, when combined with exigencies of procurement and design, can require that activities that might logically be done in series must actually go on in parallel, with irrevocable decisions postponed as long as possible. If, with philosopher Philip Rhinelander, we take wisdom to be "the art of making correct decisions on insufficient evidence, under conditions of uncertainty," then good design judgment clearly qualifies as a form of wisdom.[56]

The uncertainties connected with airfoil design in the 1930s all decreased over time, before as well as after the Davis episode. There is no need to spell out this decrease in detail; it is self-evident in our brief general history. It took place in various ways as knowledge increased with regard to wind-tunnel methods, nonviscous airfoil theory, boundary-layer phenomena, and electronic methods of computation. As a general matter, decrease of uncertainty with time is a natural accompaniment to the growth of engineering knowledge.

It would be a mistake, though, to see it as only a passive accompaniment. Decrease of uncertainty is not simply the by-product of increased knowledge caused by a desire for improved performance (as measured by speed, range, reliability, and the like in the case of airplanes). An impulse toward decreased uncertainty can itself *drive* the growth of knowledge, even at a fixed level of performance. Such action, though little evident in the Davis episode, shows up variously in aeronautics. Manufacturers of a new airplane, for example, must often guarantee a customer a given, already attained level of performance, with monetary penalties if the guarantees are not met within specified limits. This requirement puts a premium on certainty, which calls in turn for greater knowledge. The requirement to make the most reliable choice between alternative solutions to a given performance problem acts in a similar way. In both situations, the human desire of designers to feel comfortable in their decisions, on which their professional advancement may depend, reinforces the need—they want all the knowledge they can get. (This influence does appear here in the unease of the Consolidated engineers with the lack of theoretical basis for Davis's airfoil.) Though the impulse toward decreased uncertainty can occur simultaneously with a desire for increased performance, such need not be the case. Such impulse can provide a driving force for the growth of knowledge even when increased performance is not at issue. The same holds true outside aeronautics—as, for example, in the requirement of power station builders to provide performance guarantees to their utility company customers, with millions of dollars of penalties or bonuses hanging on the outcome.

The foregoing is not to deny, of course, that increased performance acts as a driving force too. Our story illustrates at least two ways in which this happens. The first, an example of what Edward Constant has called *presumptive anomaly*, is visible in Jacobs's invention of the laminar-flow airfoil. As defined by Constant, "Presumptive anomaly occurs in technology, not when the conventional system fails in any absolute or objective sense, but when assumptions derived from science indicate either that under some future conditions the conventional system will fail (or function badly) or that a radically different system will do a much better job."[57] Discussions of Constant's concept (including his own study on the origins of the turbojet engine) have focused, to my knowledge, exclusively on the first possibility. The present example qualifies for the second. Aeronautical engineers had perceived for some years preceding Jacobs's work that a "radically different" airfoil (that is, one that could maintain extensive laminar flow, if that could somehow be managed) would do "a much better job." This perception followed from theoretical estimates of friction drag derived from engineering science. Responding to this carrot of potential gain, as against the stick of expected (or actual) failure, Jacobs altered and expanded knowledge of airfoil design. Davis was apparently unaware of the presumptive anomaly. He was inspired instead by a vague notion that improvement might be possible by imitating in some way the Magnus effect.

The second way in which the move for increased performance had effect was through actual *functional failure*, which "can occur when a technology is subject to ever greater demands or . . . applied in new situations."[58] As we have seen, the conventional airfoils of the late 1930s failed to provide satisfactory aerodynamic characteristics at the higher speeds made possible during the war by other advances, most notably increases in engine power. This failure set off widespread aerodynamic research leading to a great expansion of knowledge for design of high-speed airfoils. Functional failure, presumptive anomaly, and the need to reduce uncertainty in design thus appear as three distinct, if often concurrent, driving forces (or sources) for the growth of engineering knowledge.

The historical events also show something of the means by which this growth occurs. Psychologist Donald Campbell has argued at length that all genuine increases in knowledge, from that embodied in genetic codes arrived at by biological adaptation to the theories of the modern theoretical physicist, take place by some form of a process of *blind variation and selective retention.* "Blind" is used here to connote that variations take place, not randomly, but only without complete or adequate guidance. According to Campbell, "In going beyond what is already known, one cannot but go blindly. If one can go wisely, this indicates already achieved wisdom of some general sort." But blindness, though unavoidable in any foray into the truly unknown, can vary in degree. As added by Karl Popper, "To the degree that past knowledge enters, . . . blindness is only relative: it begins where the past knowledge ends." Both the mechanisms for introducing variation and the process by which certain variations are selected for retention may be expected to differ widely depending on the type of knowledge.[59]

A variation-selection process for the growth of engineering knowledge will be set forth in detail in chapter 8. Here I want only to point out evidence for such a process in the efforts of Davis and Jacobs. For that purpose, we need to distinguish between the process at the level of a device—the airfoil—and at the level of underlying ideas. Devices depend upon ideas, of course, but in some respects the two have lives of their own.

At the level of the device, the knowledge required is the airfoil's shape. Davis's variations in shape were almost completely blind in any meaningful sense, virtually simple cut-and-try even though represented by sophisticated-looking equations. Jacobs chose his variations much less blindly on the basis of a rational theoretical concept and careful analysis of experimental experience; to the considerable extent that boundary-layer phenomena and the effects of surface roughness were uncertain or unknown, however, his variations still had a large element of blindness. Although the theoretical concept seemed sound in principle, there was no assurance at the time that any of the resulting airfoils would be worth retaining. Both Davis's and the better of Jacobs's airfoils were selected initially for retention on the basis of wind-tunnel tests and the criterion universal to technology: does it (in some practical sense) work? Davis's airfoil was soon discarded when Jacobs's sections appeared and when it was judged to no longer work as airplane speeds went up. Jacobs's sections died out more slowly for their original laminar-flow purpose when they failed to accomplish that purpose on actual airplanes in flight; they found an unintended niche of retention for a period, however, as high-speed airfoils.[60]

At the level of ideas, the knowledge is that of how to arrive at the shape of an airfoil. Davis's variations from previous thinking were ad hoc and geometrical and almost completely blind as to fluid mechanics. Only he knew what they were, and even to him they provided no general insights or rationalizations. They had no influence on the evolution of airfoil design. Jacobs's idea of designing to a pressure distribution, though a distinct variation from previous ideas, grew consciously and deliberately from accepted knowledge of fluid mechanics. When publicized in NACA reports, it provided the engineering community with a new way of thinking about airfoils and a guide for development programs aimed at specific purposes. It worked in general and was retained as a permanent part of airfoil design, even when the original laminar-flow airfoils, for which it was conceived, disappeared from the scene. Perhaps if Davis had operated less blindly—if he had at least been aware of the drag estimates behind the presumptive anomaly— he might have realized what he had found and contributed in a more substantial way to the growth of ideas. One should not conclude from a single study, however, that survival value necessarily goes up as the degree of blindness of variation in either devices or ideas goes down; I doubt if it is universally true.[61]

If growth of engineering knowledge is an evolutionary variation-selection process, technology's failures require historical examination along with its successes. Although examination of failure by scholars has been limited, the need seems accepted in principle.[62] It may be especially pertinent for epistemological study—failures are an inherent part of any variation-selection process and may be expected to be at least as numerous as successes. The Davis airfoil, however, does not fall unambiguously into either category. It succeeded relative to other airfoils in the GALCIT tunnel and was therefore adopted in what turned out to be a successful bomber. If the level of airplane performance had not been increased, it would probably have taken a useful place in the catalog of airfoils; in slightly different circumstances, it might even have led to the principle of the laminar-flow profile. At the same time, the Davis airfoil was probably no better than other contemporary airfoils would have been on the operational B-24, whose successful performance can be explained on other grounds. The aeronautical-engineering community did not see the airfoil as a particular success at the time (or later), and only Consolidated put it to use. It prompted no theoretical insights and, in fact, played no role in the evolution of airfoil design. The history of technology is undoubtedly full of such devices— devices that work well enough not to be counted as failures but not so well as to be regarded as particularly successful. Designing and building them help engineers to improve their professional competence and attain a broader frame of reference for future design judgments, that is, to add to their store of professional wisdom. Such equivocal devices too play a role in the growth of engineering knowledge.

Finally, we can observe—somewhat roughly—a progression of development in airfoil technology, which I take to comprise both explicit knowledge and methods for design. The first decades of the century saw the technology in what can be called its infancy. No realistically useful theory existed, and empirical knowledge was meager and uncodified. Design was almost exclusively by simple cut-and-try; that is, by sketching an airfoil and trying it out. No other way was possible. Today, airfoil technology has reached maturity. Using relatively complete (though not yet finished) theories, supported by sophisticated experimental techniques and accurate semitheoretical correlations of data, engineers design airfoils to specific requirements with a minimum of uncertainty. Little cut-and-try is needed by a skilled professional. Between the phases of infancy and maturity lay a half-century of growth. In this period theory provided qualitative guidance and increasing partial results, but wind-tunnel data were vital. Design was an uncertain and changing combination of theoretical thinking and calculation and cut-and-try empiricism. The most rapid changes took place precisely in the late 1930s and early 1940s, with Jacobs's new concept of airfoil design and generation of the knowledge to exploit it. At the same time, however, Davis could still come forward with what was essentially a cut-and-try airfoil and have it adopted by a major aircraft company. Perhaps we could call this decade the adolescence of airfoil technology, when rational behavior was on the increase but offbeat things could still occur. Whether or not we push the metaphor that far, we can at least see a progression of development through phases of infancy, growth, and maturity, with a characteristic relationship of knowledge and design in each phase. One wonders whether such phases and relationships appear in other technologies.

The Davis airfoil was not of itself very important for the history of aeronautics. It played only a questionable role in its one major application, and its contribution to airfoil technology was essentially nil. It constitutes no more than a footnote in the story of mechanical flight. For purposes of scholarship, however, its insignificance for aeronautics is precisely the point—a great deal goes on in engineering that is "not very important." As in all human endeavor, engineers spend a great deal of time doing things that don't come to much one way or the other. The Davis airfoil, though dramatic in its circumstances, was typical in this regard. We must give attention to such "unimportant" activities if we want to view engineering knowledge in all its richness and complexity.

Establishment of Design Requirements: Flying-Quality Specifications for American Aircraft, 1918–1943

Shortly after World War I, 1st Lt. Leigh Wade, test pilot for the U.S. Army Air Service at its engineering center at McCook Field, Dayton, Ohio, reported the Morane Saulnier training monoplane he had been testing as "tail heavy longitudinally." He went on to add that "the tail heaviness is not tiresome because of the lightness of the controls."[1] A designer, reading these words at the time, might have had some idea of what caused the tail heaviness; he would have been hard pressed to know, however, how the mitigating "lightness of controls" had been achieved—or, indeed, just what the phrase meant. A quarter century later, the situation was completely changed: in 1942–43, the military services were able to publish for the first time specifications to which their aircraft must be designed to have flying qualities acceptable to pilots. The learning process over the intervening years provides an example of how an engineering community translates an ill-defined problem, containing in this case a large subjective human element, into an objective, well-defined problem for the designer.

In the preceding chapter, the problem of concern was already well defined in a technical sense. Long before the Davis wing, engineers had learned how to specify what was required of an airfoil in a given application. The problem for the designer was how best to shape the airfoil to obtain the specified characteristics; the knowledge we examined was the knowledge required to that well-defined end. The problem in this chapter, by contrast, was initially *ill* defined—the engineering community did not know at the beginning of our period what flying qualities were needed by pilots or how they could be specified.[2] We shall examine how that community learned to identify pilots' needs and translate them into criteria specifiable in terms appropriate to the hardware; that is, so that the flying-qualities problem would be well defined in the same sense as the airfoil problem. Some learning process of this sort is necessary before any design task can be specified for everyday engineering.

Because of its subjective human element, the present problem is unique in this book, though not in engineering. Numerous mechanical devices require an operator, and vehicles—airplanes, automobiles, bicycles, and so forth—are prominent among them. Designers of such devices often have to consider the subjective as well as physical interaction of the operator with the machine. Such consideration presupposes knowledge of pertinent human opinion and of how to translate that opinion into technical criteria and specifications. Though a concern of design professionals for some time, this kind of engineering knowledge and activity has received little attention from historians of technology.

In a broad, underlying way, the history of flying-quality specifications is the history of an idea. The notion that specifications could usefully be written for something as subjectively perceived as flying qualities had itself to be realized intellectually and verified in the real world. It was not at all an obvious or obviously useful idea at the outset. Our story, though preoccupied on the surface with the airplane as a device, bears out the contention of chapter 1 that the history of engineering is, in its fundamentals, an aspect of intellectual history.

To the extent that the needs of pilots are external to the airplane, the story of flying qualities also involves one of the contextual factors that govern design. It thus provides an explicit instance of Staudenmaier's "design-ambient tension" (see chapter 1). Since flying-quality specifications are implemented mostly via detailed design of the aerodynamic surfaces and control systems, it likewise exemplifies the relatively unusual situation of contextual influence applying directly at the lower levels of design hierarchy.

The present story emphasizes, better than any other in the book, how the generation of engineering knowledge is characteristically a community activity. While a number of people play visible roles, no individual or individuals dominate our account; the protagonist must be seen as the entire flying-quality community. This community consisted, however, of at least four subcommunities having to do individually with design, engineering research, instrument development, and test flying. These subcommunities overlapped intimately, and the generation of knowledge took place—indeed, had to take place—simultaneously and interactively in all of them. The evidence thus illustrates the "community structure of technological practice" so well described by Edward Constant.[3]

This case is the longest, most detailed, and most technically demanding in the book. The length and detail are unavoidable if we want to see how an engineering community tackles a broad and ill-defined intellectual and practical problem, that is, to capture the real workings of the community. I have also attempted, more deliberately than in other chapters, to carry out the task mentioned in chapter 1 of showing how "the intellectual process confronted the participants as it came at them." To this end, I have taken special pains to include mistakes and differences of opinion in essential matters (though at the same time leaving out numerous instances of uncertain or faulty knowledge to keep the story within bounds). As emphasized in chapter 1, correction of mistakes and misunderstandings is an essential part of the learning process, in engineering as elsewhere.

Preliminaries

To discuss flying qualities in detail, we need some elementary terminology and ideas. Though some of these originated or became clear in the history to be described, our task will be easier if we know them at the outset.

Flying qualities comprise those qualities or characteristics of an aircraft that govern the ease and precision with which a pilot is able to perform the task of controlling the vehicle. Flying qualities are thus a property of the aircraft, though their identification depends on the perceptions of the pilot.[4] The control task may vary widely, all the way from maintaining constant speed and altitude over a long period of time by a bomber or airline pilot, to tracking a rapidly maneuvering target by a fighter pilot. The effort required by these tasks gives the pilot a feeling of confidence or apprehension about the airplane. Aerodynamically speaking, control of the vehicle is provided by angularly movable portions of the aircraft's surface actuated by the pilot—in most airplanes, the *elevator* on the horizontal tail, the *rudder* on the vertical tail, and the *ailerons* at the trailing edge of the wing near the tips. Deflection of these control surfaces alters the aerodynamic forces on the tail and wing. The primary function of aerodynamic control, as suggested by the aforementioned piloting tasks, is to fix or change an equilibrium condition of flight (speed, course, altitude, angle of climb, and so forth) or produce accelerated or nonequilibrium maneuvers.

A pilot's perception of flying qualities stems in large part from the forces and movements required by the cockpit devices used to move the control surfaces. (Visual cues from flight instruments and "seat-of-the-pants" reactions to acceleration also provide indications of the airplane's response, but these will not concern us here.) The usual cockpit control devices are a manual *stick* (or *wheel*) for the elevator and ailerons—moved fore and aft for the former and side to side (or rotationally) for the latter—and foot *pedals* (or *rudder bar* in older times) for the rudder. The pilot assesses the flying qualities largely via these controls. Suppose, for example, that an inordinate push on the stick is required to provide the downward elevator deflection needed for a nose-down maneuver; the pilot may then report the airplane as "stiff" to fly in this respect. British flier and writer Beryl Markham put the matter more lyrically in 1935 concerning her Avro Avian: "To me she is alive and to me she speaks. I feel through the soles of my feet on the rudder-bar the willing strain and flex of her muscles . . . . My right hand rests upon the stick in easy communication with the will and the way of the plane."[5]

Flying qualities depend, however, on more than the action of the control surfaces; they depend also on the airplane's *inherent stability*. Inherent stability has to do with the ability of an airplane, by aerodynamic action alone and *without any corrective response by the pilot*, to return to an equilibrium flight condition after a transitory disturbance, as might arise, for example, from a gust. An inherently stable airplane returns *of itself* to its original condition; an unstable airplane departs farther from it. It must not be assumed, however, that inherent stability is essential to a piloted vehicle—a normal bicycle, for example, though inherently unstable at low speeds, is ridden with ease by millions of riders who supply stability by corrective control action without being conscious they are doing so. Such control-achieved stability is likewise possible with an airplane, whether by means of a human pilot or an automatic mechanism. (This possibility is the reason for the qualifier "inherent" when purely aerodynamic stability is meant. Since the inherent situation is of almost exclusive concern here, I will usually omit the qualifier. When control-achieved stability is intended, I will say so explicitly.)[6]

Inherent stability is important to flying qualities because the stable airplane resists initiation of a change in flight condition to more or less the same degree as it does a transitory disturbance. The unstable airplane, by contrast, responds readily, even perhaps excessively, to movement of the controls. Stability and control thus work at cross purposes, and the ease and precision with which a pilot can control an airplane depend as much on its stability characteristics as on the action of the aerodynamic control surfaces. As they relate to flying qualities, stability and control are different sides of the same coin. (This explanation oversimplifies a complex situation, but it has been conventional wisdom for many years and is sufficiently correct for present purposes.)

Stability and control, and flying qualities, however, are not synonymous. Not all stability and control characteristics are important to piloting tasks, and flying qualities involve subjective pilot reactions that are outside the objective concerns of stability and control. Control and especially stability problems can, moreover, be posed in engineering terms that have little if any connection with pilots' perceptions. What airplane characteristics are important and how they can be specified effectively for the design engineer can only be decided on the basis of pilot opinion. Engineers by themselves could have solved the problems of stability and control, at least in principle; flying qualities could not have been settled without involvement of pilots.

Control of an airplane is complicated because, unlike sea, road, and rail transport, flight takes place in full three-dimensional space. Fortunately, analysis is simplified by the mirror (or lateral) symmetry of the usual airplane. By virtue of this symmetry, the vehicle's motions can be divided into two classes, longitudinal and lateral, that can be treated independently. *Longitudinal motions* take place in the plane of symmetry of the airplane; they include vertical and pitching motions such as can be brought about by deflection of the elevator. *Lateral motions* act to displace the plane of symmetry; they consist of sideways and rolling and yawing motions, such as those caused by movement of the ailerons and rudder. The various lateral motions are interrelated in a complex way and are more difficult to understand and discuss than longitudinal ones. Partly for this reason, and partly to keep the story within bounds, I shall restrict my account mostly to longitudinal flying qualities, that is, to problems of longitudinal stability and control. The systematization of lateral qualities went on simultaneously, indeed it had to, since operation of an airplane involves both kinds of motions more or less equally. The types of knowledge and methods of development, however, were much the same. Nothing is lost for our purpose by restriction to longitudinal problems.

For the airplane designer, the problem of longitudinal flying qualities, as for flying qualities generally, breaks down into two parts: (1) What stability and control characteristics should an airplane have to possess desirable flying qualities? and (2) How is an airplane to be arranged and proportioned to obtain those characteristics? As indicated in the introductory paragraphs, this chapter deals only with the first question; it is this question that had to be answered to transform flying-quality design from an ill-defined to a well-defined problem. Unfortunately, restricting discussion in this way may leave the reader wondering how some of the characteristics to be described were practically achieved. The history of this second question, however, of how to design the wing, tail, and control surfaces and how to dispose them relative to one another and to the airplane's center of gravity to achieve the desired end, is another and even larger story. It could profitably constitute a book, though the matters involved would be nearer to the historiographically and methodologically familiar sort. As pointed out earlier, the first question involves elements that have not frequently been examined.

The State of Knowledge circa 1918

By the end of the 1910s, when the story of flying qualities begins per se, the aeronautical community more or less recognized that something needed to be done concerning the relationship between pilot and machine. What that should be, however, was not obvious, and the tools for the problem were still incomplete. How this situation arose, in particular the development of ideas in the amorphous period preceding, has yet to be examined completely. Some background is essential, however, to understand the narrative that follows. As in all such engineering circumstances, growth of knowledge could begin in earnest only when the problem had become sufficiently recognized.[7]

What appears to have been the majority view in 1918 concerning the tension between stability and control was expressed in that year's *Annual Report* of the National Advisory Committee for Aeronautics (NACA). This government agency had been set up by Congress three years earlier "to supervise and direct the scientific study of the problems of flight, with a view to their practical solution," and was still orienting itself among the problems. As part of a general review, the committee wrote as follows under the heading "Stability and Control:"

> We are confronted with one of those situations so frequently encountered in scientific and technical work, where a choice must be accepted on some middle ground between wide extremes, and where the attempt to secure some desirable quality in high degree may lead to a limitation of desirable qualities in other directions. . . . If stability is carried to an extreme, then mobility and quickness of maneuvering are reduced and control in the sense of ready response is lacking. . . . For machines of the fighting type, where mobility is of the highest importance, this would be a serious shortcoming, and hence such machines can not be given too much stability in the ordinary sense of the term. On the other hand, for heavy machines of the bombing type, where mobility of evolution is not so vitally important, the margin of stability may be greater.

This view—reasonable in retrospect and the view that would ultimately prevail—took it for granted that some amount of inherent stability was necessary; the question was how much. As we shall see, it was not a view corroborated by all aircraft at the time or shared by all pilots. Of the twelve members of the committee, most knew about airplanes theoretically and at second hand; only two or three representatives from the military are likely to have had piloting experience.[8]

The NACA's view, however, was consistent with that appearing in the technical literature. In a magazine article in 1917, Harlan Fowler had written, "The view now generally held is that it is necessary that the machine should be stable, but not *too* stable." John Rathbun, in an elementary aeronautical textbook of 1918, subscribed to the same philosophy; he also warned against excessive stability, but for a reason not mentioned by the NACA: "An aeroplane can be too stable and therefore difficult to steer and control in gusts because of its tendency toward changing its attitude with every gust in order to restore its equilibrium."[9]

The situation of 1918 was the legacy of two contending schools of thought. The contention had been noted in 1910 by Charles Turner in an early British book on "aerial navigation":

> The question of how to obtain . . . equilibrium has developed into a controversy dividing aviators into two schools. One school holds that equilibrium can be made automatic [i.e., inherent] to a very large degree; the other, known as the American school, following the methods of the Brothers Wright, claims that equilibrium is a matter for the skill of the aviator, who, with practice, acquires perfect control of his machine.[10]

The history of this controversy has been perceptively traced (more in its practical than its ideational aspects) by aeronautical historian Charles H. Gibbs-Smith. Gibbs-Smith calls the adherents of the two schools the *chauffeurs* and the *airmen*.[11]

The chauffeur philosophy conceived that the airplane should be a highly stable vehicle, a kind of winged automobile, that simply required steering by the pilot. This philosophy, which dominated in Europe in the first decade of the 1900s, stemmed from a century of off-and-on thinking and experimentation, mostly with hand-launched models, by a number of Europeans. Most influential of these were Sir George Cayley (employing gliders in England from 1799 to 1809 and 1843 to 1852), Alphonse Pénaud (with a rubber-powered, propeller-driven model in France in the 1870s), and Frederick W. Lanchester (with model gliders, again in England, in the 1890s). Working as they did almost entirely with unpiloted models with no possibility of control, these men "rapidly discovered that they had to have an inherently stable system for success." (They also learned how to achieve such stability, though that does not concern us here.) Although this approach was initially fruitful, it had the effect of focusing European thinking on stability, largely to the exclusion of control. The resulting chauffeur mentality proved counterproductive for later attempts at piloted flight, where the problems of controlled maneuvering were crucial.[12]

The airmen, by contrast, thought primarily in terms of control. Realizing that active control under the myriad of possible conditions would be essential for safe human flight, they sought to ride the air in controlled, piloted gliders. The preeminent airmen were Otto Lilienthal in Germany in the 1890s (though with little effect on the European chauffeur tradition) and, as Turner emphasized in the above quotation, Wilbur and Orville Wright in America. Using the mastery of control obtained with their gliders from 1900 to 1902, the Wrights achieved the first powered, piloted flight in 1903 in an airplane that was, like almost all their gliders, inherently unstable both longitudinally and laterally. Disagreement exists about whether this instability was deliberate.[13] Despite the instability, the Wrights as pilots learned to control their machine to maintain an effectively stable system. Although the inherent instability of the Wright machines may have been extreme, the airmen's insistence on control was essential for the attainment and improvement of human flight. As pointed out by Gibbs-Smith, "Until the pilot has experienced and can anticipate and control the behaviour of his machine in the air, he cannot decide what is necessary (or possible) to leave to automatic mechanisms and built-in qualities."[14]

When word of the Americans' achievement reached Europe, the first reaction was to modify the Wrights' glider configuration—specifically, by Ferdinand Ferber in France—to provide inherent longitudinal stability. This was done, however, with little appreciation of the importance of control. Only after Wilbur Wright's masterful demonstration of controlled flight in France in 1908—not only to achieve effective stability but to initiate maneuvers—did the European philosophy begin to change. Turner's statement of 1910 shows that the ideological controversy persisted, but the long, hard task of understanding and marrying stability and control was under way.[15]

How rationally and rapidly the marriage proceeded is difficult to know. Different writers take different views, pointing to either instability or stability in the early designs.[16] There now seems little doubt, however, thanks to recent flight experience with restored aircraft and accurate replicas built for motion pictures, that several successful airplanes at the beginning of the 1910s were highly unstable longitudinally as well as laterally. The fact that the earliest automatic pilots (1912) were intended as aircraft stabilizers also suggests a lack of inherent stability.[17] Whatever the situation, the practical problems must have seemed confusing at the time, when designers and fliers were feeling their way dangerously into the unknown.

To the extent that early designers thought analytically about longitudinal stability, it was probably in terms of simple physical ideas. Such

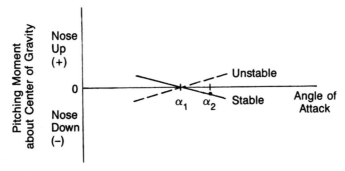

Fig. 3-1. Variation of pitching moment with angle of attack in straight steady flight for stable and unstable airplanes.

ideas had been put forth by a number of people, perhaps most directly by Lanchester in his *Aerodynamics* of 1907, in which he described his earlier experiments with hand-launched model gliders. In present-day language, these ideas can be stated as follows: Suppose an airplane to be in straight equilibrium flight at angle of attack α_1. (The *angle of attack* is the angle between some arbitrary fore-and-aft reference axis in the airplane and the direction of flight.) For this to be so—in particular, for there to be no tendency for the vehicle to rotate in pitch—the net, or resultant, of all the aerodynamic forces must pass through the center of gravity, or "c.g.," of the vehicle. Put another way, the pitching moment about the c.g. (that is, the resultant force times the displacement of its line of action from the center of gravity) must be zero. Suppose now that the angle of attack is by some external cause momentarily increased by a small amount from α_1 to α_2 (figure 3-1), the elevator angle remaining unchanged. For the airplane to be stable, its design must be such that the resultant force moves aft of the c.g., producing a negative or nose-down moment, thus tending to restore the condition of equilibrium (solid line). Were the resultant force to move ahead of the c.g., the moment would be positive or nose-up, the angle of attack would tend to increase further, and the airplane would be unstable (dashed line). The criterion for longitudinal stability is thus that the slope of the curve of pitching moment versus angle of attack at the equilibrium condition must be negative. This requirement on the *moment-curve slope* appears as the basic criterion of stability in elementary textbooks of aeronautics to the present day. Whether the early designers attempted to use such ideas is difficult to know. Though how best to satisfy the pitching-moment criterion was far from obvious, the criterion itself was the kind of thing many analytically minded people, given the recent demonstration that mechanical flight was possible, might have figured out for themselves.[18]

To trace the history further, we need to know more about stability. Consider, for concreteness, the transitory disturbance provided by quick movement of a control surface followed by return to its original position. Suppose, in particular, that an airplane is in steady level flight at some given speed. The pilot now deflects the elevator just long enough to put the machine into a shallow dive at a slightly higher speed and then returns the surface as rapidly as possible to its original position. In the main, two *initial tendencies* in the airplane's response are possible. If the airplane, left to itself, tends to return toward its original condition of level flight at the given speed, it is said to be *statically stable*. If it tends toward a steeper dive at even higher speed, it is *statically unstable*. (The airplane could also remain permanently in the shallow dive at a constant speed, in which case it would be neutrally stable. This situation is rare.) As indicated in its derivation, the requirement on moment-curve slope pertains to such initial tendencies following a disturbance; it is in fact a criterion for *static* stability. Though "static" (or "statically") suggests fixity rather than motion, it must be added to distinguish the foregoing from still another aspect of the problem.

The additional aspect arises because even the statically stable airplane, although it tends initially toward the original condition of flight, might not in the end return to that state. In most cases, it will start to oscillate about the original condition, similar to the way in which a deflected pendulum oscillates about the vertical. If these oscillations die out (or "damp"), as with the pendulum, with the result that the vehicle returns to its original state, it is said to be *dynamically stable*. If the oscillations build up continuously, resulting usually in a violent stall or dive (a situation that has no analogy with the simple pendulum), it is *dynamically unstable*.

The ideas here, both static and dynamic, have been explained in terms of a specific disturbance causing movement in the vertical plane; similar statements can be made about other kinds of disturbance and other types of motion (e.g., banking and turning). Whatever the details, the topic of *static stability* has to do with the tendency of an airplane to return toward or diverge from an equilibrium flight condition immediately following a disturbance. *Dynamic stability* concerns what actually happens in the resulting motion. The important point to remember is that an airplane that is statically stable can be dynamically either stable or unstable. A machine that is statically unstable, however, cannot be dynamically stable.[19] In other words, static stability is a necessary but not sufficient condition for dynamic stability.

Analysis of static stability is reasonably simple. Not so for dynamic stability, which calls for sophisticated mathematical tools. Fortunately for early researchers in aeronautics, such tools lay at hand in the discipline of rigid-body dynamics, developed by scientists and applied mathematicians in the nineteenth century. Following partial efforts by a number of people (again most notably Lanchester, who relied on ingenious methods of his own), the British applied mathematician George H. Bryan set the problem on a rigorous foundation. In his book *Stability in Aviation*, published in 1911, Bryan established mathematically the separation of an airplane's motion into longitudinal and lateral components as described earlier. He then obtained, among other things, the important result that the oscillatory longitudinal motion following a disturbance comprises two modes: (1) a heavily damped oscillation of short period, typically one to two seconds (the *period* is the time for one oscillation), and (2) a weakly damped or undamped oscillation of long period, typically twenty to forty seconds. Bryan reasoned that the first of these, being short and heavily damped, would disappear so quickly as to be of no practical importance. He—and most investigators who followed his lead—therefore dismissed it out of hand and analyzed in detail only the long-period mode (what Lanchester, its discoverer, had termed the *phugoid oscillation*). As we shall see, this omission would prove misleading with respect to flying qualities.[20]

Prompted by Bryan's study and by the general uncertainty about the proper levels of stability, the new community of aeronautical research workers turned to experiments on scale models. Two groups were important here. The first, under the direction of Leonard Bairstow and B. Melvill Jones, tested a model of a Blériot monoplane in 1912–13 in the new 4-foot wind tunnel at Britain's National Physical Laboratory; the second, led by Jerome C. Hunsaker at Massachusetts Institute of Technology in 1915–16, employed models of a Curtiss JN-2 biplane and a specially designed variant thereof. Hunsaker had visited Bairstow and Jones in England in 1913, and his tunnel duplicated the one at the NPL. At both MIT and NPL, the models were tested at varied fixed orientations and in restrained oscillations. Both groups used the test results to help evaluate the aerodynamic constants in Bryan's equations and then used the equations to calculate the dynamic motion of the vehicles. The NPL group found the calculated long-period mode of the Blériot to be stable at the one speed analyzed. The MIT investigators deduced that both their designs were stable in this mode in the upper part of their speed range; both became unstable, however, as speed was reduced. Only the British gave attention to the short-period mode, and that only briefly.[21]

Besides these results and others dealing with lateral motion, the investigations provided a great deal of valuable understanding of the mechanics and mathematics of stability generally. They became an important source of instruction to the aeronautical community, or at least to that part of it that could understand the advanced ideas involved. The study of dynamic stability, however, provided no simple criterion of stability for the designer. (The negative moment-curve slope appeared as a necessary but not sufficient mathematical condition for dynamic stability. Whether a design was dynamically stable could be found only from a complex analysis based on this and other aerodynamic properties.) Since the work did not involve movable control surfaces, the results of both the NPL and MIT studies said little directly about control.[22]

Though studies of this kind could reveal the level of stability of a given design, they could not indicate what level was desirable. Without coordinated flight tests and meaningful pilot opinion, the investigators could only recommend what they thought sensible in light of their theoretical reasoning. In their report, Bairstow and Jones wrote, "While some may still be found to dissent from this view, it will at least be conceded that instability is clearly a disadvantage; that a machine flying in still air should tend continually to depart from the condition of normal flight, and require the constant attention of the pilot to bring it back to the correct attitude, will hardly be urged as a characteristic in general desirable." Hunsaker thought it "conservative to conclude that airplanes should not be unstable" and that "it is likely that the most satisfactory aeroplane will be only slightly stable and that this aeroplane will in any possible attitude be easily controlled by the pilot."[23]

These opinions helped foster the view expressed by Fowler, Rathbun, and the NACA in 1917–18. Fowler's earlier quoted statement that "the view now generally held is that . . . the machine should be stable, but not *too* stable" was in fact word for word (without attribution) from Bairstow and Jones's report. And the official opinion from the NACA's *Annual Report* was preceded by acknowledgment that understanding of stability and control was "due largely to the splendid theoretical and experimental investigations initiated by British scientists, and to which certain workers in the same field in the United States may have contributed something." The NPL and MIT studies, tentative as they were, were by far the best available, and their influence spread accordingly.[24]

Even in 1918, however, and despite intensive development in World War I, the practical situation did not always correspond to the "view now generally held." As observed by Courtland Perkins, a later engineering authority on stability and control, "The perplexing fact, for [the] researchers, was that when airplanes encountered the instabilities directly forecast by the mathematicians, the airplanes got along quite well and flew anyway." As with the machines of the early 1910s mentioned earlier, flying of restorations and replicas shows that a number of successful aircraft from the war—the Sopwith Camel in Britain and the Curtiss JN-4 and Thomas-Morse S-4C in the United States—were longitudinally (and laterally) unstable. The words of Brian Lecomber, a pilot who tested a restored Camel in 1978, are especially illuminating:

> The Camel is mildly unstable in pitch and considerably unstable in yaw, and both elevator and rudder are extremely light and sensitive, with very little feedback pressure. The ailerons, on the other hand, are in direct and quite awe-inspiring contrast. . . . All this results in an aircraft which, initially, feels horribly wrong to a present-day aeroplane driver. . . . Once the initial shock has worn off . . . [however,] the machine becomes not so much difficult to fly as merely different.

The instability of the JN-4, in fact, had been established experimentally as early as 1919, and a prominent design engineer recalled in 1925 that "many unstable airplanes [he mentioned in particular the Camel and the JN-4] were not only much used but were very popular as well." Clearly, various solutions of the stability-control problem are possible and even acceptable. The question to be answered was not what is correct, but what is best.[25]

Of course, the instability of actual aircraft may have come, not from designers' intentions, but from lack of knowledge about how to achieve stable designs—conclusive evidence is hard to find and understanding of aerodynamics was still elementary. It may also have come, however, from remaining uncertainty about what was in fact desirable. Even Hunsaker, writing in 1916, moderated his recommendations for stability with the observation that "it is well known that the French monoplane pilots demanded at one time a neutral aeroplane with no stability whatever against pitching, on the ground that 'stable' aeroplanes were too violent in their motion in gusty air." He also concluded near the end of his report that "safety in flight may well depend more upon ease of control than upon stability. The almost universal prejudice among accomplished fliers against so-called 'stable aeroplanes' appears to have a rational foundation."[26] (Why he chose to put stable in quotation marks is not clear.) Apparently, the "generally held" view of stability in relation to control circa 1918, like the situation regarding airfoils in the 1930s in the preceding chapter, had its full share of uncertainties.

As Hunsaker's statements implied, the opinion of pilots was crucial. Bairstow and Jones in their 1913 report had said explicitly that "the knowledge [of the desirable degree of stability] can only be obtained from the experience of pilots in flight." By 1918, military test pilots, a profession brought into being by the war, were recording such experience routinely, though hardly in a form useful to designers. The words of Capt. Rudolph W. ("Shorty") Schroeder, test pilot at McCook Field, are typical. Reporting on August 15, 1918, on "Trials of Flying Qualities" of the Packard–Le Père LUSAC-II biplane fighter, Schroeder wrote as follows under the conventional headings:

EFFICIENCY OF CONTROLS

The elevators, ailerons, and rudder respond very readily in climbing, diving, banking, and in horizontal flight.

PRESSURE OF THE CONTROLS

The pressure exerted on the controls for the movement of the elevator is normal. . . .

REMARKS ON STABILITY

The longitudinal, lateral, and directional stability of this machine is good.

Such subjective, qualitative remarks did little for the needs of the designer. The last, in particular, bore no relation to the engineering criterion for static longitudinal stability.[27]

This generally accepted criterion remained the moment-curve slope described earlier. D. R. Husted of the Curtiss Engineering Laboratories, writing in an aeronautical magazine in 1920, advised designers that in his company's experience "it is usually safe enough to assume that when the model test gives a pitching moment-angle of incidence [i.e., angle of attack] curve having a negative slope of the proper magnitude, the actual machine will have sufficient pitching stability."[28] Husted, who thus subscribed to the view that some level of stability was required, did not specify what the "proper magnitude" should be. The moment-curve slope, in any event, affords no correlation with the pilot's experience in flight. It provides, as suggested by Husted's words, a useful criterion in terms of a model test, since pitching moment as a function of angle of attack is readily measured in a wind tunnel. In flight, however, neither moment nor angle is easily measured or sensed by the pilot, and the criterion loses its utility. What was missing in 1918 was a stability criterion in terms of control forces and movements felt by the pilot and measurable in flight.

Other things were missing as well. Little research consideration, theoretical or experimental, had been given to control, and no analytical criterion whatsoever for control was available. And even if suitable criteria for stability and control had been known, means to assess them in flight were lacking. As stated by a pair of modern engineers, "There were no data recording systems to show the designer what the airplane characteristics actually were in comparison to what he thought they might be"—or, one might add, to what the test pilot said they were.[29]

In summary, by the end of the 1910s the aeronautical community realized, at least implicitly, that flying qualities posed a problem (witness existence of the term in the title of Captain Schroeder's report). The controversy between the chauffeurs and airmen had largely subsided. Designers and researchers agreed, on the whole, that some sort of compromise between stability and control was necessary but had little idea just what. Doubts lingered, however, that inherent stability was desirable for certain types of aircraft, and we now know—and some people apparently realized at the time—that some successful airplanes were in fact unstable. In the attempt to handle the problem, ways for analyzing stability (but not control) had been developed, but they provided no connection with the realities of piloting experience. As a result, though the need for such experience was recognized, no one knew how to quantify it and express it in terms useful to designers. As stated by Perkins, "Stability had to be better related to control and ultimately to the [pilot] who was to fly the system."[30] Until these research tasks could be accomplished, the design problem remained perplexingly ill defined.

Developing Understanding and Capability, 1918–1936

Seen in retrospect, the quarter-century-long process by which the problem became well defined divides into two stages. The first stage, from 1918 to 1936, established the basic analytical understanding and practical capability needed for the job. Both capacities were prerequisite to the idea that specification of flying qualities might be desirable and feasible. The second stage, from 1936 to 1943, put the understanding and capability to work to show that such desirability and feasibility were practical, that is, to establish concrete, workable specifications. The stages were not recognized as such at the time, but the division is evident historically. As we shall see, it is marked by the appearance of the first suggested flying-quality specifications in 1936.

The narrative in this and the next section deals only with events in the United States. A more or less parallel development took place in Britain; the British, in fact, led by a year or two at the outset. The learning process was essentially similar in both countries, however, and it is the process that is of primary concern here. I have not examined the situation on the European continent, but similar events presumably took place there; each national engineering community, no matter how much information it may acquire from outside, needs a certain amount of hands-on learning in matters of this kind. In the United States, the work in both stages came predominantly from the NACA's newly established laboratory at Langley Field, Virginia.[31]

The years after the war saw a vigorous program of flight experiments at the Langley laboratory. These studies, which appeared in

numerous NACA publications from 1920 to 1923, were the work mainly of three men: Edward P. Warner, Frederick H. Norton, and Edmund
T. Allen. Warner, who commuted from Hunsaker's group of engineers
at MIT to work part-time at Langley, was the first to have the (misleading) title of chief physicist at Langley in the years 1919 to 1920. Norton,
also an MIT product who had been hired at Langley in 1918, succeeded to the position when Warner returned to MIT full-time to join
the faculty in 1920. Allen, a test pilot who had acquired experience in
England and at McCook Field, had also attended MIT; he performed
most of the flying, mainly with the Curtiss JN-4H "Jenny" (figure 3-2),
though other planes were tested as well. These men, with the instrument engineers and others in the Langley group, soon led in flight
research in the United States.

The Langley work quickly provided the needed criteria for stability
in terms of control variables apparent to the pilot. This knowledge
arose out of tests, conducted in the summer of 1919, of two examples of
the JN-4H and one De Havilland DH-4. The results for the Jennies
were meant for comparison with Hunsaker's wind-tunnel measurements of the aerodynamic characteristics of the closely similar JN-2, to
see in part how accurately wind-tunnel tests could predict conditions in
flight. Tests of both designs included, among other things, measurement of the elevator angle and stick force required for straight steady
flight throughout the range of flight speeds at various throttle settings.
These were evidently the first quantitative flight tests of stability and
control in the United States.[32]

In their lengthy and detailed report, published in 1920, Warner and
Norton explained how properties of the observed curves of elevator
angle and stick force versus speed provide a measure of an airplane's
static longitudinal stability. Such relationship is plausible in light of the
earlier considerations of pitching moment versus angle of attack (figure 3-1). In straight steady flight, angle of attack is a function of speed:
the lower the speed, the higher the angle of attack required to generate
the necessary lift. An airplane flying at a given speed will therefore, at a
lower speed, need to fly at a higher angle of attack; if the airplane is
statically stable and the elevator angle were unchanged, it would then
experience a nose-down pitching moment (see figure 3-1, solid line). At
a higher speed, it will need to fly at a lower angle of attack and would
experience a nose-up moment. To maintain steady flight at the new
speeds, the pilot will find it necessary, therefore, to deflect the elevator
to supply a counterbalancing nose-up moment at the lower speed and
nose-down moment at the higher speed. The converse argument and
results hold for an unstable airplane. The necessary control action to
maintain steady flight thus provides a convenient way to assess inherent

Fig. 3-2. Two Curtiss JN-4H "Jennies" tested at NACA's laboratory at Langley Field, 1919. The horizontal booms extending forward from the interplane struts on the near machine support aerodynamic measuring instruments at their forward end. The bright spot on the far machine is apparently a blemish on the negative. From archives, NASA Langley Research Center.

static longitudinal stability; to wit, fly the airplane at some speed of interest and note the control actions required to hold it in steady flight above and below that speed.[33]

To apply this idea in practice, distinction must be made between two conditions of stability: (1) stability with control stick and hence elevator held fixed (*stick-fixed stability*) and (2) stability with control stick released and elevator left free to trail up or down in the airstream (*stick-free stability*). Detailed analysis shows that an airplane has stick-fixed stability at a given speed V_1 if the pilot must move the stick back and raise the elevator to maintain flight at reduced speed and move it forward and lower the elevator at increased speed (figure 3-3, upper graph, solid curve). It has stick-free stability at that speed if the pilot must pull on the stick at reduced speed and push on it at increased speed (figure 3-3, lower graph, solid curve). For instability, the reverse is true in each case. For both the stick-fixed and stick-free situations, the *gradient* of the appropriate curve thus gives a measure of the corresponding static stability.[34] Both kinds of stability are important; though pilots usually hold onto the control stick, occasionally they release it to take care of other duties. Warner and Norton arrived at their criteria by physical reasoning too technical to include here; equivalent explanation is provided in an appendix at the end of the chapter.[35] According to the criteria, the JN-4H was unstable, both stick fixed and stick free, at high speeds but stable at low speeds; the DH-4 was stable in both respects throughout its speed range. (Stick-fixed and stick-free stability usually go together at a given speed, but not always.)

Warner elaborated the work on longitudinal properties in two more reports in 1920. These reports included stick-force results for three other aircraft tested, presumably at NACA request, by the Army at McCook Field—a Vought VE-7 and a LUSAC-II, both single-engine fighters, and a Martin twin-engine bomber. They also provided a certain amount of mathematical analysis and a great deal of discussion, much of it exploratory and tentative. Clearly, a difficult learning process was taking place. The lasting intellectual result, however, was knowledge of the significance of the elevator-angle and stick-force curves. The gradient of these curves provided at the same time a criterion for static stability and a measure of controllability, the latter by indicating the degree of control action required to change the steady flight speed. This was a major conceptual advance. Warner realized this fact when he wrote in one report that "the actual measurement of the forces at several different speeds and the plotting of a curve is far more accurate and satisfactory as a means of determining longitudinal stability than is the customary method of recording the pilot's impressions on the subject, as it . . . gives a definite quantitative result in place of

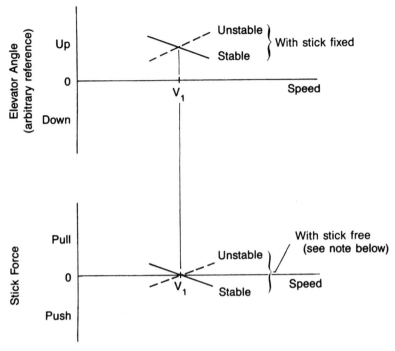

Fig. 3-3. Variation of elevator angle and stick force with speed in straight steady flight for stable and unstable airplanes. The stick-free criterion has meaning only at zero stick force, since only there can the airplane maintain steady flight with the stick released.

such vague phrases as: 'Stability very good'; 'Stability poor'; 'Stick pushes strongly against the hand at low speed.'"[36] No one attempted, however, to correlate such subjective impressions with the measured, quantitative results. Such research would come much later.

Following Warner's departure from Langley in mid-1920, Norton, Allen, and others explored additional areas of stability and control. These studies, published over the years 1921 to 1923, included measurements of linear accelerations in maneuvering flight—first along the vertical axis of the airplane and then along all three axes—and of angular velocities about the three axes. Of particular importance were pioneering studies by Norton and Allen of control positions and forces in steady turns and by Norton and William G. Brown of the unsteady response of an airplane to sudden, given movements of the controls. The first of these extended the stick-fixed and stick-free criteria to horizontal turning flight; it indicates how quickly these criteria became routine at Langley. The second attempted (not completely successfully) to distinguish between "controllability" and "maneuverability" and to define suitable measures for these properties. In the course of the work, Norton in 1921 wrote a summary discussion of stability and control for designers (though it has signs of being done in haste, and one wonders how useful it could have been), and in 1923 he and Brown published measurements of the long-period dynamic oscillations of the JN-4H and VE-7. These and the earlier reports, taken together, constituted a broad-ranging exploration of the field of stability and control. Much of it was pertinent to the problem of flying qualities. Most importantly, however, it provided the first extensive experience and capability in the United States concerning research (as against simple testing) in flight.[37]

The capability at Langley depended crucially on new instrumentation. Warner and Norton's first instruments were primitive: standard altimeter, tachometer, and airspeed meter, plus a graduated sector attached to the rocker arm of the control stick for elevator angle and a spring grip on the control column for stick force. The pilot or observer hurriedly recorded the readings on a knee pad. Within two years, however, the airspeed meter and control instruments had been replaced by more sophisticated, photographic-recording devices. The investigators at first designed these mostly themselves and had them built at Langley, though with some assistance initially from the National Bureau of Standards. Recording was achieved, as in succeeding NACA instruments, by having the sensor in the device rotate a small, mechanically movable mirror, which reflected a light beam onto a revolving film drum. Langley instrument engineers, most notably Henry J. E. Reid, also developed the specialized equipment required to measure and record linear accelerations and angular velocities. Records from the multiplicity of instruments, of course, needed synchronization to show how the different control movements and vehicle responses related in time. This was accomplished by a special, motor-driven chronometer that illuminated a light intermittently in each instrument, putting simultaneous timing marks on the films. Detailed descriptions of these mechanical instruments (which today would be replaced by electronic devices) appeared either in the research reports or in a series of specialized papers.[38]

The Langley pilots—Thomas Carroll, William H. McAvoy, and others besides Allen—also had to develop new skills and techniques. Research flying had to be invented virtually from scratch. Standard routines were needed to measure the various quantities and to obtain the most from limited flight time. These required flying the airplanes level and steady for one to two minutes to take the necessary data; to do this with the accuracy needed for research called for considerable skill. The matter-of-fact description by Warner and Norton of the pilots' problems in flying the Jenny steadily for extended periods with wings stalled (where the lift decreases instead of increases with angle of attack and the required elevator action is reversed) makes hair-raising reading to anyone familiar with flying. Flying an airplane so as to obtain repeatable readings of the various quantities also required infinite patience and practice. Eddie Allen, who became the most renowned test pilot of his generation, wrote some years later, "Once, when I was a test pilot for the N.A.C.A., I spent three months flying one plane in a study of 'control in circling flight' before I was able to separate and record [the] frequently overlapping forces to the satisfaction of myself and Edward P. Warner." Warner and Norton recognized these problems when they wrote in their report, "Test flying is a very highly specialized branch of work, the difficulties of which are not generally appreciated, and there is no type of flying in which a difference between the abilities of pilots thoroughly competent in ordinary flying becomes more quickly evident." The existence of the subcommunities of engineering research, instrument development, and test flying appears clearly in the work at Langley; so too do their essential overlap and interdependence.[39]

From 1923 till the end of the decade, flight research on flying qualities mostly subsided at the NACA. Flight research continued and even increased in quantity, but on problems such as aerodynamic loads, scale effects, spin recovery, and propeller performance. The Langley group also did a good deal of performance testing of aircraft for other government departments. This work, and the instrument development and refinement that went with it, contributed to the capability in flight testing generally.[40] So far as flying qualities were concerned, however, the pertinent first-order questions about stability and control appear to have been answered for the time being, at least to the satisfaction of the research community. A factor in this shift at Langley may have been the departure in mid-1923 of Frederick Norton, who was evidently the driving spirit (following Warner) in the vigorously productive work on stability and control in the preceding years. In the aeronautical community as a whole, there appears to have been no felt need—or perhaps the time was not ripe—for a concentrated attack on flying qualities. The years until 1936, however, including some relevant flight work at Langley in the early 1930s that I shall describe below, saw an accumulation of experience and sifting of knowledge that would be useful later.

Designers, of course, did not have the luxury to put their practical problems aside. To help them, as well as the increasing number of students of aeronautics, writers included discussion of stability and control in textbooks and magazine articles. Some of these—notably the general books on aerodynamics by Charles N. Monteith and Edward Warner (the latter a massive work of nearly 600 pages written at MIT) and an extended series of articles on stability and control by the design engineer Boris V. Korvin-Krukovsky—reproduced the Langley results and discussed the elevator-angle and stick-force criteria.[41] Most, however, including well-known books by Walter S. Diehl, Virginius E. Clark, Alexander Klemin, Edward A. Stalker, Clifton C. Carter, and Karl D. Wood, treated the problem mainly in terms of moment-curve slope, with little or no attempt to relate this to quantities sensed by the pilot.[42] As we have seen, this approach, which predominates in elementary textbooks to the present day, is simpler to explain and understand (as well as to implement by the designer). One wonders, however, if a resulting lack of awareness of the control criteria for stability and of the ideas and reasons behind them may not have contributed to the slowness in coming to grips with flying qualities.

Any doubts that inherent longitudinal stability was desirable disappeared in the 1920s. This important fundamental judgment took place with little explicit deliberation and is therefore difficult to trace. Despite the advice of Husted quoted earlier, one still finds in the first years of the decade statements like the following from Norton and Allen's study of circling flight: "Generally speaking, a pilot does not know a stable from an unstable machine, and if the forces on the controls are small he is just as well satisfied with the unstable one as the other. . . . the beginner learns to fly as quickly in an unstable machine as in a stable one." In 1925, Korvin-Krukovsky still pointed out the instability of some popular airplanes of World War I and added: "It is true that an unstable machine will not fly very long with hands off, but even in stable machines the pilots do not leave the controls for long, simply because the stick is the best place to keep the hands on if not for any other reason." By 1928, however, Clark, an experienced practical engineer, was stating categorically in his book that "instability in any sense is useless and unsafe." And two years later, Klemin wrote similarly, "With the technical knowledge now available . . . there is no excuse for designing a machine of longitudinal instability."[43]

By the mid-1930s the shift was complete. The young Clarence L. ("Kelly") Johnson, who would become one of the most famous of American airplane designers, put the matter clearly in a 1935 article on wind-tunnel tests of the projected twin-engine Lockheed Electra:

> The reasons why an airplane must be stable are more or less obvious. A moderately unstable airplane may be flown safely if the pilot continually manipulates the controls. However, this procedure is very nerve-wracking and tiring. Landing or take-off is dangerous in an unstable ship. In flying

blind, it is a great comfort to know that if the controls are released, the ship will continue on its course safely even in rough air.[44]

Thus, what in 1910 had been controversial and in 1920 at least arguable had by 1935 become "obvious."

The reasons Johnson cites came into being in the intervening years. Throughout the period, increasing range and duration of flights increased pilots' exposure to fatigue when they were required continually to manipulate the controls. Higher landing speeds, especially with the introduction of more streamlined aircraft after 1930, added to the danger from that source, and improvements in cockpit instrumentation made blind flying more commonplace. At the same time, pilots were having to give increased attention to peripheral duties, such as radio operation, that required them to remove their hands from the controls. Certain military operations, such as improved precision bombing and photographic mapping, also required a steady platform. And experiences by pilots in the air races of the early 1930s showed the potentially (and sometimes fatally) dangerous effects of instabilities in the high-performance racing aircraft of that day. From a variety of causes, the art of flying had grown up and become more complex. How the various influences became translated into statements in books and articles is not the kind of thing that appears in the record. It must have involved pilots, designers, research engineers, and academics talking to each other and participating to some extent in each other's activities. However it took place, it reflected a complex and widespread learning process by a considerable engineering community.[45]

The alleged "view now generally held" at the end of World War I had thus by 1935 become truly general. In this view, a highly maneuverable airplane, such as a fighter, should have a slight amount of static stability and a bomber or transport somewhat more—much as we saw stated in the NACA's *Annual Report* of 1918. A few of the writers cited even attempted to specify quantitative values for the negative moment-curve slope or for some essentially equivalent criterion. Stalker, in his book of 1931, made tentative mention of such values and of how they should vary with weight and speed; and Kelly Johnson, who had been a student of Stalker at the University of Michigan, recommended in his report on the Electra a value for transports that he claimed "has been found from practice" to give "satisfactory longitudinal stability." Walter Diehl, in a 1936 revision of his earlier book, defined a relatively complicated coefficient and gave ranges of recommended values for three types of aircraft of increasing stability: (1) highly maneuverable (i.e., fighters); (2) moderately maneuverable (trainers and observation planes); and (3) very stable (bombers and transports). Commander

Diehl, a key engineering officer with the navy's Bureau of Aeronautics, did not say how he arrived at his figures; he presumably did so by analyzing the wide range of wind-tunnel and flight data and pilot opinion available to him in his navy post. How generally Diehl's values were used is difficult to say. His book was well known and practically oriented; and Max M. Munk, in an article of 1942 analyzing the physical significance of Diehl's coefficient, said the use of "Diehl's rule" by designers was "wide-spread." On the other hand, Alexander Klemin and J. G. Beerer in 1937 criticized the coefficient and proposed a new one in its stead. Whatever use it may have had for the designer, however, any quantitative coefficient based solely on moment-curve slope provided no connection with the flying qualities sensed by the pilot.[46]

The concentration on static stability by practical men like Diehl was accompanied by lessened concern for dynamic stability. At the same time, the relevance of the topic for flying qualities remained confused. In his 1923 report on longitudinal oscillations in flight, Frederick Norton remarked that "the practical man has never considered dynamic stability very seriously," and experiences in the 1920s appeared to confirm this attitude. Machines that had been made statically stable almost always proved dynamically stable; and, in the exceptional cases where the long-period mode—so painstakingly studied by Bryan, Bairstow, Jones, and Hunsaker—was unstable, the dynamic instability was mild and readily controlled by the pilot. Norton from his study was forced to conclude that "while dynamic stability is interesting from a scientific point of view, the designer may entirely disregard it unless the airplane is such a radical departure from the usual practice as to make an investigation of this property advisable."[47]

Aeronautical textbooks of the 1920s and 1930s nevertheless incorporated treatments of dynamic stability, and theoretical studies extending and refining the work of Bryan continued, including extensive charts to facilitate application. In 1932 Langley engineers also published a careful comparison between flight results of long-period longitudinal oscillations of a Doyle O-2 light monoplane and theoretical predictions derived from theory. Results of such analyses, however, probably saw little use in design. In 1936 one of the theoretical workers, Arthur G. B. Metcalf of Boston University, still complained that the designer "is usually only vaguely aware of 'dynamic' stability as 'something you don't have to worry about,'" though he admitted that, given the complexity and difficulty of the subject, this attitude was understandable. In fact, a more enlightened attitude would have availed the designer little concerning flying qualities. Analyses still concentrated on the long-period mode, which was just at that time (as we shall see) being found irrelevant to such concerns.[48]

Simultaneously with these engineering developments, test flying became increasingly professionalized, mainly at the hands of the military. Military pilots in the process adopted the quantitative techniques of the NACA. James H. ("Jimmy") Doolittle, an air corps officer in the 1920s, is a notable, if atypical, example. While detailed to study under Warner at MIT, where he received the D.Sc. in 1925, Doolittle for his master's thesis made flight tests (carried out at McCook Field) in which he measured accelerations of a Fokker PW-7 pursuit airplane in a wide range of maneuvers. The tests, which employed a recording accelerometer copied after Norton and Warner's at Langley Field, were reported in an NACA publication. More typical of test pilots in the air corps was attendance at a training course prepared by Lt. Eugene Barksdale at McCook Field in 1926. The object of the course, according to its manual, was "to develop the ability of analyzing the behavior, possibilities, and limitations of aircraft and equipment, as distinguished from performance testing." The following year, William F. Gerhardt and Lawrence V. Kerber, both of whom had worked at McCook, also issued a flight-testing guide through the Department of Engineering Research at the University of Michigan. This extensive manual gave detailed plans and procedures for both production and research testing, though the stability-and-control tests were all qualitative and dependent mainly on pilot observation and opinion. In 1929, however, the air corps, through an information circular, adopted essentially the Langley Field method for measuring elevator angle and stick force as a function of speed. These developments did not aid directly in solving the flying-quality problem. The heightened professionalism, however, contributed to awareness and knowledge of the problem in the flying community.[49]

In the early 1930s, flight work at Langley Field turned back to topics more closely related to flying qualities. Besides the study of dynamic stability of the Doyle O-2 already mentioned, these included comparative measurements of the maneuverability of three navy biplanes conducted at the request of the Bureau of Aeronautics. These studies in effect extended the work of Norton and Brown of the early 1920s and used the recording instruments developed at Langley in that period. (Development of new instruments had not stopped, however; the dynamic-stability work produced a new angle-of-attack recorder.) The Langley group also carried out intensive flight measurements of the effectiveness of a variety of lateral-control devices on a Fairchild 22 light monoplane. The experimental methods, including an especially devised standard series of maneuvers, were "widely copied by industry." Most important for present concerns, these flight studies supplied valuable experience and learning for the flying-quality investigations to follow.[50]

A crucial part of the experience was formation of a small, mutually sympathetic group of engineers and pilots. This involved more than simply cooperation. A pilot and aircraft, taken together, form a single dynamic system, with feedback loops to the pilot via the feel of the cockpit controls plus cues from instruments and from vehicle orientation and acceleration. The pilot is a dynamic part of this *closed-loop system* (to use modern control terminology) and senses him- or herself as such. The engineer, on the other hand, views the system from outside and tends to focus on the airplane, the part that can be designed. Engineers of the 1930s, as a result, tended to see the airplane as an *open-loop system*—though they didn't use the term—with the pilot as an external agent who supplied whatever more or less quasi-static actions were required. (The preoccupation with moment-curve slope for specification of longitudinal stability reflected this view.) The difference was one of viewpoint rather than pilot-aircraft reality; it gave (and still gives) rise to subtle and troublesome differences between pilots and engineers, not only in how problems are defined and solutions attempted, but in psychology and language as well.

The group at Langley learned to transcend these differences. In working together through the early 1930s, pilots William H. McAvoy and Melvin N. Gough and engineers Hartley A. Soulé and Floyd L. Thompson (plus others who joined them later) learned to understand each other's view and to combine them into an essentially common approach. In so doing, they turned themselves into *research* pilots and *flight-research* engineers. The process was probably an unconscious one that took place naturally out of the demands of the job. With the addition of instrumentation engineers such as Howard W. Kirschbaum, it gave the Langley group a powerful potential for tackling the flying-quality problem in earnest.

That the problem needed tackling could hardly be doubted. The practical situation for aircraft of the mid-1930s, even successful ones, was not always satisfactory. Some showed adequate flying qualities, some did not. Three quotations illustrate the situation.

(1) In 1935, Pan American Airways, operating in the Pacific, pioneered transoceanic commercial flying with the highly successful Martin M-130 four-engine flying boat. Thirteen years later, Scott Flower, a longtime Pan Am pilot, confessed as follows to a symposium on "flight handling characteristics" at the Annual Meeting of the Institute of Aeronautical Sciences:

> This flying boat had extremely heavy control forces, and, as a consequence, its flight characteristics were compromised. We have been criticized at times for not placing our transpacific operations on a 24-hour basis sooner than we did. I'll let you in on a secret; we couldn't! Because after 10 to 12 hours of flying the M-130, a stop was necessary to revive the crew![51]

(2) On July 12, 1939, Charles A. Lindbergh flew from Washington, D.C., to Langley Field piloting a Seversky P-35 pursuit plane. The P-35 was a 1935–36 design that was by 1939 in routine service with the air corps. In his journal, Lindbergh described the takeoff and landing thus:

> No sooner had I opened the throttle than the load on the stick became excessive. I had to push forward on it with many times the normal pressure, and still the plane seemed to jump into the air. When I could spare my attention for a few seconds, I looked down at the stabilizer adjustment and found that I had set it in the center of the quadrant, whereas the neutral point was a little farther forward. I moved it slightly, and the load on the stick disappeared instantly—exceptional sensitivity. My fault, of course, but what a delicate balance that plane was designed to. How carefully she would have to be watched and coached and held in check—no margin of reserve there. . . .
>
> Langley Field was full of activity when I approached. . . . I circled around several times before getting a clearance to land. . . . I lowered the landing gear and flaps and felt as though I was balanced on a pin point, ready to fall quickly in any direction if I let myself be off guard for an instant. I thought of the captain who said, "It's a feat every time you land in a P-35." I came in with the engine idling rapidly and let the wheels touch first—a two-point landing—taking no chance. The plane landed slowly for a pursuit . . . , but I did not feel at ease until I stopped rolling, had lifted the flaps, and had turned in toward the line.[52]

(3) In April 1941, Capt. Robert S. Hatcher, an officer at the navy's Bureau of Aeronautics in Washington, D.C., replied to an inquiry from the head of the bureau concerning the advisability of issuing specifications for flying qualities. In an exchange of penciled intraoffice memoranda, Captain Hatcher let down his hair with a candor that rarely appears in more official documents:

> At present we simply specify that the airplane shall be perfect in all respects and leave it up to the contractor to guess what we really want in terms of degree of stability, controllability, maneuverability, control forces, etc. He does the best he can and then starts building new tails, ailerons, etc. until we say we are satisfied.[53]

Unsatisfactory airplane characteristics were doubtless due in part to the inability of designers to design for what was wanted. As indicated by Hatcher's memo, however, what was wanted was far from clear.

By the mid-1930s, then, engineers had learned to relate stability to control in terms of measurable quantities sensed by the pilot. The elevator-angle and stick-force criteria were probably little used by designers (who favored moment-curve slope); the aeronautical community agreed, however, that some degree of inherent stability was desirable, whatever the criterion. At the same time, a new breed of flight-research engineers and research pilots had achieved growing skill and sophistication; working together, they had familiarized themselves with a wide range of stability-and-control problems. Accurate and reliable instruments also existed to record the needed quantities continuously in flight. (Although we haven't had reason to go into the matter here, a great deal of wind-tunnel work on how to design for given levels of stability and control had also contributed to the general awareness.) The basic analytical understanding and practical capability required to deal with flying qualities had thus been brought into being. Notwithstanding, all was far from well in design. As two flying-quality experts put it looking back from 1956, "There were good and bad aeroplanes, and pilot opinion could make or break a machine without the designer understanding why."[54] Means and need thus both existed for a concerted attack on the problems of stability, control, and flying qualities.

Still, much of what has just been said could also be said about the situation in 1923. By that time, thanks to the work at Langley Field, the necessary rudiments were at hand and more could have been done. Why, we may justifiably ask, did twelve years elapse before the aeronautical community attempted the next logical step? The answer appears in a contemporary lecture "On the Results of Aerodynamic Research and Their Application to Aircraft Construction" delivered by Professor Clark B. Millikan of the California Institute of Technology to the Lilienthal-Gesellschaft on October 13, 1936. Introducing his section on stability and control, Millikan told his German audience, "In the last year or so, performance, especially of commercial planes, has become more or less stabilized, the result being a much increased interest in stability and controllability problems."[55] That is to say, control of an airplane is, in a sense, secondary to its speed, range, ceiling, or carrying capacity, attributes usually lumped together under the term *performance*. With the essential proviso, of course, that the airplane can be reasonably and safely flown by the pilot, its utility depends crucially on these other attributes. In the late 1920s and early 1930s, great potential appeared for advance in these areas, and researchers and designers focused their attention accordingly. These efforts bore fruit in the streamlined, metal-structure configurations exemplified most conspicuously by the Douglas DC-1, -2, and -3 of 1933 to 1936. Only when the performance gains had been at least in part realized did concentration on problems of stability and control become advantageous. A similar situation occurred earlier when the performance increases that took place under the wartime pressures of 1914 to 1918 were followed by the Langley Field work on stability and control of 1919 to 1923. In both cases, intellectual and practical resources were limited, and first things had to be put first. That is to say, in the face of limited resources, the purposes of engineering devices set priorities in engineering knowledge.[56]

Establishing Specifications, 1936–1943

With the increased interest in stability and control came explicit concern for flying qualities. Work aimed deliberately at the problem began about 1935–36, more or less independently at first, at Langley Field and at United Airlines and the Douglas Aircraft Company. The sustained—and soon related—effort that followed would lead to the flying-quality specifications of 1942–43.

The initial work at Langley focused on the relation between dynamic longitudinal stability and pilots' observations of flying qualities. In research published in 1936 (and begun, presumably a year or two earlier, because the people at Langley saw a need for it on their own), Hartley Soulé and his co-workers carefully measured the period and damping of the long-period oscillation of eight single-engine airplanes, ranging from a 1,400 pound civil monoplane to a 6,000 pound military biplane. They then attempted to correlate these quantitative results with qualitative rankings by two pilots with respect to overall "stiffness," stick force and movement, and pitching motion in rough air. The investigators hoped such correlation would provide an indication of the degree of dynamic stability desirable for design. The results did not fulfill this hope. To Soulé's seeming surprise, the considerable range of opinions of the pilots concerning flying qualities appeared to be entirely independent of the equally considerable range of dynamic properties. He was forced to conclude that "the dynamic longitudinal stability characteristics, as defined by the period and damping of the [long-period] oscillation, are not apparent to the pilot and, therefore, cannot be taken as an indication of the handling characteristics of airplanes."[57]

The Langley work thus brought into doubt the twenty-five-year preoccupation with the long-period oscillation, at least so far as flying qualities were concerned. The following year (1937), Robert T. Jones, a theorist at Langley Field, carried the matter further in a letter to the editor of the *Journal of the Aeronautical Sciences*. Jones's letter was in reply to the 1936 article by Arthur Metcalf, quoted earlier, which had discussed only the long-period mode. Citing Soulé's results, Jones suggested that the out-of-hand dismissal of the short-period mode by Metcalf and most theoreticians before him was probably a mistake. In particular, this mode, with its very short period and heavy damping, meant that "there is an effective restraint of the machine against movements of the airplane in gusts. . . . If we wish to make more practical

Fig. 3-4. Douglas DC-4E. The airplane first flew in June 1938. From United Airlines.

use of the dynamical theory . . . to analyze and predict the response to control and the motions in gusty air, we cannot neglect this factor which is of primary significance in these questions." Several others were coming to the same realization, and Soulé himself had remarked that the possible relationship between the short-period mode and behavior in rough air "should perhaps be investigated." This relationship would, in fact, turn out to be the significant one with regard to flying qualities.[58]

The primary catalyst for flying-quality research, however, appeared in a set of specifications prepared for the Douglas DC-4E (figure 3-4). A group of five U.S. air carriers, spearheaded by United Airlines, ordered the DC-4E from the Douglas Aircraft Company of Santa Monica in March 1936. The airplane, Douglas's first venture into four-engine aircraft, was intended (though not in the end adopted) as an increased-capacity follow-on to the two-engine transports then in use. Specifications for the airplane stemmed from thinking at United about the company's expanding needs and from a study conducted for the airline in 1935 by an outside group directed by Jerome Hunsaker of MIT and George J. Mead of United Aircraft Manufacturing Corporation. The company later that year also engaged Edward Warner, now a consulting engineer, to evaluate proposals from three aircraft manufacturers for designs of the type indicated by the study.[59] The final specifications to Douglas, drawn up in conjunction with the four other airlines and agreed to in 1936, were in part Warner's work. Like most airplane specifications, they included the airlines' requirements for performance, plus the usual particulars on strength, workmanship, interior arrangements, and so forth. They went beyond the usual specifications, however, by incorporating a section on flying qualities.[60]

This section, evidently proposed and written by Warner, represents probably the first attempt in the United States (perhaps the world) to set down such requirements for a new design. The idea behind them, however, went deeper than just another set of ordinary specifications. They embodied for the first time the notion that desired subjective perceptions of pilots could be attained through objective specifications for designers. This was an idea not in evidence before. It was not unique with Warner, however. According to Scott Flower, the Pan American pilot quoted earlier, his colleague Harold Gray produced in the same period a similar set of criteria (though not for a specific design) from their company's experience with the M-130.[61] Other people may well have had the same thought. Though the idea was apparently new, it was the logical product of what the aeronautical community had learned about stability, control, and pilots' reactions. Whether or not it was feasible, of course, remained to be seen.

The flying-quality specifications to Douglas aimed, in Warner's words, "to produce an airplane suitable for all the purposes of transport operation—possessing adequate stability, responsiveness to the controls, lacking in eccentricities or sudden changes of behavior, and generally satisfactory to the pilot"—in itself an excellent characterization of flying qualities. The ambitiousness of this goal, given the state of knowledge, was implied by the further statement that "there is the probability of an accumulation of . . . information bearing on the subject of flying-quality tests, beyond any that now exists, within the period of construction of this airplane." "Probability," however, did not fully describe the situation. As we shall see, Warner was at that same time engaged in activities to promote such accumulation. According to a later NACA report, Warner arrived at his specifications by consulting "a considerable number of air-line pilots, engineers connected with both the operating and the manufacturing companies, and research men, including members of the N.A.C.A. staff." Unfortunately, no record of this consultation appears to have survived.[62]

The specifications, though hedged with qualifications because of admitted lack of knowledge, were highly detailed. They addressed both longitudinal and lateral characteristics for the full range of flight speeds and load distributions and with flaps and landing gear retracted and extended. Regarding longitudinal characteristics, the specifications called for qualitative static stability both stick free and stick fixed (as indicated by negative stick-force and elevator-angle gradients, respectively). They stipulated, however, that the degree of stability should be "small." To help keep stick forces from being excessive,

Warner also placed a quantitative upper limit on the absolute value of the stick-force gradient. (The *absolute value* of a negative number is the value with the negative sign removed.) In all the foregoing, he made no reference to moment-curve slope. Concerning longitudinal dynamic stability, Warner held with current beliefs by placing quantitative requirements on the damping and period of the long-period oscillation; he made no mention of the short period. Concerning effectiveness of the elevator in maneuvers, the specifications required that it be able to pitch the airplane either up or down by 5 degrees in 1.5 seconds at the lower end of the speed range and that the stick force needed to do this or to turn the airplane with 45-degree bank at 140 mph should not exceed 75 pounds. (Banking an airplane in a steady level turn tilts the lift of the wing toward the inside of the turn, providing the horizontal component of force required to make the turn. To maintain the vertical component needed to balance the weight requires at the same time that the magnitude of the lift be increased. Such increase, in a stable airplane, calls for upward deflection of the elevator and hence a pull on the stick.)

Warner's specifications undoubtedly went into more detail than was warranted by existing knowledge. As we shall see, however, they helped focus attention on flying qualities in a way that had not occurred before. That the topic was a live concern was shown by the fact that Millikan, after the statement quoted earlier, went on to give extensive coverage to Warner's specifications in his talk to the Lilienthal-Gesellschaft.[63]

As implied above, Warner (figure 3-5) had impact in more ways than one. After resigning his faculty position at MIT in 1926, he had spent three years as the initial assistant secretary of the navy for aeronautics, five years as editor of *Aviation* magazine, and a one-year term as vice-chairman of the Federal Aviation Commission. He was also from 1929 to 1945 a member of the main committee of the NACA and from 1919 to 1941 of the NACA's Committee on Aerodynamics, of which he became chairman in 1935. This last appointment put him in a key position to influence NACA research on flying qualities, of whose importance he was keenly aware from his consulting with United and Douglas.[64]

Reacting to Warner's apparently vigorous leadership, the Committee on Aerodynamics, on December 9, 1935, issued to the Langley Laboratory an official research authorization under the title "Preliminary Study of Control Requirements for Large Transport Airplanes." This authorization had as its purpose "to obtain data for determination of the requirements as to flying qualities, particularly maneuverability and stability, of transport airplanes, and evolve a technique for making

Fig. 3-5. Edward Pearson Warner (1894–1958). From National Air and Space Museum, Smithsonian Institution.

tests to determine these qualities." In the research that followed, Warner functioned as a kind of de facto intermediary between the airline and aircraft industries and the NACA research community at Langley Field, which he visited periodically. He affords a classic example of what Hugh Aitken calls a "translator" between social subsystems (or communities). Aitken sees such people as playing a key role in "translating the information generated in one [community] into a form intelligible to participants in others and organizing the movement of resources between them."[65]

Though prompted by Warner, the committee's action gave the people at Langley authority to move energetically in a direction their research already inclined them to go. The required program, as projected by Soulé and his immediate supervisor, Floyd Thompson, in the spring of 1936, would comprise three steps: (1) tentative identification of the factors that can be used to define flying qualities quantitatively and that can reasonably be measured in flight, plus an outline of routines for such measurement; (2) tests on one or two airplanes to evaluate the feasibility of the scheme; (3) accumulation of data and pilot opinion from existing airplanes to provide a base for quantitative specifications for future designs. They expected that experience from the second and third steps would require revision of the requirements and procedures laid out in the first.[66]

To accomplish the first step, Thompson during the summer of 1936 prepared a schedule of "Suggested Requirements for Flying Qualities of Large Multi-Engined Airplanes." He patterned these requirements in most respects after Warner's, whose specifications were available to him. Thompson's differed in some matters, however, and went generally into less detail. The lesser detail came from the fact that Warner's requirements were for a specific airplane, whereas Thompson's were intended as a basis for broad research. In preparing his schedule, Thompson limited himself to factors that could be measured by existing NACA instruments or by instruments that might readily be designed and developed. With respect to longitudinal requirements, Thompson, like Warner, called for stick-free and stick-fixed stability in steady flight but omitted Warner's stipulation that the degree should be small. Though he also omitted Warner's quantitative upper limit on the absolute value of stick-force gradient, he introduced a lower limit to assure that control forces would be at least sufficient to overcome friction in the control system. Surprisingly, despite Soulé's recent negative findings concerning the importance of the long-period oscillation, Thompson continued to place requirements on the period and damping of that mode, while making no mention of the short period. Perhaps the Langley people were not yet convinced of the implications of their findings, or perhaps old habits simply died hard. The most marked difference from Warner's work, however, appeared in the requirements on stick-force in maneuvering flight, a property for which there was no generally accepted criterion. Where Warner had put an upper limit on stick force required to achieve a banked turn, Thompson set both upper and lower limits on the force needed to pull up, from a prescribed dive, when using a wing lift of 0.8 of the maximum structurally designed value. Thompson looked on his schedule as essentially a thinking exercise "to crystallize ideas about what items are important and [to indicate] wherein data are lacking concerning quantitative values for various items." To the schedule he appended a "General Program of Tests" outlining flight routines needed to check compliance with the requirements.[67]

The next step was to evaluate the requirements and procedures in flight. For this purpose, Soulé, who supervised the tests, employed a Stinson SR-8E single-engine high-wing monoplane (figure 3-6), which belonged to the NACA and was readily available. That it was not a large multiengined craft didn't matter, since the tests were intended only to assure that the items on Thompson's schedule could be measured with adequate precision and that they related to flying qualities as intended. The investigators employed two alternative sets of instruments: (1) the NACA's own special recording instruments, with the stick-type sensor for control force replaced by a device suitable to the Stinson's control wheel; and (2) standard indicating instruments normally available in a

Fig. 3-6. Stinson SR-8E used in tests at Langley Field. From H. A. Soulé, "Preliminary Investigation of the Flying Qualities of Airplanes," *Report No. 700*, NACA (Washington, D.C., 1940), fig. 2(a).

transport airplane (plus a stopwatch and suitable force and position indicators for the controls). The standard instruments were grouped together on the airplane's instrument panel and photographed with a small motion-picture camera. This set showed what might be done by aircraft manufacturers and other agencies who wanted to measure flying qualities but did not have NACA instruments available. The tests took about twenty hours of flight time over three months in the winter of 1936–37. Soulé reported the results in September 1937 in a classified report with restricted circulation in the American aircraft community.[68]

Soulé's report took up in order the seventeen requirements suggested by Thompson, eight for longitudinal qualities and nine for lateral. In each case, he listed the requirement (usually with a number of subdivisions), laid out the pertinent flight routine, and presented and discussed the results. To help others who might want to make similar tests, he also described the experimental difficulties that were encountered. Problems had arisen, for example, in photographing the stopwatch, thanks to an indistinct dial and "because the face was white with black markings rather than black with white markings." Such mundane matters can spell the difference between success and failure in physical measurements. Soulé concluded that Thompson's list of requirements required minimal revision as a basis for tests of other airplanes and that "the tentative schedule of flying qualities and flight tests is generally satisfactory." The few suggested revisions included elimination of the condition on the damping (but not the period) of the long-period oscillation. Reassessment of dynamic stability thus proceeded slowly.[69]

A sidelight on Soulé's report is instructive. In February 1938, when engineers at Chance Vought Aircraft in East Hartford received the report, they were puzzled by a seeming inconsistency in curves of elevator angle versus speed for different angular settings of the fixed portion of the horizontal tail. C. J. McCarthy of Chance Vought wrote to the NACA asking if the supposed "elevator angles" had perhaps been assessed from the position of the control wheel (which was not clear one way or the other from the report). If so, the inconsistency might be explainable on the basis of the stretch of the cables connecting the wheel to the elevator—such stretch would be expected to vary with control force and hence with elevator angle, and this variation could affect the shape of the nominal elevator-angle curves. Robert R. Gilruth, a young engineer who had recently taken over the flying-quality program when Soulé moved to wind-tunnel duties, measured the stretch under applied static loads and found that Chance Vought's supposition was in fact correct. The unrestrictedly available version of

Soulé's report, published in 1940, presented both the apparent elevator angles as measured at the control wheel, which condition the pilot's
impression of stability, and the true elevator angles (more precisely, the
foregoing values corrected for cable stretch), which reveal the true
stability. In tests of later airplanes, elevator angles were measured directly at the elevator. Such matters seem obvious in retrospect, but they
have to become known somehow.[70]

Following success with the Stinson, the Langley group was eager to
try its instrumentation and methods on a larger, multiengine airplane.
To this end, the NACA managed the loan for two months in mid-1937
of a Martin B-10B twin-engine bomber from the army air corps. Flight
tests took place in June, while Soulé was preparing his report on the
Stinson, and Thompson summarized the results in a memorandum
soon after. As with the Stinson, the schedule of flying qualities and
flight routines checked out satisfactorily, and Thompson judged "the
procedure has been fairly well perfected." Gilruth wrote a complete
report, which was sent to the air corps in confidential form in January
1938.[71]

Warner, in his position as chairman of the Committee on Aerodynamics, learned promptly of the findings for the two airplanes.
Though the tests had been primarily to evaluate the NACA scheme,
the quantitative results caused him to reconsider portions of his original specifications for the DC-4E. To help guide tests of the airplane,
which had begun following initial flight on June 7, 1938, Warner in July
revised the specifications "to represent what seems to be desirable, and
reasonably attainable in the present state of the art." Throughout the
revision, he eliminated many of the qualifications that hedged the original. Some of his quantitative limits he altered in light of the NACA
results, and a few of the requirements he rewrote in varying degree,
explaining in each case why he did so. In regard to stick-force gradient,
for example, he added the lower limit suggested by Thompson to allow
for friction. For the stick force required in maneuvers, he gave up his
requirement in terms of a banked turn and replaced it, following
Thompson but differing in specifics, with a limit on the force needed to
attain specified loads in a pull-up. While he could not cite the findings
for the B-10B because of their military classification, he supported the
many revisions by liberal appeal to the results from the Stinson. Thus,
as Warner had intended in pressing for the NACA work, the research
guided by his original specifications had fed back into their improvement. As Stephen Kline has emphasized, such iterative interplay between research and practice is an essential feature in the development
of engineering knowledge.[72]

Though such work was not part of the central plan, the Langley group studied pilots as well as airplanes. In early 1936, even before the flying-quality work got under way, Melvin Gough and A. Paul Beard authored a report titled "Limitations of the Pilot in Applying Forces in Airplane Controls." Beard was an engineer and Gough a senior test pilot who played a key role in the flying-quality program. (Though he had an engineering degree, Gough's point of view was mainly that of a pilot.) Gough and Beard constructed a representative cockpit in which the location of the control stick and rudder pedals could be adjusted relative to the seat and measurements made of the maximum forces exerted by the pilot. The cockpit could also be rotated so that tests could be made with the pilot headup, on the side, and headdown. Besides graphs of quantitative results, the report included general conclusions such as "the ability to pull on the elevators was greater than the ability to push" and "the forces required on aileron, elevator, and rudder should be in the ratio of approximately 1 : 3 : 10 for the application of equal efforts by the pilot." The next year, William McAvoy, also a senior test pilot, extended the study to wheel-type controls. Results of such studies helped establish the limits on control forces called for in flying-quality requirements.[73]

Several peripheral events, taken together, also illustrate the context within which engineering research workers typically function: (1) On October 7, 1937, Gough gave a talk on "The Handling Characteristics of Modern Airplanes from the Pilot's Standpoint" to two squadrons of pilots aboard the aircraft carrier *Yorktown*. Drawing from the NACA's growing research experience, he warned the navy pilots of flying-quality problems they might encounter in an impending shift from biplanes to higher-performance low-wing monoplanes. (2) In March 1938, Lawrence Kerber, now with the Bureau of Air Commerce, wrote to the NACA regarding stalling tests of the Douglas DC-3 conducted at Langley Field in late 1937. In his letter, Kerber raised general questions about the relation between stalling behavior and longitudinal stability. Soulé prepared a detailed reply, answering some questions on the basis of Langley's flying-quality experience and saying of others that "this is exactly the sort of thing that is now being studied." (3) Later the same month, Walter Diehl of the navy's Bureau of Aeronautics, in a visit to Langley Field, requested the staff's opinion on tentative flying-quality specifications prepared at the bureau for a four-engine flying boat to be built by Consolidated Aircraft. Commander Diehl had modeled these specifications after those written by Warner for the DC-4E. Soulé and Thompson, in reply, wrote comments and recommended changes based on their experience with the Stinson and B-10B. They questioned (among other things) Diehl's requirement for complete stability of the long-period dynamic motion.[74]

These incidents illustrate how engineering research typically takes place within a context of practical demands. Since the demands usually will not wait, research engineers are often called on to give advice when their knowledge is still at a formative stage. Doing so requires them to articulate and sometimes alter and improve their developing ideas. Such constructive exchange between the generators and users of engineering knowledge is an essential element in the learning process. It forms another aspect of the interactive interplay between research and practice mentioned above.

With success on the Stinson and B-10B behind it, the Langley group went on to the third step of the program—collection of data and opinion on a number of existing airplanes. From 1938 to early 1941, when Gilruth prepared the overall report, sixteen airplanes received full tests; a number of others were also studied in part. Figure 3-7 shows fifteen of the fully tested aircraft, ranging from the four-engine Boeing XB-15 of 65,000-pound gross weight (no. 1) to two single-engine light planes of about 1,000 pounds (nos. 14 and 15). Most came from the military, but some were supplied by private companies at the request of the Civil Aeronautics Board. Occasionally an airplane would be sent to Langley for solution of some specific aerodynamic problem and, after the problem was solved, the opportunity would be taken to test it generally for flying qualities. The tests followed the routines proved on the Stinson and B-10B, modified by the evolution and improvement to be expected from experience. Instrumentation, which used the NACA recording devices, also evolved and improved, in the direction, for example, of greater recording time and of new instruments for angles of attack and yaw.[75] Throughout the program, lateral qualities (which I do not go into) proved typically more intractable, conceptually and experimentally, than longitudinal ones.

It is interesting to note that, whereas the title of the original research authorization of 1935 had referred to "Large Transport Airplanes" (and that of Thompson's schedule to "Large Multi-Engined Planes"), a second authorization issued in 1938 to cover the third step mentioned simply "Various Airplanes."[76] This raises the question, Did the people involved expect from the outset that the same criteria might apply to all types and simply not say so, or did they widen their ideas in the course of the tests? Did, for example, the results on two such dissimilar craft as the Stinson and B-10B open their eyes to possible generalization? The evidence I can find is mute on the point. Whenever it arose, of course, the idea would have to be checked out in practice. The assortment of types in figure 3-7, with more than one example of each type, implies the investigators were doing just that. As things turned out, the flying-qualities problem was seen as a general one subject to broadly

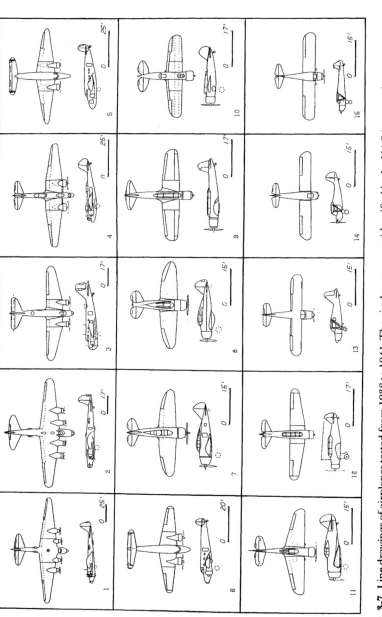

Fig. 3-7. Line drawings of airplanes tested from 1938 to 1941. The airplanes are not identified in the NACA report, but are, as nearly as can be judged, as follows: 1. Boeing XB-15; 2. Boeing B-17; 3. Douglas B-18; 4. Martin B-10B; 5. Lockheed 12-A; 7. Republic P-43; 8. Seversky P-35; 9. Vought-Sikorsky SB2U-1; 10. Brewster XSBA-1; 11. Curtiss P-36; 12. North American BT-9B; 13. Stinson 105; 14. Aeronca 65-C; 15. Taylorcraft BC-65. From R. R. Gilruth and M. D. White, "Analysis and Prediction of Longitudinal Stability of Airplanes," *Report No. 711,* NACA (Washington, D.C., 1941), fig. 1.

applicable specifications. In engineering knowledge, as in all knowledge, generalization of ideas often occurs in the course of knowledge growth.[77]

Starting from small beginnings, the flying-quality group at Langley grew rapidly in the period of the tests. Maurice D. White, who joined Gilruth to make up the engineering staff in June 1938, recalls that by 1941 the full-time crew had increased to about six engineers, three or four test pilots, two or three instrument specialists, and a like number of operations personnel who kept the airplanes flying. Support for instrument development also came from the laboratory's general instrumentation group. Gilruth, according to White, pretty much "ran the show" and kept track of the complete picture; solution to the many individual problems came from "the people in the trenches," who were occupied with their immediate work and had little opportunity to know the overall program. Management exercised little detailed supervision, however, and productivity was high and self-generated. As usual in any learning process, a great deal of "fumbling and head scratching" took place.[78] The situation exemplifies the kind of fruitful melding of personal and group ambition and interest that can arise when talented technical people join in what they see as a demanding and worthwhile task. The whole was more than the sum of the parts.

The input of pilot opinion came from unrecorded conversations between the engineers and pilots, especially Gilruth and Gough, the chief pilot of the program. (The formalized pilot-evaluation scales and methodologies that are standard today had yet to be devised.) The finally published requirements were thus based on pilots' judgment, but the process does not appear in the record. The professional relationship between Gilruth and Gough has been described by the former in an interview:

> Mel Gough . . . was very much interested in handling qualities. He didn't know too much about why an airplane did what it did, but he knew what he liked. He used to take me up and show me what airplanes would do. When he got an airplane that was a rogue, he'd say "I want you to come up because I want to show you this. Maybe you can tell me how you can fix it." And so we did a lot of flying together. In fact we even flew one time in a single seater. I had a foot on one pedal and he had a foot on the other.[79]

Conclusions were arrived at by the combined judgment of Gilruth, who understood thoroughly the engineering criteria, and Gough, who knew his own and his fellow pilots' opinions.

Results and analysis for each airplane appeared in a classified report as the program progressed. The report by Gough and Gilruth on the twin-engine Lockheed 14-H, a popular transport (no. 5 in figure 3-7), was typical. The airplane, tested in early 1939 at the request of the Air Safety Board, exhibited elevator-angle and control-force gradients that were small but in the "correct" (i.e., stable) direction. In the authors' opinion, however, "the degree of longitudinal stability . . . is believed inadequate in many respects. The variation of control position and force is too small to give the pilot an indication of a change in speed, particularly at low speeds with power on." Dynamically, a short-period oscillation of less than usual damping also proved "very objectionable when excited by gusts." The oscillation appeared to arise from a coupling between the inherent pitching motion of the airplane and an oscillatory movement of the elevator about its hinge (resulting from interplay between the aerodynamic forces on the elevator and the elasticity of the control system). Because of the unusually short period (about one second), the pilot could not react effectively to control the motion. The authors suspected that the same sort of oscillation, excited when a 14-H became prematurely airborne by striking a rut during takeoff, probably figured in an accident reported in airline operation. In addition to the reports on specific aircraft, Gough also gave at least two public presentations on the progress of the program. In both, he expressed the view, consistent with findings on the Lockheed, that the static longitudinal stability of current low-wing monoplanes was generally not sufficient for adequate pilot "feel."[80]

By 1941 the accumulated information and experience enabled Gilruth to publish a refined set of requirements. To arrive at his results, Gilruth reviewed the mass of flight data and pilot opinion to see what measured characteristics proved significant; he also considered what was reasonable to require of an airplane and its designers. The requirements appeared in classified form in April of that year and for unrestricted circulation in 1943. As the first generally accepted set of flying-quality requirements, the work is an oft-cited milestone in its field. The requirements addressed essentially the same items as had Thompson and Soulé, but in simplified and more logically organized format and with considerable change in detailed content (see below). Looking back after forty-five years, Gilruth says, "I boiled that thing down to a set of requirements that were very straightforward and very simple to interpret. [The] requirements went right back to those things you could design for." Thanks to the extensive testing, quantitative limits could now be stated for the designer with some certainty. The seemingly remarkable result that the same requirements could be applied (with minor exceptions) to all current types and sizes of airplanes, however, was nowhere explicitly mentioned. If anyone had ever debated the point, by now it seemed taken for granted.[81]

Gilruth's requirements gave comprehensive coverage to longitudinal stability and control (with seven subdivisions), lateral stability and control (nine subdivisions), and stalling characteristics. Each subdivision stated a requirement, usually in a number of parts, gave a simple explanation of the reasons for it, and briefly discussed the design considerations necessary to its achievement. Under longitudinal stability and control, in addition to the three fundamental subdivisions I have been following (static stability, dynamic stability, and maneuverability), he set requirements on elevator control in landing and takeoff, on the effects of engine power and wing flaps, and on the characteristics of longitudinal trimming devices (aerodynamic devices used to adjust to zero stick force at a desired flight speed).[82] I shall discuss only the three subdivisions as before. Gilruth headed them, more descriptively for his purpose, (1) elevator control in steady flight, (2) uncontrolled longitudinal motion, and (3) elevator control in accelerated flight.

(1) As in Thompson's tentative schedule, the manifestations of elevator control in steady flight—the gradients of elevator angle and stick force versus speed—were both required to show static stability (i.e. negative values). Gilruth felt justified, however, in being more quantitative, though elsewhere than in the requirements themselves. In a separate publication in 1941, which Gilruth cited in the requirements, he and Maurice White examined the analysis and prediction of longitudinal stability. In this study the authors recommended a quantitative lower limit on the absolute value of elevator-angle gradient for gliding (i.e., power-off) flight. Absolute values over this limit could be expected, according to flight experience, to insure adequate static stability and pilot feel under power-on conditions, where the effects of power tend to be destabilizing. This recommendation obviously stemmed (though Gilruth and White did not say so) from the inadequacies in current aircraft that Gough had complained of in his talks. As to absolute values of stick-force gradient, neither the requirements nor the Gilruth-White report recommended an upper limit to keep control forces within the pilot's capacity; elevator requirements for accelerated flight (see below) were stated to be more stringent here. The requirements did include, however, a minimum-force provision sufficient to overcome friction and maintain adequate pilot feel. Statement of some of the quantitative requirements, here as well as elsewhere, was deliberately loose—the numbers and reasoning, as White recalls, were not always firm enough to commit to print. Gilruth and others, however, discussed them frankly with designers who came to Langley seeking advice on specific designs.[83]

(2) The requirements for uncontrolled longitudinal motion (i.e., dy-

namic stability) made a firm break with the past. Gilruth disregarded completely the long-period oscillation on grounds that Soulé's findings of 1936 had shown it to be irrelevant to pilots' opinions and that "subsequent tests have not altered this conclusion."[84] He required instead that any short-period oscillation with elevator free should be sufficiently damped to have vanished after one cycle.

Conjecture of the need for such requirement went back at least to Robert Jones's letter of 1937 in reply to the article by Arthur Metcalf. Events had proved Jones correct about the importance of the short-period mode but for reasons he had not foreseen. Although the fact had gone largely unnoticed, the theory of dynamic motion, unlike that of static stability, had focused almost entirely on fixed controls. The Langley tests of a number of airplanes besides the Lockheed 14-H gave unmistakable evidence, however, of severe rough-air oscillations accompanied by uncontrolled movements of the elevator (again associated with elastic stretch of the control system). To understand this troubling behavior, Jónes, in collaboration with Doris Cohen, extended the theory of dynamic stability to the detailed study of free controls. The results, published the same year as Gilruth's classified report, revealed not one but two short-period oscillations, both shorter than the single short-period mode with elevators fixed. One was of such extremely short period and high damping as to be unidentifiable with the observed oscillations. The other, of somewhat longer period, involved reinforcement of the pitching motion by flapping of the elevator, leading to the possibility of marginal damping or even instability. This oscillation correlated with those observed in flight, and it was against it that Gilruth's requirement was directed. (Fortunately, the Jones-Cohen analysis also guided the designer in how to meet the requirement.) Thus, in the end, the short period replaced the long period as the focus of concern for the designer; the importance of the two modes had become completely reversed. The background for this reversal (that is, the quarter century in which the aeronautical community was blinded by concern for the long-period oscillation) provides a cautionary example of how preconceived, uncritical use of mathematical theory can mislead in practice. The received tradition had turned into a barrier that had to be laboriously dismantled. (Subsequent research has shown that, for subtle reasons, complete dismissal of the long-period oscillation is not really justified and minimum requirements need to be set for damping of that mode.)[85]

(3) To specify elevator control in accelerated (or maneuvering) flight, Gilruth adopted a criterion that was just at that time becoming current—"stick force per g," where g is the symbol for the acceleration

of gravity (see below). This was a perceptive choice; from the extensive testing in World War II, it would quickly prove the most meaningful in relation to pilot opinion. The criterion seems to have been arrived at independently and more or less simultaneously by Gilruth and Sidney B. Gates of Britain's Royal Aircraft Establishment. Gilruth used it in his requirements without rationale; Gates explained his reasoning in a published report in 1942.[86]

Gates's arguments, based on sophisticated theoretical reasoning, boil down roughly to this: Logic would suggest for the criterion for maneuverability some ratio between an action on the part of the pilot and a significant effect in the resulting maneuver. Gates considered several alternatives. For the pilot's action, he recommended stick force (customarily in pounds) "because the pilot's strength is fundamental to the problems both of manoeuvrability and safety."[87] For the effect he took the normal acceleration (acceleration at right angles to the flight path) in a simple curved-path maneuver such as a steady turn or a pull-up from a shallow dive. The normal acceleration is significant because it is this acceleration, directed toward the center of curvature of the path, that causes the path to deviate from a straight line. Such acceleration derives from an increase in lift on the wing and is therefore limited by the maximum load the wing can sustain; it thus at the same time relates to the structural strength of the vehicle. Acceleration in flight is customarily measured in multiples of the acceleration of gravity, g, that is, the vertical acceleration experienced by a body falling freely in a vacuum.[88] The ratio of stick force to the resulting acceleration in multiples of g—the stick force per g—is thus a rational criterion for maneuverability. Though reasoned from simple steady-state maneuvers, it has the advantage of a generalized final form. Unlike the criteria of Warner and Thompson, which were stated in terms of specific flight routines, it retains meaning in a general context.

Gilruth put the new criterion to good use. To avoid heavy controls and pilot fatigue, he required on the basis of his tests that its value measured in a steady turn should be less than six pounds per g on fighter-type airplanes. For bombers and transports, which are not called on to maneuver continuously, values up to fifty pounds per g were allowed.[89] (He also argued on physical grounds that, under comparable conditions, stick forces are less in steady than in accelerated flight; an upper limit therefore had to be set only for the accelerated case.) To insure both types of airplane against inadvertent overload of the structure by small actions, Gilruth added an overriding requirement: attainment of the maximum number of gs the structure was designed to withstand (a higher number for fighters than for bombers and transports) should call for a steady pull of not less than thirty pounds.

Aeronautical engineers today express amazement that any maneuverability criterion besides stick force per g ever existed and that the idea took so long to develop. They consider it obvious on the face of it. The fact is, however, that five years were needed for the aeronautical community to grapple with the problem and come up with the answer. No matter how simple it may look now, the solution required considerable analytical and experimental effort from the undoubtedly competent people who developed it. Wallace Stegner's analogy with the Doppler effect, quoted in chapter 1, was clearly at work.

The three foregoing subdivisions were typical. Of the requirements as a whole, Gilruth wrote that they "require characteristics that have been demonstrated to be essential for reasonably safe and efficient operation of an airplane. They go as far toward requiring ideal characteristics as present design methods will permit. Compliance with the specifications should insure satisfactory flying qualities on the basis of present standards."[90]

Gilruth's requirements quickly attracted attention. Immediately upon receiving the classified report, officers at the Bureau of Aeronautics exchanged the penciled memoranda mentioned earlier, in which they discussed rewriting their existing specifications on stability and control. After some disagreement about whether or not to do so, Commander Diehl drafted a proposed specification that "includes all of [the NACA requirements] worth including." Gilruth also recalled some years later that, when his report was received in England, "with whom we were working very closely," the English "thought it was a great report and . . . sent a mission over to America to talk to me about it. And they used many of our criteria." Designers in the United States likewise found the requirements useful. Irving L. Ashkenas, who was a design engineer at North American Aviation at the time, has written that "the author can personally recall his own problems with the sizing of the B-25 aileron . . . and how delighted he was to find his problem solved by Gilruth's very early set of requirements for satisfactory flying qualities." Reception was not without its skeptics, however. Waldemar Breuhaus, then a young engineer at Vought-Sikorsky (successor to Chance Vought), remembers his department head, Paul Baker, flinging a copy of Diehl's draft on his desk with the words, "Read this and see if we can live with it. If it's adopted we'll be stuck with the damned thing forever!"[91]

Gilruth's NACA "requirements" had only the force of effectiveness behind them. Many of their features became mandatory for military aircraft, however, when the navy's Bureau of Aeronautics (1942) and the army air forces (1943) incorporated them into their specifications for stability and control. The services had issued stability-and-control specifications before, but these were the first to give comprehensive consideration to the demands of flying qualities. The military specifications differed from each other and from the NACA requirements in organization and detail; they followed Gilruth's approach, however, and took over many of his provisions. They adopted, for example, the short-period mode as the focus of concern in dynamic stability and stick force per g as the criterion for elevator control.[92]

Regulations of government agencies overseeing civil transport aircraft showed less influence from Gilruth's work. Maneuverability is generally less critical for civil than for military airplanes, whose performance in combat may require rapid aggressive or evasive action. Civil regulations therefore tend to be limited to the minimum necessary for safety, with the finer points of flying qualities left to the judgment of operators and designers. The airworthiness requirements of the Civil Aeronautics Board in 1946, for example, while they put considerable emphasis on longitudinal control near stalling speed and with flaps and landing gear extended, made no mention of stick force per g in accelerated flight. Airlines could and did specify non-safety-related flying qualities for airplane manufacturers (as Warner and the airlines had done for Douglas), but this remained for them to decide.[93]

Thus, by the early 1940s, the problem of flying qualities, though far from completely solved, was at last well defined. The road from the recognized but ill-defined problem of 1918 had been a long and complicated one. The idea that subjective pilot preferences could be embodied in objective design requirements, itself the product of a decade and a half of learning, had been validated by producing a set of requirements that accomplished that job. From here on, the problem of flying qualities was conceptually a different ball game. Research engineers could now devote themselves to refining and extending the requirements with confidence that the idea was useful. Designers at the same time had a greatly improved understanding of what was wanted in flying qualities and explicit specifications at which to aim. They didn't always succeed, of course; knowledge of *how* to design to a given requirement still left much to be desired. (Control forces on the DC-4E, for example, turned out at first to be "unduly heavy.")[94] Their problem now, however, was mainly one of designing (i.e., proportioning) the airplane rather than deciding at the same time what to design *for*. The early 1940s thus marked a watershed in the evolution of the flying-qualities problem. Paul Baker's skeptical statement a few paragraphs

back conveys the sense that contemporaries realized something fundamental and irrevocable was taking place.

Work on flying qualities and flying-quality specifications has advanced enormously in subsequent years. Largely under pressure of
World War II, the NACA tested from 1939 to 1949 approximately sixty
airplanes of all types. On the basis of experience from this and later
sources, the military services have revised their specifications periodically to cope with the flying-quality effects of increases in aircraft performance and changes in aircraft configurations and piloting tasks.
The work has been fostered by an assortment of new research tools: (1)
variable-stability airplanes, whose response characteristics can be varied deliberately (by means of power control systems and computer
electronics) for systematic exploration of a wide range of pilot-aircraft
interactions; (2) ground-based simulators that achieve the same goal in
the laboratory; (3) conceptualization and analysis of pilot and airplane
as a dynamic closed-loop system, including mathematical modeling of
the pilot; (4) standardized terminology and definitions for research
pilots and engineers and standard rating scales for quantification of
pilot opinion. The variable-stability airplane, pioneered in the late
1940s at the Cornell Aeronautical Laboratory in Buffalo and the
NACA's Ames Aeronautical Laboratory at Moffett Field, California,
had particularly dramatic impact. Based on the foregoing tools, military specifications have become greatly enlarged in scope and highly
differentiated with regard to phase of flight and level of acceptability of
compliance. Four types of airplane are also distinguished, with differences between types now appearing in numerous places. Evolution
continues today, and flying qualities remain a highly active field.[95] The
initial specifications arrived at by the community of engineeers and
pilots in the 1920s and 1930s, however, pointed the way.

Specifications, Communities, and
Traditions of Practice

The story of flying-quality specifications illustrates how a community
of engineers translates an amorphous, qualitative design problem into
a quantitatively specifiable problem susceptible of realistically attainable solution. In this sense I have called the former problem "ill defined" and the latter "well defined." Ill-defined problems and the specifications that render them well defined, however, are not all of a kind.

This fact becomes evident if we compare the sections on performance and flying qualities from the specifications for the DC-4E. The
former, headed "Performance Guarantees," gave quantitative requirements on such things as gross weight, useful load, and fuel capacity,

plus speed, rate of climb, and ceiling under various power conditions. A typical requirement of the last sort stated, for example, that at a gross weight of 54,000 pounds the airplane "shall be capable" of a cruising speed of 175 miles per hour at 1,800 horsepower at 10,000 feet. The performance requirements represented a compromise between the sometimes divergent wants of United and the other participating airlines. Edward Warner played an important and difficult role in resolving these differences. The performance specifications, however, were not peculiarly his work in the same way as the flying-quality requirements.[96]

The performance specifications, like those for flying qualities, gave directions to Douglas engineers about what to design for; the problem of how to meet the specifications was left to the designers. By giving directions, both sets of specifications embodied initially ill-defined goals in well-defined form. The president of United, William A. Patterson, and his board of directors had initiated this process for the performance specifications by asking for a four-engine airplane that could handle the increased traffic expected in the late 1930s and that would enable United "to develop passenger and express business at a cost which we hoped would make us less dependent on any one particular source of revenue than exists today."[97] The airplane would, of course, have to be compatible with the distances and airport elevations in United's route structure. United's staff, the Hunsaker-Mead study group, and, finally, Edward Warner and engineers from the various cooperating airlines then translated these qualitative commercial goals into quantitative performance requirements. The process involved consultation with Douglas engineers and drew on past experience together with projections of what might reasonably be expected from recent and anticipated technical advances. In terms of the design hierarchy set forth in chapter 1, the process was that of "project definition." Warner, perceiving the need and that the time was ripe, added the specifications for flying qualities. Here, as we saw, he drew again on existing understanding, his own and that of the pilots and engineers with whom he consulted. In the end, both sets of specifications laid down quantitative requirements for the designer, expressed in terms of the hardware. These requirements constituted the well-defined, objective means whereby the originally ill-defined ends—a commercially viable airplane in one case and piloting ease, precision, and confidence in the other—were to be attained. Both sets of specifications had utility only on the assumption that their requirements could realistically be met. Such assumption is basic to formulation of any practical engineering specification.[98]

Despite these similarities, the two kinds of specifications have important differences. The differences surface in the way requirements are put—the performance specifications give precise figures (such and such speed at such and such power and altitude), while flying-quality specifications only set limits (stick force per *g* of not more than such and such or less than such and such). This contrast reflects the fact that the parameters and criteria for performance are relatively simple and obvious and can be related fairly directly to operating costs and hence to the objective economic (i.e., bottom-line) goals of the airline. Once such relationship has been worked out in a given case, precise specifications can be set down for the designer. Analogous though less categorical observations can be made about performance specifications for military aircraft. For flying qualities, the parameters and criteria are complicated and problematic and not so directly related to pilots' reactions, which are themselves subjective. Here it is not warranted to set down precise requirements, hence the recourse to limits within which the designer has freedom to work. As a result, flying-quality specifications, though written as specifications, have something of the air of a guide for the designer instead of hard and fast requirements.

The hard and fast nature of performance specifications has significant implications. Though such specifications are the means to an end for the customer, satisfying their objective, precisely stated requirements becomes an end in itself for the designer. The final test of whether the specifications are met can be made by unambiguous measurements of speed, power, altitude, and so on in flight. As in the agreement between the airlines and Douglas, both parties may even call such specifications "performance guarantees." Though conformity with the quantitative flying-quality specifications can also be measured in flight, the final test there remains the pilots' subjective reactions. The flying-quality specifications retain their function as means—a design guide—and resist becoming an end. In the words of a veteran flying-qualities engineer, "We aren't smart enough to be sure that strict compliance with the requirements of the flying-qualities spec will insure a 'good' airplane."[99] Thus, *for the designer*, the quantities set down in performance specifications are themselves *objective ends*; the quantities prescribed in specifications of flying qualities are *objective means* to an associated *subjective end*.

Other related differences exist. Performance specifications, being highly precise, are particular to particular airplanes or to several airplanes in a design competition (e.g., two different fighter designs competing for a given military order); flying-quality specifications, to the extent they differ at all, tend to be generic for airplanes of a given class (e.g., trainers, fighters, or bombers and transports). Performance specifications, by the same token, may also shift more drastically with time, as, for example, when the steep increase in fuel prices in the 1970s put suddenly increased emphasis on fuel economy for commercial aircraft. Flying-quality specifications likewise change with changes in piloting tasks or airplane configurations; pilot opinion being as subjective and imprecise as it is, however, and major configuration change being a slow process, changes here tend to be gradual modifications. Finally, consistent with their status as design ends, performance specifications are enforced unquestioningly in their own right when a new airplane is tested and evaluated. With flying-quality specifications the situation is not so clear. Once they are written, a temptation exists, despite their relative looseness of statement, to see them as objective ends in the same way as performance figures and desire to enforce them accordingly. This view is at odds with the nebulous character of the flying-qualities problem and its original subjective goal. This tension can cause problems between customers and designers. In the opinion of the flying-qualities engineer quoted previously, "It will be a black day if conformance with specification requirements is permitted to override the considered judgement of the pilots."[100]

The matter of configuration change in the preceding paragraph deserves special note. As shown in figure 3-7, the airplanes that led to Gilruth's requirements all had lateral symmetry, a straight wing, and a horizontal and vertical tail at the rear. The data and pilots' opinions were limited to this normal configuration. The resulting specifications could be expected to apply with assurance only to designs that did not depart significantly from such arrangement. Limitation of this kind is fundamental to any set of specifications intended as means to achieve certain qualities in a device and based on empirical evidence from a cross section of such devices. When flying-quality specifications came to be applied to airplanes with swept-back wings in the 1940s and 1950s, they had to be reassessed and modified. Similar reassessment has taken place in response to other configuration changes. It is going on currently (mid-1980s)—by means necessarily of ground-based simulators, since flight experience with such airplanes is still very limited—for proposed designs with lateral asymmetry.

Not only is the nature of flying-quality specifications complex, so also was their attainment. In the detailed history we can distinguish at least seven interactive elements, all essential to the story; together they show the complex epistemological structure of the learning process. I shall list the elements and then discuss them only briefly. As should be apparent, they did not emerge in sequence or in simple relation to the divisions in our historical study. Some obviously had to come before others, but mostly they interacted concurrently and iteratively throughout the history. They are as follows (boldfaced words will provide an identifying shorthand phrase for the subsequent discussion):

1. **Familiarization with** vehicle and recognition of **problem.**
2. **Identification of** basic variables and derivation of analytical concepts and **criteria.**
3. **Development of instruments and** piloting **techniques** for measurements in flight.
4. **Growth** and refinement **of** pilot **opinion** regarding desirable flying qualities.
5. Combination of partial results from 2, 3, and 4 into deliberate, practical **scheme for** flying-quality **research.**
6. **Measurement of** relevant flight **characteristics** for a cross section of aircraft.
7. **Assessment of results** and data on flight characteristics in light of pilot opinion to arrive at general specifications.

The elements and their interaction were conditioned throughout—in some instances, required—by the ill-defined, subjective nature of the initial problem. They have implications beyond the present study.

Familiarization with problem is evident in the events before 1918, though it continued to grow thereafter. Activities of the early pilots and designers, including the intensive experience of World War I, provided the essential basis for learning "the true nature of the airplane."[101] Pilots found, sometimes fatally, that some airplanes could be flown and others could not. In this, the point of view of the birdmen was crucial. But, while even Lilienthal and the Wrights undoubtedly possessed some feeling for flying qualities, the problem had to become explicitly recognized by a considerable community before the widespread cumulative action needed for its solution could begin. The matter-of-fact appearance of the term *flying qualities* in the title of Shorty Schroeder's test report in 1918 suggests such recognition was commonplace by that time.

Identification of criteria, development of instruments and techniques, and **growth of opinion** provided the basic analytical, experimental, and evaluative capability, respectively. **Identification of criteria** included such things as stick force and elevator angle as pertinent variables, the long- and short-period modes as concepts in dynamic stability, and stick force per g as a criterion of control. The issue here was how to represent flying qualities quantitatively in engineering terms. This intellectual element interacted iteratively with the empirical elements all through the story, the dynamic modes appearing near the beginning and stick force per g not till nearly the end. Which variables, concepts, and criteria survived as significant ones was decided on the basis of such interaction. The equally essential **development of instruments and techniques** took place rapidly in the early 1920s, but improvement and refinement continued throughout. **Growth of opinion** occurred cumulatively and in part subconsciously over three to four decades in the general flying community. Pilots gradually learned the nature of their confidence or apprehension about the airplanes they flew, so that in time an Eddie Allen or a Mel Gough "knew what he liked" and could say so.

Identification of criteria, development of instruments and techniques, and **growth of opinion**, though interactive in an ad hoc way from the start, came together in a deliberate **scheme for research** at Langley Field in the mid-1930s. This scheme started with Thompson's tentative schedule of requirements and test routines, inspired in part by Warner's attempt at specifications for the DC-4E; essential verification and revision came from Soulé's and Thompson's flight evaluations with the Stinson and B-10B. (The schedule and evaluations constituted the first and second steps, respectively, in the Langley master plan.) This analytical and methodological synthesis supplied the essential tool for the organized research that followed.

Measurement of characteristics and **assessment of results** mainly carried through that research, though they had operated in a less organized way for some time. Measurement of relevant flight characteristics had gone on increasingly since the early 1920s in research, military, and industrial establishments, but Gilruth's program (the third step in the Langley plan) was the first to collect consistent, comprehensive data. Such collection, however, was hardly routine; as exemplified by the short-period oscillations of the Lockheed 14-H, it also brought to light phenomena that caused revision of knowledge under **identification of criteria. Assessment of results**—in particular, of just *what* about airplanes made pilots confident or apprehensive—culminated in Gilruth's requirements, based on the comprehensive Langley data from **measurement of characteristics** plus ongoing **growth of opinion**. Such assessment had already gone on tentatively, however, when Warner produced his specifications for the DC-4E and—still earlier and in more amorphous fashion—when the aeronautical community achieved consensus on the balance between stability and control in the 1920s.

The early formulation of performance specifications, though I don't propose to examine it in detail, would presumably show different structure. Basic performance variables—weight, speed, altitude, and so forth—must have been obvious to the earliest designers, and no complicated concepts and criteria like those for flying qualities had to be found. Moreover, nothing equivalent to pilot opinion entered the picture; military officers, executives, and engineers who translated the military services' and airlines' ill-defined needs into well-defined specifications had to exercise judgment, but the subjective component was of an order smaller than that involved here. For these (and perhaps other) reasons, the aeronautical community learned to write performance specifications without the lengthy experience and deliberate research needed for flying qualities.

The process that brought flying-quality specifications into being was thus conceptually, analytically, and experimentally complex. It entailed complicated interplay between numerous elements having both objective and subjective components. The latter entered explicitly wherever pilots' perceptions were involved (**familiarization with problem, growth of opinion**, and **assessment of results** above), though they were constantly in the background elsewhere. Throughout the story, the nature of the flying-quality problem and the means for its solution unfolded simultaneously, always within an urgent context of practical design demands. For a quarter century, the aeronautical community engaged, to use the words applied by Donald Schön to the design process itself, in a "reflective conversation with the materials of the situation."[102]

Other devices with a human operator can be expected to require a similar approach. Most of the elements observed here find counterpart, for example, in the investigation of automobile handling qualities, which followed and was influenced by experience with aircraft. The similarity is also not limited to vehicles. Neuromuscular perceptions of the operator influence the design of stationary devices such as powered telemanipulators (used for remote handling in hostile environments) and—while it is not usually thought of as a mechanical device—the piano. Concert pianists such as Mozart (demonstrably) and generations of anonymous piano tuners and technicians (presumably) expressed their opinions to piano designers and, though without producing formal specifications, thus affected their designs. In all these cases, the forces and movements of the controls, in relation to the perceived response of the device, engender in the operator a feeling of ease or difficulty in performing the required tasks. Treatment of such problems is a logical part of what in the last forty years has come to be called human-factors engineering (or ergonomics), which encompasses all aspects of human interaction with complex technical systems. The instances here, however, have occurred largely outside explicit activity in that field.[103]

As mentioned at the outset, the socially complex nature of our story also illustrates the complicated structure of "communities of technological practitioners" discussed by Edward Constant. Over time, a loosely but clearly identifiable community appeared, made up of people with professional commitment to the problem of flying qualities in one way or another and communication with and dependence on each other. Following the hierarchical and overlapping pattern noted by Constant, this flying-quality community was a subset of a broader stability-and-control community, itself contained within a much larger community of practitioners of aerodynamics generally. The flying-quality community in turn included at least four functionally differentiated subcommunities (most of whose members had concurrent commitments to other problems as well). Three of the subcommunities—engineering research, instrument development, and test flying—appear explicitly in the story; the fourth—design—is less specifically evident in, for example, the engineering organizations with which Warner dealt at United Airlines and Douglas, the design engineers such as Husted and Korvin-Krukovsky, who wrote articles for magazines, and the military design monitors such as Walter Diehl at the Bureau of Aeronautics. These subcommunities interacted through numerous means, including people who belonged to more than one subgroup: for example, Warner (research and design) and Gough (research and test flying). The overall community, in fact, actively encouraged the latter combination; in doing so, it learned that "it is easier to make a test pilot out of an engineer than an engineer out of a test pilot." (The Wright brothers, it should be noted, embodied all four functions at a time before division into subcommunities was practicable.) At the same time, one sees obvious evidence of institutional communities intersecting and overlapping the functional subcommunities; these are identified variously with manufacturing companies, airlines, research laboratories, universities, and the military. In focusing usefully on the functional groups, the present work confirms Constant's emphasis on the utility of "a community of practitioners as a primary unit of historical analysis." This confirmation comes within an overlying focus on, to use Aitken's phrase (see chapter 1), "the flows of information." Historians of technology may do well to concentrate more deliberately than they have on the generation and movement of knowledge in and between communities.[104]

Certain people—Norton, Allen, Soulé, Thompson, Gilruth, and, most of all, Warner—obviously played key roles within and between the subcommunities. It in no way downgrades their achievements to suggest, however, that if they had not made them, others surely would have. Even Warner's activities as a translator could well have been supplied by several people working in concert. In fact, parallel development did take place simultaneously and for the most part independently in Britain. The context of aeronautical development generally in the 1920s and 1930s clearly called for the codification and specification of flying qualities. What was indispensable was the community of research and design engineers, instrumentation specialists, and test pilots within which the learning process took place. One can imagine the story with a different cast of characters but not without a flying-quality community.

The history of flying qualities also illustrates what Constant calls "traditions of practice" in technology, which he defines as "conventional system[s] for accomplishing a specified technical task." Such a tradition is "proximately tautological with the community that embodies it; each serves to define the other." In the 1940s, the conventional system employed by the airplane-design community was quickly extended to include specification of flying qualities. From the military and commercial regulations of the period, in fact, such inclusion became mandatory. Today everyone in the design community simply takes it for granted that flying qualities, like numerous other things, must be specified before design of a new airplane can begin. Such specification thus forms part of the continually expanding tradition of practice in airplane design.[105]

With the foregoing development came another kind of tradition, also a component in Constant's model of technological change. Our period and the years following saw the devising by engineers and pilots of prescribed routines to test aircraft to see if they met their flying-quality specifications. These methods, essential to customer and airplane designer alike, have become part of well-established proof-testing procedures used across the industry to evaluate prototype aircraft. Such procedures warrant Constant's term of *tradition of testability*. In Constant's view, such traditions, which embody the crucial evaluative element that testing supplies for technology, constitute "the central mechanism both for the incremental improvement of conventional systems and for the comparative evaluation of alternative systems." Both incremental improvement and comparative evaluation can be served by flying-quality testing, though the comparative evaluation here is between variants of a given conventional system (the airplane) rather than between markedly different systems as Constant had in mind.[106] In the present instance, however, the main goal is the still additional one of assuring that a new variant of the given system simply meets design specifications. Whatever the goal, the tradition of testability in the case of flying qualities lies entirely within a larger, single tradition of practice, that of airplane design. Such relationship contrasts with Constant's detailed example of dynamometer testing of prime movers, which provides a tradition of testability separate from and cutting across a number of different traditions of practice (water turbines, electric motors, internal-combustion engines, etc.). Both types are clearly possible.[107]

Finally, some remarks about the consensus on stability in the 1920s, which epitomizes the essentially engineering nature of the flying-quality work. The conclusion that airplanes ought to be inherently stable but not too much so was a judgment rendered, not from careful deliberation by an individual or individuals, but more or less instinctively by the flying-quality community as a whole. It was arrived at gradually in the first three decades of the century by relating the growing, objective knowledge of stability and control to the similarly growing body of subjective piloting experience with expanding flying tasks. The outcome was a balance, or tradeoff, between conflicting requirements—control versus stability—of a kind that engineers often find necessary. The necessity, here, however, did not arise, as engineering tradeoffs often do, for economic reasons. Neither did it derive, of itself, from purely theoretical requirements—it came into being because of the practical needs and limitations of the human pilot. The balance therefore could not have been achieved on purely intellectual grounds and without extensive flight experience. It summarized a practical design judgment (based in this case on subjective opinion) of a sort that cannot be avoided in engineering. Achieving such judgment—and these remarks apply equally to the flying-quality work as a whole—involved elements that can properly be regarded as science (e.g., the analysis of dynamic response by applied mathematicians). Predominantly, however, it called for a great deal more than simply the application of scientific knowledge and principles. It required a complex interaction of intellectual and experiential elements reacting to and going on within a context of practical design demands. This "great deal more" provides evidence, if such is still needed, of the epistemological difference between applied science and engineering.[108]

Thus, an engineering community, by characteristically engineering activity, gained knowledge of how to translate a generalized, ill-defined problem into a relatively well-defined problem for engineering design. The initial problem contained an essential, subjective human component and arose out of the different but related needs of airplane designers and pilots. The twenty-five-year learning process proved socially as well as technically complex, involving numerous intellectual and practical elements and the interaction of a number of subgroups within a growing flying-quality community. In the end, the concept of flying-quality specifications (that is, the idea that desired subjective ends for the pilot could be attained through objective requirements for the designer) led to establishment of such specifications for the aircraft industry. These specifications, and the test procedures required to support them, have become incorporated into the tradition of practice in airplane design. The story of flying qualities, perhaps more than any other in the book, thus illustrates the intellectual and practical richness of engineering knowledge.

Criteria for Stick-Fixed and
Stick-Free Static Longitudinal Stability

Stick-fixed stability: Assume an airplane with the measured variation of elevator angle as a function of angle of attack for straight steady flight shown in the upper part of figure 3-8. (As pointed out in the text, flight speed decreases as angle of attack increases. The horizontal axis of this figure is therefore inverted from that of figure 3-3.) We do not know at this point whether the variation shown is stable or unstable.

Suppose the airplane to be flying steadily at angle of attack α_1 *with elevator locked* (stick fixed) at the corresponding elevator angle δ_1. Suppose further that the angle of attack is suddenly increased to a slightly higher value α_2 by some chance, transitory disturbance, such as an upward gust. The pilot now decides not to allow the airplane to respond on its own with controls locked but to establish steady flight at the new angle of attack α_2. To accomplish this, he or she must (according to

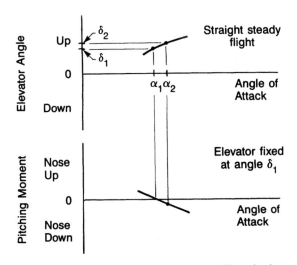

Fig. 3-8. Demonstration of stick-fixed stability criterion.

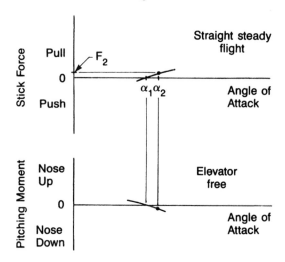

Fig. 3-9. Demonstration of stick-free stability criterion.

the figure) unlock the elevator and raise it to the corresponding elevator angle δ_2, which imposes a downward aerodynamic load on the tail and hence a nose-up pitching moment on the airplane. This means that the elevator-*fixed* moment that is being counteracted to establish steady flight at the new angle must have been nose-down; that is, the pitching-moment curve of the airplane *with elevator fixed at angle* δ_1 must be as shown in the lower part of figure 3-8. But we already know from figure 3-1 that an airplane having a moment curve of this type (i.e., with negative slope) is statically stable. It follows that the measured variation in the upper part of the present figure is that for elevator- or stick-fixed stability. When the horizontal axis of the upper part of figure 3-8 is inverted to plot elevator angle versus speed, we obtain the situation labeled "Stable" in the upper half of figure 3-3. Analogous reasoning can be used to arrive at the situation marked "Unstable."

Stick-free stability: Assume the airplane has the measured variation of stick force versus angle of attack for straight steady flight shown in the upper part of figure 3-9. Again, we do not know whether such variation is stable or unstable.

Suppose the pilot has the airplane adjusted to fly steadily at angle of attack α_1, now with *hands off* (stick free) and therefore with zero stick force. Suppose again the angle of attack is increased to α_2 by a chance disturbance. To establish steady flight at this new angle, the pilot would now have to grab hold of the control stick and apply a pull of amount F_2. But existence of a pull indicates the elevator has been raised above its otherwise freely trailing angular position. As before, such upward rotation causes a downward load on the tail and a nose-up pitching moment. Thus, the elevator-*free* moment being counteracted must have been nose-down; that is, the moment curve of the airplane *with elevator free to trail* must have the inclination shown in the lower part of figure 3-9. (This moment curve is quantitatively different from that in figure 3-8.) This again is a stable moment curve, and the measured variation of stick force versus angle of attack is therefore stable. With the horizontal axis of this variation inverted, we obtain the situation marked "Stable" in the lower half of figure 3-3. Analogous reasoning will lead to the situation marked "Unstable."

CHAPTER 4

A Theoretical Tool for Design:
Control-Volume Analysis, 1912–1953

The statement is sometimes made, usually in the abstract, that engineers think about their problems differently from scientists. This chapter identifies a specific instance of that difference, traces its history, and analyzes the reasons behind it. I shall argue that the difference flows in the present case primarily from a combination of differences in the physical problems and economic constraints of science and engineering. The findings say something about engineering science in relation to science, about the sense in which engineering knowledge is cumulative, and about the importance of ways of thinking in engineering. A basic determinative throughout is the fundamental difference between engineering as the creation of artifacts and science as the pursuit of understanding.

The instance in question appears most strikingly in the way modern textbooks of engineering and physics treat problems in thermodynamics, a subject of vital concern to both professions. Engineering texts speak frequently of a so-called control volume, an imagined spatial volume having certain characteristics and introduced for purposes of analysis. These texts are replete with diagrams showing—conventionally by dashed lines—the imaginary "control surface" enclosing such a volume. Textbooks on thermodynamics for physicists, on the other hand, are notable for the complete absence of such ideas and diagrams. The same difference appears in textbooks dealing with fluid mechanics, at least to the relatively limited extent that physicists have concerned themselves with this subject in the past fifty years. These differences in the literature provide unmistakable evidence of a difference in thinking between science and engineering.[1]

As the foregoing suggests, the original focus of this chapter arose when my concern was still for the relationship of science and technology. The topic has pertinence here because the way of thinking in question and the mathematical formalism that embodies it stemmed directly from the demands of design. Engineers require analytical tools to carry out design calculations; the control volume and the equations that go with it serve that purpose for the wide variety of fluid-flow devices with which aeronautical, mechanical, civil, and chemical engineers must deal. (It was belated appreciation of the role of this connection that redirected my attention to design.) Such tools sometimes draw, as did this one, from existing science. Even so, the scientific knowledge must usually be reformulated to make it serve for engineers. Engineering knowledge, even when closely related to science, thus exhibits its own goals and characteristics.

Nature of Control-Volume Analysis

Figures 4-1 and 4-2 show typical examples of control-volume diagrams from two well-known textbooks in engineering thermodynamics. As exemplified in these figures, the *control volume* is an arbitrarily chosen volume, fixed in space and with fluid flowing through it. An imaginary *control surface*, shown in each case by the closed dashed line, separates the control volume from its *surroundings* (i.e., everything outside the volume). In some applications (figure 4-1), heat Q may also flow through the control surface over certain regions, or mechanical work W_x may be transmitted through the surface by a moving shaft. In others (figure 4-2), significant forces F_x (representing in this case the resultant of distributed pressures) may act on the fluid inside the control surface. In all cases, however, fluid (either liquid or gas) must by definition flow through the control surface at some location, as for example where the surface intersects the steam pipes in figure 4-1 or cuts across the entrance pipe and nozzle exit in figure 4-2. This requirement stems from the fact that the control volume is designed specifically to analyze problems involving the flow of fluids, as occurs in engines, rockets, boilers, turbomachinery, and a multitude of other devices with which engineers must cope.

The importance of the control volume is that it provides, for flow devices, a convenient framework in which to apply the physical laws governing mass, momentum, energy, and (for an occasional problem) entropy. Such application leads to general integral equations involving changes inside the volume, transport of various quantities through the control surface, and forces acting on the material within the volume. These equations provide a prescription for doing the bookkeeping, so to speak, on the physical quantities going in and out of the control volume. The engineer chooses the volume in a given case both to facilitate application of the equations and with an eye toward the quantities he knows and those he needs to find. In the advice of one text, the boundaries should be put "either where you know something or where

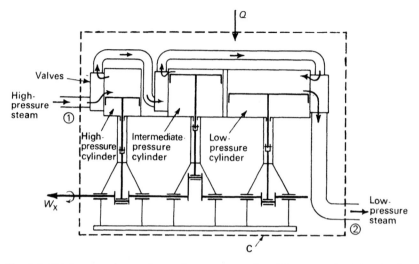

Fig. 4-1. Triple-expansion reciprocating steam engine. The control surface is the dashed line C. From D. B. Spalding and E. H. Cole, *Engineering Thermodynamics*, 3d ed. (London, 1973), p. 139, reproduced by permission of Edward Arnold Ltd.

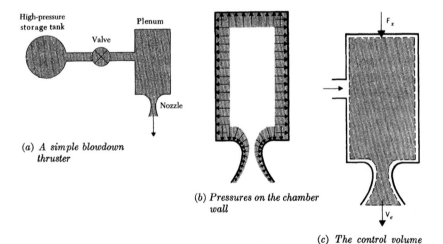

(a) *A simple blowdown thruster*

(b) *Pressures on the chamber wall*

(c) *The control volume*

Fig. 4-2. Blowdown thruster. From W. C. Reynolds and H. C. Perkins, *Engineering Thermodynamics*, 2d ed. (New York, 1977), p. 348, reproduced by permission of McGraw-Hill Book Company.

you want to know something."[2] The choice may require some ingenuity and may lead to a control surface either independent of or closely following the contours of the device (figures 4-1 and 4-2, respectively). Control-volume analysis thus comprises two elements: (1) the control volume (or control surface) and (2) the control-volume equations. With these elements the engineer has a systematic, explicit method for thinking about and making design calculations for a large and important class of engineering devices.[3]

An alternative frame in which to apply the physical laws is the *control mass*, which is an arbitrary collection of matter of given identity.[4] Here, by definition, and in contrast to the situation for the control volume, no material, fluid or otherwise, may flow through the imaginary boundary that separates the matter from its surroundings.[5] Unlike the control volume, the control mass is a common feature of thermodynamics texts for physicists as well as engineers. Almost all thermodynamic problems of interest to physicists (for example, the physical, chemical, and electrical properties of matter) do not involve fluid flow in any essential way and are thus amenable to control-mass analysis. So also is a smaller proportion of engineering problems, as, for example, the compression and ignition strokes of an internal-combustion engine. Problems that do involve fluid flow can in principle also be treated by control-mass analysis. In all but the simplest cases, however, the treatment is hopelessly complicated by the fact that the boundary of a mass of fluid of given identity changes shape and location as the fluid moves. Ascher Shapiro, an engineer who will figure prominently in the historical account, has characterized the situation as follows: "Fluids are extremely mobile, and it is therefore difficult to identify the boundaries of a [given mass of] fluid . . . for any appreciable length of time. This is particularly true in the interior of turbomachines, where complex processes occur and where different particles of fluid passing through the machine experience different histories. With fluids in motion it is, therefore, simpler to think in terms of a given volume of space through which fluid flows than it is to think in terms of a particular mass of fluid of fixed identity."[6] (Notice the verb "to think.") The situation is rather like that of traffic engineers in a large city who would become impossibly confused if they tried for long to track the cars from an initially well-defined group as a method for doing their accounting. Instead they take a simpler course and count the rate of flow of cars past a fixed boundary.

But feasibility of thought and analysis is not the only advantage of the control volume. Another, equally important, has been described in these words by Ludwig Prandtl, the central figure in the historical development: "The undoubted value of theorems [of control-volume analysis] lies in the fact that their application enables one to obtain results in physical problems from just a knowledge of the boundary conditions. There is no need to be told anything about the interior of the fluid or about the mechanism of the motion. These theorems are usually helpful where equations of motion cannot be written down, or at least cannot be integrated, and they give knowledge of the general flow without going into details."[7] That is to say, engineers frequently must deal with flow problems so complex that the underlying physics is not completely understood or the differential equations that describe the phenomena point by point cannot practically be solved throughout the flow. In such situations control-volume analysis, by working with information only on boundaries and ignoring the interior physics, can often supply limited but highly useful results of an overall nature. In figure 4-1, for example, the efficiency of the engine can be calculated without dealing with the complicated processes inside the cylinders; in figure 4-2 the force provided by the thruster can be found without knowing the complex details of the flow through the nozzle. The advantage of ignoring the interior details applies, of course, to the control mass as much as to the control volume. When such advantage is crucial in a flow problem, however, the greater simplicity of the control volume for such problems makes it the useful choice.

The utility of control-volume analysis thus depends on two requirements. The first is the need to deal with a problem involving fluid flow. This is basic—no flow, no control-volume analysis. The second, once the first is met, is a concern, usually because of some element of difficulty in the problem, for overall (as opposed to detailed) results. When these requirements are present together, as they often are, thinking in terms of a control volume becomes almost essential. Prompted by this situation, the modern engineer has developed control-volume analysis into an analytical method of power, rigor, and considerable elegance. The historical question therefore arises: When and through what process did control-volume analysis emerge as an explicit, systematic way of thinking in engineering? And are there perhaps reasons for its emergence and adoption beyond those already mentioned?

The words *explicit* and *systematic* are crucial to these questions. As we shall see, control-volume ideas were used implicitly in the ad hoc solution of individual problems long before they were organized into anything worthy to be called a method. Uses of this kind obviously involved control-volume thinking to a degree. They did not, however, constitute control-volume analysis as I have defined it and as it appears in the modern engineering literature.

In the account that follows, care in the distinction between thermodynamics and fluid mechanics is essential. Thermodynamics starts from the physical laws governing thermal energy and related energy transfers and embraces any and all problems in which these laws find use. Fluid mechanics (except for the relatively minor area of fluid statics) sets out to solve problems of fluid motion and employs whatever physical concepts and laws are needed for that purpose in the given situation. It follows that the two disciplines overlap in problems of fluid motion where thermal energy and the laws that govern it are essential. This shared area can sometimes make the disciplines seem identical. In our story both the distinction and the overlap play a role.

Development of Control-Volume Analysis

Control-volume analysis as an explicit systematic method appears to have originated in fluid mechanics in the early 1900s. That it should have its origins in fluid mechanics, instead of thermodynamics where it is so visible today, is understandable. In thermodynamics until well into the twentieth century the problems of both scientific and engineering concern did not for the most part involve fluid flow, the sine qua non of the control volume. Control-mass ideas were therefore sufficient. In fluid mechanics, on the other hand, there had been almost two hundred years of experience with problems that by definition entailed fluid flow and thus invited the control-volume approach. Moreover, these problems involved only incompressible fluids (i.e., liquids, or gases at low speeds) which could be treated without bringing in the concept of thermal energy. Fluid mechanics could therefore proceed independently of thermodynamics and, insofar as energy relations were concerned, on the basis of only the various forms of mechanical energy. Thus at the time when control-volume analysis was established, the distinction between fluid mechanics and thermodynamics was still clear, and only fluid mechanics had need for control-volume ideas.

Motivated by this need, the founders and practitioners of hydraulics and hydrodynamics of the eighteenth and nineteenth centuries had frequent recourse to control-volume thinking. Mostly they did so, however, in an implicit way in the ad hoc solution of specific, isolated problems. (As usual, the solution of individual problems was a prerequisite for systematic generalization.) The use in this period was intimately related to—indeed part of—the growth and application of what Clifford Truesdell has called "Euler's cut principle," which dates from the mid-1700s and which is fundamental to all continuum mechanics. This principle states, in Truesdell's words, that "the way to discuss the motion of a body is to cut it in the imagination into two parts, an inside and an outside, and to represent the entire action of the exterior on the interior by fields defined on the boundary." Although, according to

Truesdell, Leonhard Euler was probably not the first to state this principle, "he used it again and again with ever increasing ease, generality, and directness, to the point that his hydrodynamical research at mid-[eighteenth] century laid it bare for everyone to see, once and for all."[8] However this may be, the cut principle does not of itself distinguish between the control volume and the control mass. Furthermore, to the extent that it was applied to the former, such application before 1900 had little resemblance to the systematic method of analysis observed today.

An explicit, isolated example of the control volume, though not of the control-volume equations, does appear in Daniel Bernoulli's *Hydrodynamica* (1738), which laid the foundations of the field for which its title supplied the name. In an analysis to show that the force of a jet impinging perpendicularly onto a plate is equal to the reaction experienced by the vessel from which the jet flows, Bernoulli wrote as follows with reference to figure 4-3: "It is permissible to assume that the plate *EF* is fixed to the vessel and that the stream is surrounded by the lateral surfaces *CHDGLM*, so that the water can be assumed to flow out from the vessel *ABCHDEFGLM* through the circular opening *DEGF*."[9] The imaginary control surface *ABCHDEFGLM*, though not given its later name or indicated by a dashed line, is thus unmistakable.

A limited search of the eighteenth- and nineteenth-century literature following Bernoulli reveals no one else so conscious or explicit in this regard. Instances of the implicit use of control-volume thinking (or its equivalent) in the solution of individual problems, however, are not difficult to find. A few prominent examples will suffice: Euler's pioneering treatment in 1754 of the reaction turbine; Jean Charles de Borda's analysis in 1766 of the contraction of a jet leaving a tank through a certain type of exit (the so-called Borda mouthpiece); Robert Froude's derivation in 1889, the first correct one, of the momentum theory of marine propellers.[10] Solutions to problems thus accumulated that brought control-volume ideas slowly into the consciousness of workers in hydraulics and hydrodynamics, though in an unorganized and mostly inexplicit way. To trace this accumulation in detail through the fluid-mechanics literature of the eighteenth and nineteenth centuries would be a formidable task that I have not attempted.

Fortunately it is not essential for present purposes. The important thing here is that by early in the twentieth century control-volume ideas were prevalent implicitly in the technical literature, and there existed a collection of practical problems for which such ideas provided a solution. Moreover, these solutions had entered the engineering textbooks of the period. Typical examples were the texts on hydraulics by Henry Bovey in the United States (1895) and F. C. Lea in Britain (1911).[11]

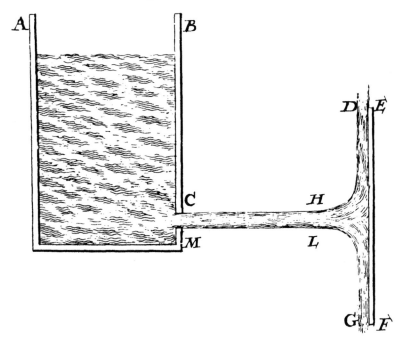

Fig. 4-3. Liquid jet flowing from a vessel and impinging onto a flat plate. From D. Bernoulli, *Hydrodynamica* (Strasbourg, 1738), fig. 84.

These books included treatment of most of the problems already mentioned—jet reaction, jet impact, the reaction turbine, the Borda mouthpiece—along with a few additional ones, such as the energy loss in a sudden pipe enlargement and the force exerted by a fluid flowing through a pipe bend. These problems all involve overall as opposed to detailed results. They also have the common feature that their solution depends crucially on application of the law governing momentum; indeed, it is the need for careful consideration of momentum changes in fluid flow that appears to have given the main impetus to control-volume thinking. The energy law appeared in Bovey and Lea, in conjunction with implicit control-volume ideas, only for derivation of the *Bernoulli equation*, an important but straightforward application that relates the various forms of energy in frictionless flow along a streamline (a path followed by a tiny portion of fluid). Both writers classified their solutions according to problem rather than method (which was still not evident), with the result that the treatments were scattered throughout their books; neither author attempted a unified treatment or presented general equations. Bovey also used control-mass instead of control-volume ideas in some of his solutions, with no apparent realization of the difference.[12] The various problems had yet to be pulled together under the common method that was implicit in their individual treatment.

The synthesis into a common method first appeared clearly in the work of Ludwig Prandtl in Germany in the 1910s and 1920s. Prandtl (figure 4-4) was an engineer by training with a doctorate from the Technical University of Munich. His research and teaching at the University of Göttingen from 1904 to 1946 have gained him recognition as the most important single contributor to the development of modern fluid mechanics. His most illustrious student, Theodore von Kármán, described him as "endowed with rare vision for the understanding of physical phenomena and unusual ability in putting them into relatively simple mathematical form."[13] To "simple" he might have added "well-organized," for Prandtl's codification of control-volume ideas showed this characteristic as well.

Interestingly enough, the first published example of the control-volume approach from Göttingen was not in Prandtl's writing but in Kármán's famous theoretical work of 1912 on the rows of vortices that appear in the flow behind a bluff body (the *Kármán vortex trail*). Kármán, who had obtained his doctorate with Prandtl in 1909 and in 1912 was working as a graduate assistant, made explicit and sophisticated use of a control volume and momentum considerations to relate the drag of the body to the spacing and speed of translation of the vortices. In a more extensive paper of the same year, written by Kármán and H. Rubach, a doctoral candidate, the control surface was delineated clearly in one of the figures (though in neither paper do the authors use the terms *control volume* or *control surface*). This advanced ad hoc application to a specific problem suggests that self-conscious use of control-volume ideas was in the intellectual atmosphere at Göttingen.[14]

Certainly it was in Prandtl's thinking. His article on fluid motion published the next year in the *Handwörterbuch der Naturwissenschaften* gave the first example I have found of the application of control-volume ideas systematically to a collection of problems. Prandtl began his discussion with a general treatment of the momentum of the fluid entering and leaving an arbitrary length of stream-tube (a more or less small, imaginary tube surrounding a streamline and within which fluid is conceived to flow) and gave a general mathematical procedure for calculating the corresponding force reactions. In a sentence that both defined the control volume and introduced the terminology and convention that have since become standard, he wrote: "In order to apply [this momentum procedure] correctly, the mass of fluid must be surrounded by a suitably closed surface, which we shall call the . . . 'control surface' [*Kontrollfläche*] (marked by a dotted line in some of the

Fig. 4-4. Ludwig Prandtl (1875–1953). From National Air and Space Museum, Smithsonian Institution.

following figures), and the reactions arising from all the stream tubes entering or leaving it must be calculated." On this basis he then applied the general procedure successively to five of the problems scattered through books like those of Bovey and Lea and to one additional problem—the aerodynamic lifting of heavy loads (as by a helicopter). One of his drawings is reproduced in figure 4-5. Nowhere in this article, however, did Prandtl set down explicitly the control-volume equation governing momentum, and the treatment left something to be desired as a systematic presentation of the control-volume concept. Its main virtue was that it established control-volume ideas explicitly and correctly in a general way and gathered a number of problems together under the same theoretical frame. Prandtl reproduced the same material without essential change in a series of influential publications in both German and English over the next forty years.[15]

Prandtl employed control-volume ideas with increasing clarity in his publications and lectures in the 1910s and 1920s. In 1917, in an appendix to a practical discussion of aircraft propellers by two of his associates, he gave a clear exposition of the momentum theory of propellers, using two different control volumes for different parts of the analysis (figure 4-6). In 1925 he used control-volume methods to correct an error in the work of another writer by analyzing the drag of a so-called half body inside a channel (figure 4-7). More importantly, he treated these problems together with those in the *Handwörterbuch* in improved fashion in his lectures, which were written down by his assistant Oskar Tietjens and published in two volumes in 1929 and 1931. In a specialized chapter in the first of these volumes, which were issued in English translation in 1934, Prandtl and Tietjens began by deriving the general control-volume equation governing momentum by mathematical transformation from a control mass to a control volume. (Newton's law concerning momentum is stated for a mass of given identity, and such transformation is logically necessary for a rigorous deriva-

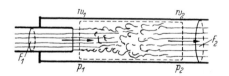

Fig. 4-5. Flow in a sudden pipe enlargement. From *Ludwig Prandtl Gesammelte Abhandlungen*, 3 vols. (Berlin, 1961), vol. 3, p. 1441, reproduced by permission of Springer-Verlag.

 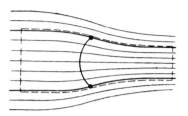

Fig. 4-6. Control volumes for momentum theory of propellers. The observer is stationary relative to the propeller, and the flow is from left to right. The effective disk of the propeller is the curved line concave to the right. From *Ludwig Prandtl Gesammelte Abhandlungen*, 3 vols. (Berlin, 1961), vol. 1, p. 303, reproduced by permission of Springer-Verlag.

tion.) This material provided a more complete and satisfactory presentation than before of the control-volume concept. The authors then went on in the same chapter to apply the general equation to six problems, five of which were in the *Handwörterbuch*. In the second volume they used the same equation to treat the half body and the Kármán vortex trail, plus two additional problems: the zero drag of a body of arbitrary shape moving in an ideal fluid (the so-called d'Alembert paradox) and the calculation of the drag of a body in a real fluid from measurements of the flow in the wake behind the body. With this systematic treatment of ten diverse problems, the power and rigor of control-volume analysis were evident. Students and practicing engineers now had a clear method and example to follow in tackling other problems.[16]

At this point we may ask, What led Prandtl to make his synthesis? The answer is that we do not know, though a number of influences can be examined. Prandtl's doctoral professor at Munich, August Föppl, is recognized as having had important influence on the application of theory to design in German engineering generally. Yet his highly popular *Vorlesungen über technische Mechanik*, published in six volumes from

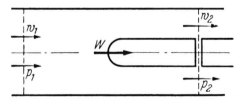

Fig. 4-7. "Half body" in a channel. From *Ludwig Prandtl Gesammelte Abhandlungen*, 3 vols. (Berlin, 1961), vol. 2, p. 618, reproduced by permission of Springer-Verlag.

1897 to 1910, contained even less of the control-volume approach than did the books of Bovey and Lea. Textbooks dealing with technical hydrodynamics were numerous in Germany in Prandtl's time, but the content of these books suggests that he influenced other writers more than they influenced him.[17] A spur to many developments in fluid mechanics in the early 1900s, which obviously did have influence here, was the demand from aeronautics for analysis of the forces on bodies moving through a surrounding fluid (in contrast to the earlier problems from hydraulics of flow inside pipes, orifices, jets, and turbines). Influence from this source appeared clearly in the new problems to which Prandtl and Kármán applied the method—the lift of a helicopter, the thrust of a propeller, and the drag of various moving bodies. Moreover, with these new problems plus the problems already accumulated by others, the material almost cried out for integrated treatment. Such treatment, in addition, coincided perfectly with a category of contribution that Hunter Rouse and Simon Ince in their *History of Hydraulics* see coming from Prandtl's institute: "methods of analysis that were neither mathematical abstractions nor empirical formulations, but practical applications of perceptive physical reasoning."[18] By its generality and organizing power, control-volume analysis provided a useful theoretical framework for just such reasoning. But the same complex of circumstances and motives could have influenced others besides Prandtl. (I shall return to a general consideration of such influences in the discussion.) In the end, if control-volume analysis was to be systematized, a development that would certainly have come sooner or later, someone had to supply the necessary insight and synthesis. For whatever immediate reasons (and I know of no autobiographical information on the point), Prandtl was that person.

Though Prandtl's lectures marked the emergence of control-volume analysis as a systematic method, his treatment dealt only with an incompressible fluid and focused almost exclusively on the equation governing momentum. In his limited concern with the energy law, only mechanical energy needed to be included. These limitations to an incompressible fluid were accurate for liquids and an acceptable approximation for gases at sufficiently low speeds. To extend Prandtl's treatment to a compressible fluid (that is, to gases at high speeds), energy considerations still had to be addressed in a more inclusive way by bringing in the concept of thermal energy from thermodynamics. This generalization of control-volume analysis took place following World War II in response to two influences: (1) the increase of flight speeds in aeronautics to the point where the effect of compressibility on aerodynamic forces was no longer negligible and (2) a growing need for the analysis of shock waves and other inherently compressible phenomena that occur in atomic blasts and in the flow around ballistic projectiles. These influences gave rise to the overlapping area of fluid mechanics and thermodynamics known as *gas dynamics*.

At the time of generalization to a compressible fluid, control-volume analysis as a systematic method had not appeared within thermodynamics itself or within the application of thermodynamics to heat-power engineering.[19] (We need consider here only engineering thermodynamics, since the control volume has not appeared in thermodynamics for physicists even today.) Prandtl himself contributed to an article on technical thermodynamics in 1905 for the *Encyklopädie der mathematischen Wissenschaften*, and, though he dealt with fluid flow, he apparently felt no need to introduce control-volume ideas in this predominantly thermodynamic treatment. More surprising in reference to the German literature is the absence of explicit control-volume analysis from such an advanced work as Aurel Stodola's volumes from 1903 to 1924 on steam and gas turbines. Stodola, professor at the Federal Technical University in Zurich, would certainly have known of Prandtl's developments by the time of his later volumes, and the control volume would have been a useful concept for him. Nonetheless, it nowhere appeared in his essentially thermodynamic approach. In the United States, textbooks in engineering thermodynamics up to the 1940s used control-volume ideas implicitly—and sometimes in rather curious and questionable ways—to derive the energy equation for steady flow in a stream-tube, but for little else. Paul Kiefer and Milton Stuart in their *Principles of Engineering Thermodynamics*, a standard American text in the 1930s, gave a statement of what is in effect the general control-volume approach for energy (including, of course, thermal energy). They did not go on, however, to develop the associated equations in any comprehensive way. *Thermodynamics* by Joseph Keenan of the Massachusetts Institute of Technology, published in 1941 and the most advanced and influential engineering text on its subject for some time, contained explicit and careful use of a control volume (though not by that name) to derive the steady-flow energy equation. Keenan's derivation, however, was an extension of an approach published by L. J. Gillespie and J. R. Coe in 1933 based on procedures from chemical thermodynamics, an approach that is awkward and obscure compared with Prandtl's transformation from a control mass. Wherever one looks, the level of analysis of flow situations in thermodynamics was primitive compared with Prandtl's synthesis in fluid mechanics. Although thermodynamics contributed important ideas to gas dynamics, control-volume analysis was not one of them.[20]

To this point the story has required discussion of events primarily in Germany. In what follows I shall limit my examination to the United

States. Although similar generalization took place in Germany, the details in America will be sufficient for the points I want to make.[21]

In the United States the generalization of control-volume analysis to its completed form was the work of engineering professors at the Massachusetts Institute of Technology. The first appearance was in the introductory fluid-mechanics text by Jerome Hunsaker and Brandon Rightmire in 1947. Hunsaker was a professor of aeronautical engineering, Rightmire of mechanical engineering, and their book showed the influence of both specialties. Shortly after, in 1953, their colleague Ascher Shapiro published a more finished version in the first of his two volumes on compressible flow. Shapiro had taken his doctorate in mechanical engineering at MIT with Keenan in 1946 and was a professor in the mechanical engineering department by the time of publication of his volumes. Despite his formal identification with mechanical engineering, the contents of his volumes also showed a clear influence from aeronautics. The texts by Hunsaker and Rightmire and by Shapiro both used the transformation from a control mass to a control volume to derive in a uniform way the control-volume equations governing mass, momentum (both linear and angular), and energy, all for a compressible fluid. Shapiro also set down for the first time the control-volume relation for entropy. Both texts used the terms *control volume* and *control surface* and followed Prandtl's convention of showing the latter by a dashed line. Both gave not only the typical applications to specific problems but also an assortment of exercises that required students to apply the concepts in new and unfamiliar situations. Control-volume analysis was now in essentially the completed form in which it appears in modern engineering texts.[22]

Further evidence of activity at MIT appeared in the textbook *Mechanical Engineering Thermodynamics* by David Mooney, published like Shapiro's work in 1953. Mooney, an engineer with the consulting firm of Jackson and Moreland in Boston, had been an assistant professor of mechanical engineering at MIT from 1945 to 1952. His treatment of the control volume, while clear and explicit, was less careful and comprehensive (only energy being included) than that of Hunsaker and Rightmire and especially Shapiro. Though less definitive, Mooney's book indicates how thoroughly the ideas of control-volume analysis had taken hold at MIT.[23]

The influences on developments at MIT came more from Prandtl's work in Germany than from anything previous in the United States. In the United States in the 1930s, several writers in fluid mechanics had made embryonic attempts at control-volume analysis, and the book with the most satisfactory treatment (by M. P. O'Brien and G. H. Hickox) appeared in a bibliography in the text of Hunsaker and Rightmire. Even at best, however, these attempts were crude and incomplete compared with that of Prandtl. Shapiro remembers that his ideas about the control volume "had been pretty well formed and developed" by the time Hunsaker and Rightmire's book appeared and that they "almost certainly came from the same source, namely the German development of this concept." Heinrich Peters, who had been an assistant at Göttingen, was on the MIT faculty from 1931 to 1940, and Shapiro remembers hearing him lecture clearly on the control-volume equation for momentum. The preface by Hunsaker and Rightmire acknowledged help in teaching their course in fluid mechanics not only from Peters but also from Richard von Mises, a member of the faculty at Harvard since 1939. Mises, who had been professor of applied mechanics at the University of Berlin from 1920 to 1933, had carried on a long correspondence with Prandtl and would certainly have known of his control-volume work; in lectures of his own at Brown University in the summer of 1941 he made careful use of a control surface, calling it by that name. Shapiro also says, in reference to the exposition of the control volume in the English translation of Prandtl and Tietjens, that his copy is "well thumbed at this spot in the text." Numerous channels thus existed for transmission of Prandtl's ideas to MIT. Recalling the circumstances, Shapiro considers that his book "came at a time when it was possible to put the control-volume concept as it applied to both fluid mechanics and thermodynamics in a coherent and unified form." The impetus for translating the possible into the actual, however, came from the changing demands of aeronautics and ballistics.[24]

From MIT control-volume analysis spread rapidly in the United States. The books of Hunsaker and Rightmire, Shapiro, and Mooney served to propagate the idea. Graduates from MIT who went into teaching at other universities used it in their lectures and reproduced it in turn in their own textbooks. Universities such as Stanford then became secondary sources for the same diffusion process.[25] The process observed in these paragraphs is, of course, not peculiar to engineering; it typifies diffusion of all scholarly knowledge.

By the foregoing events, the modern engineering professions in which flow situations play a significant role—mechanical, aeronautical, civil, and chemical—were supplied with a powerful method of analysis and a way of thinking about a broad class of problems.[26] Most important, the method is not limited to the few problems from which it grew and which are reproduced in the texts; it can be applied equally to new problems as they arise. Such applications indeed show up frequently in the literature of engineering research and design.[27] The control volume also appears in working publications dealing with standard problems of the engineering profession: for example, in the code established by the American Society of Mechanical Engineers for the testing of gas-turbine power plants.[28]

Since the mid-1950s control-volume analysis has been evident not only in fluid mechanics, where it originated, but in engineering thermodynamics as well. The situation today is thus very different from that before 1940, when such ideas, though prominent in fluid mechanics in the writings of Prandtl, were almost unknown in the literature of thermodynamics. This change followed in part from the convergence of the two fields as engineering thermodynamics became more concerned with difficult flow problems while fluid mechanics dealt increasingly with compressible flow. A second reason, however, is that engineers working with thermodynamics soon learned that use of an explicitly defined control volume helps in keeping track of the energy flows in complex systems, even when the fluid-flow aspects are elementary. Moreover, this lesson learned from the control volume—that explicitness and care in the specification of the boundary of the thermodynamic system have practical benefits—has carried over even to the control-mass analysis of nonflow problems by engineers. Regarding the explicit use of both the control volume and control mass, D. B. Spalding and E. H. Cole state in their text on engineering thermodynamics that "the extent to which they assist in clarifying thought and separating the relevant from the irrelevant can be fully appreciated only by those who were taught thermodynamics without being enjoined to relate the fluxes across and the interactions at a boundary to the changes within."[29] For this reason the use of explicitly bounded regions in general and of the control volume in particular is today, if anything, even more characteristic of engineering texts in thermodynamics than in fluid mechanics.

In retrospect, the development of control-volume analysis can be divided into four stages, dated roughly as follows: 1740 to 1910, accumulation of ad hoc solutions to individual problems in the mechanics of incompressible flow by implicit use of control-volume ideas; 1910 to 1945, synthesis and dissemination, first by Prandtl and later by others, of an explicit control-volume methodology for incompressible flow; 1945 to 1955, generalization and completion of control-volume analysis by extension to compressible flow, involving the introduction of concepts from thermodynamics; and 1955 to present, diffusion of control-volume methods in engineering education and practice in both fluid mechanics and thermodynamics. As we have seen, the impetus for the second and third stages came in large part from aeronautics, as was the case with many engineering advances at the time. The third and fourth stages were part of a general convergence of fluid mechanics and thermodynamics that has taken place in mechanical and aeronautical engineering.

Purposes, Goals, and Engineering Science

Control-volume analysis obviously has served a valuable purpose for engineers. As observed at the outset, it does not appear to have served a compelling purpose for physicists. The circumstances in physics cannot be traced historically, since it is hardly possible to write the history of something that did not happen. The situation in recent decades, however, is nicely epitomized by the widely used series of volumes on thermodynamics by Mark Zemansky, professor of physics at the City College of New York from 1925 to 1966. Zemansky's early editions (e.g., 1943), though intended for engineers as well as physicists, contained only two pages about flow processes and only elementary and implicit use of a control volume. Later (1966) Zemansky and an engineering co-author, Hendrick Van Ness, produced a derivative book designed, according to its preface, specifically "for use in . . . engineering curricula." In this version the authors expanded the treatment of flow problems to a chapter under the title "Applications of Thermodynamics to Engineering Systems." Here they stated that "in deriving the energy equation for steady-state flow processes, we found it useful to introduce the concept of a control volume." Two years later Zemansky, writing as sole author, again revised his earlier work, now with an audience only of physicists in mind. Here, in contrast to the version for engineers, all reference to flow processes was eliminated, and there was no evidence, implicit or explicit, of control-volume ideas. That there is a difference in thinking between engineering and physics could hardly be more plain.[30]

As indicated in my opening description of control-volume analysis, the source of this difference lies basically in differences in the problems confronted by engineers and physicists. To understand in more detail how and why the problems differ, we must return to the two requirements mentioned earlier for the use of control-volume analysis: (1) a need to deal with fluid flow, and (2) given such need, a concern for overall rather than detailed results.

For the physicist the two requirements are seldom present together in either thermodynamics or fluid mechanics; the reasons are different, however, in the two disciplines. The basic goal of the physicist, as with physical scientists generally, is knowledge about the physical world, but this goal manifests itself differently in the two situations. In their use of thermodynamics, physicists are interested for the most part in understanding and predicting the properties of matter. This fact is illustrated by the titles of the chapters on applications in Zemansky's volume for physicists, which include such phenomena as phase transition, paramagnetism, and superconductivity.[31] Such phenomena do not involve fluid flow in any essential way, and the basic requirement for the control volume is missing. They can be analyzed readily by examining a fixed quantity of matter that may be treated as uniform in space and time, and for this purpose the control mass suffices. In fluid mechanics, on the other hand, physicists must, by definition, deal with fluid flow. All nontrivial flows, however, vary either in space or time (or both), and a physicist in his quest for knowledge characteristically wants to know the point-by-point details of this variation. For this end the overall results of control-volume analysis are not enough. In fluid mechanics physicists therefore concentrate on solution of the differential equations of motion and have little or no need for control-volume ideas. On the infrequent occasion when such need does arise, they proceed in ad hoc fashion and without the formal methods of the engineer.

In engineering the situation is very different. Engineers, unlike physicists, are after useful artifacts and must predict the performance of the objects they design. Many of these—turbines, propellers, airplane wings—involve the flow of fluids, and this fact is essential to their analysis whether it derives from thermodynamics or fluid mechanics. Moreover, the problems that arise often present serious difficulties in the underlying physics or in the solution of the differential equations, so that more than overall results are not feasible.[32] Of the problems mentioned earlier, the turbulent flow in a sudden pipe enlargement is of the former sort, the aircraft propeller of the latter. In some problems overall results may be all the engineer needs for design, and to go into detail would be a waste of time and money—calculation of the fluid force to be resisted by the structure supporting a pipe bend is an obvious example here. For reasons of this sort, both requirements for the control volume are present in a vast number of problems in mechanical, aeronautical, and hydraulic engineering. This widespread need has made it worthwhile for engineers to develop control-volume thinking as an organized method.

But the direct requirements of engineering problems do not explain fully the eagerness with which engineers have developed and embraced control-volume analysis. After all, they could have continued, as in the books of Bovey and Lea of the early 1900s and the occasional practice of physicists today, to use control-volume ideas in ad hoc fashion as the need arose; no imperative existed in the problems themselves to develop an organized method. That is to say, the requirements of the problems constituted a necessary but not sufficient condition. That Prandtl and his successors felt prompted to develop the method as they did must have involved additional considerations.

One set of possibilities comes from the fact that Prandtl and those who generalized his work were teachers as well as engineers. Although engineering research and practice generate the relevant problems, teaching strongly motivates a search for improved ways to organize the resulting knowledge. For the teacher, organization of different flow problems under a common theoretical framework as initiated by Prandtl is clearly superior to the scattered treatment according to problem as used by earlier writers. For one thing, possible problems are numerous and diverse, whereas the physical laws embodied in the equations of control-volume analysis are few and unique. Organization according to control-volume ideas is thus not only simpler but brings clearer understanding of the physical principles common to otherwise disparate situations. Moreover, such organization allows the sometimes confusing application of the basic physical laws to be done once and for all, so that all that remains is the specialization to the case at hand. These attributes of control-volume analysis aid the teacher greatly in transmitting knowledge and understanding to new generations of engineering students. As I shall point out, the organization of knowledge helps the practicing engineer as well. The teacher, however, is more likely to perceive the need and foster the development.[33]

Whatever their role in the origin of control-volume analysis, the demands of education have certainly been important in its diffusion. Considerable numbers of students are awarded first degrees in engineering each year—47,303 bachelor's degrees in the United States in 1975, for example, compared with 3,655 in physics.[34] Engineering classes in basic subjects such as fluid mechanics and thermodynamics are often large, with correspondingly large variations in student temperament and aptitude. This situation puts a premium on organization and clarity of knowledge. Such implications of the size of the audience can hardly have influenced Prandtl's initial synthesis for his presumably small classes at Göttingen; his interest was probably more in pedagogical methods per se. They are a clear factor, however, in the proliferation of control-volume analysis in the flood of fluid-mechanics and thermodynamics texts for engineers in recent decades.[35]

Still, organization of knowledge and pedagogical clarity are as much valued by physicists as by engineers. They might equally well have influenced the physicist toward control-volume analysis if his problems had called for it. Moreover, even with engineers, whose problems do call for it, one can question whether the organization of knowledge alone would be decisive.

In the end the requirements that have tipped the scales in favor of control-volume analysis lie in the goal or mission of the engineer—to design and produce useful artifacts. The scientist, by contrast, is after knowledge and understanding of the workings of nature. This difference, of course, is not always clear-cut in practice and is often a matter of priority rather than exclusivity. It is nonetheless real. As we have seen, it translates in the present situation into crucial differences in the problems that confront the engineer and the physicist. It gives rise in engineering, however, to at least two additional requirements of critical importance.

The first of these concerns economy. Engineers, because they produce things on order in a finite world, are usually hard pressed by limited resources of time, money, and manpower. Engineering is an economic as well as a technical activity, and engineers, unlike physicists, work in an environment where cost constraints are central. Engineers are therefore trained to be economical in their use of resources, and, if they are not, their industrial employers soon see that they become so. Hence practicing engineers are always on the lookout for more effective tools with which to think and do. Control-volume analysis fills this need by saving the design engineer's time in various ways in problems of fluid motion. By organizing knowledge according to physical laws rather than known problems, it aids in recognizing a control-volume problem when met in an unfamiliar disguise. By applying the physical laws to a generalized volume once and for all, it saves time in translating the problem into mathematics. Most important, by providing an explicit conceptual framework, it facilitates clear thinking about the complexities of fluid flow. (This virtue is cited in much the same words by Spalding and Cole in the quotation given earlier.) A currently active mechanical engineer summarizes the situation thus: "I use control-volume analysis because it saves me very large amounts of time, as well as because it clarifies my perception of what is important in a problem. But even in the latter, I am mostly concerned with quick and accurate perception because it allows me to do a better job with the expenditure of fewer resources."[36] A physicist would be unlikely to write in this vein.

A second requirement, which could be subsumed under economy but which deserves mention in its own right, is freedom from error. In the words of Samuel Florman, an experienced civil engineer, "Human error, lack of imagination, and blind ignorance. The practice of engineering is in large measure a continuing struggle to avoid making mistakes for these reasons."[37] Control-volume analysis is a weapon in this struggle, a weapon the physicist does not require. The point is not that the physicist is less prone to mistakes than the engineer; it is simply that the costs of a mistake in engineering are often so much greater. In most fields of physics that do not themselves require extensive engineering projects (such as particle accelerators in high-energy physics), failure of an experiment from an error in design results in loss of a certain amount of time and money and perhaps reputation. Failure of an engineering project, besides harming substantially the professional standing of the engineer (and his employer), may lead to large losses of time and money and to considerable loss of life. Engineers, like certified public accountants, must therefore adopt standardized procedures of thought and practice to protect them from themselves—that is, to make the chance of human error as small as possible initially as well as to facilitate checking later. Control-volume analysis, by setting up an explicit method of bookkeeping for the various flow quantities, provides such a procedure for the many engineers who must deal with fluid-mechanical devices. Again, this could hardly have been a reason for Prandtl's initial work, but it undoubtedly has been a factor in the adoption of the method in engineering teaching and practice.

The special nature of engineering problems, the need of engineering teachers to organize knowledge, and the engineering demands for economy and accuracy are interrelated requirements. Although I have discussed them separately, the discussion of each has invoked one or both of the others. Taken together they supply sufficient reason for the origin and development of control-volume analysis. Underlying both the problems and the demands for economy and accuracy is the mission of the engineer to design and produce useful artifacts.

These requirements together also illustrate the fundamental role of ways of thinking in engineering. Although engineering activity produces artifacts, conceiving and analyzing these artifacts requires thoughts in people's minds; the clearer these thoughts, the more likely it is that the artifacts will be successful. Control-volume analysis is useful precisely because it provides a framework and method for thinking clearly about a large class of the often confusing problems that arise in engineering design. One wonders if other aspects of engineering knowledge might not have significance in the same way.[38]

As an example of engineering knowledge, control-volume analysis has the unusual feature, evident from its history, that it is independent of any particular contrivance. Although it is by definition intended for use in flow problems, within this restriction it applies to a broad range of items—to household refrigerators as much as to jet engines. That is to say, control-volume analysis provides a general method for a class of problems rather than information on a specific device. It thus differs from much of the technological knowledge that has been studied historically, which tends to be specific to a given machine, material, or process. In this sense, control-volume analysis is an example of *pure* technological knowledge. It offers an opportunity to view such knowledge free of the complications that attend a particular application.[39]

Control-volume analysis also affords an opportunity to examine the part of engineering knowledge called *engineering science*. Prandtl and the professors at MIT—and this statement in no way diminishes the importance of their accomplishment—did not produce anything that could be described as new science in a basic sense. The fundamental laws they used, Newton's law governing momentum and the laws of thermodynamics concerning energy and entropy, were well established from scientific inquiry. The crucial transformation they employed to go from the control mass, in which the scientific laws were expressed, to the control volume was a well-known mathematical procedure for differentiating an integral with variable limits (although they presented their own derivation of it ab initio in a physical context). However, in putting these elements together in an organized and purposeful way, which was far from a trivial matter, they generated a new and coherent body of knowledge of importance to engineers. Such specialization and extension of scientific knowledge to meet the needs of engineers is an essential aspect of engineering science.[40]

Engineering science in general is both similar to and different from science itself, and it is instructive to inquire further into this relationship in the present context. First, as to similarities:

(1) Engineering science and science both conform to the same natural laws, although one deals ostensibly with artifacts and the other with nature. This fact is almost trivially obvious in control-volume analysis, where the control-volume equations are simply transformed statements of basic laws from physics. In areas of engineering science further removed from first principles, the fact is less obvious but nonetheless real.[41]

(2) Engineering science and science diffuse through the same mechanisms. The mechanisms are visible in our account: textbooks, encyclopedia articles, journals, classroom and research teaching, and movement of people from one institution to another. The same elements could be found in almost any study of the growth of scientific knowledge. The institutional remodeling of engineering in the image of science that Edwin Layton sees taking place in the nineteenth century is complete in this area.[42]

(3) Engineering science, like science, is cumulative in the sense of one piece of knowledge building on (or being derived from) another. The stages in the development of control-volume analysis illustrate the cumulative process: Prandtl built on the earlier ad hoc solutions, the people at MIT built on Prandtl's work, and the writers of present-day texts continue to build on the work accomplished at MIT. Similar situations abound in the history of science. As pointed out by Layton, the cumulative process is especially clear-cut in science "precisely because the end product of science [as an activity] is normally an increment of knowledge" and "knowledge builds upon knowledge." The end product of engineering, on the other hand, is artifacts, whose cumulative

nature is questionable. "For artifacts cannot build upon artifacts, except through the mediation of mind." The *immediate* product of engineering science as a subactivity of engineering, however, is also knowledge, and the cumulative process is again clear.[43] It is especially obvious in control-volume analysis because, as pointed out, no specific artifact is involved.

Despite the common characteristics, differences between engineering science and science must exist—engineers have developed control-volume analysis and use it, physicists have not and do not. As with the reasons underlying the control volume, the differences arise out of a difference in purpose. In scientific knowledge the purpose is understanding of nature; in engineering science the ultimate goal (ultimate in contrast to the immediate production of knowledge as discussed above) is the creation of artifacts.[44] In the case at hand the artifacts involve fluid flow, and control-volume analysis is tailored explicitly to their design and analysis; only secondarily, if at all, does it contribute to the understanding of how they work. As in many aspects of engineering knowledge, in engineering science the purpose of design is determinative. This influence is evident here in the design origins of the problems that control-volume analysis serves, in the emphasis on overall results of concern to the design engineer, and in the demands for economy and accuracy in the design process. The overall results in particular, though of little interest for the scientist, may be the difference between success and failure for the engineer. Matters of this kind make engineering science as a body of knowledge different from science "in both style and substance." They show up in the problems to which the knowledge is adapted, the way the knowledge is formulated, the kinds and range of phenomena the knowledge encompasses, and the depth of detail the knowledge is designed to provide.[45] To the person from outside, these differences may be obscure, and the bodies of knowledge may appear much the same. Engineers and scientists, however, though they rarely think about it consciously, have little trouble making the distinction in their daily work—or, as we have seen, in the content of their textbooks.

The ideas of this discussion also say something about the aim or direction of the cumulation of engineering science. While one thing clearly built on another in the development of control-volume analysis, this cumulation was more than a simple piling up of information. The synthesis and generalization from 1910 to 1955, which produced control-volume analysis as such, resulted in much more than the addition of individual solutions to those already amassed. The result in fact was an organized way of thinking that enabled engineers to solve problems in the design of fluid-flow devices more effectively than they had been able to without it; that is, with clearer definition of the relevant parameters and principles, over an increasingly broader range (e.g., for compressible as well as incompressible fluids), and with more economy of time and less likelihood of error. (They have also been enabled to transmit this ability more effectively to succeeding generations of engineers.) Such increased effectiveness in problem solving seems to me to define the direction of cumulation in engineering science and indeed in engineering knowledge generally. Henryk Skolimowski has described technological progress broadly as "the pursuit of effectiveness in producing objects of a given kind."[46] Whatever the truth generally, the idea appears valid within the sphere of engineering knowledge.

As always, objections can be raised. My example is without doubt untypically clean and clear-cut, even within engineering science. But perhaps simplicity has its virtues in affording something firm to build on. Whether the cumulation of engineering knowledge is in the direction of solutions that are "better" in some social sense, and why Prandtl and his followers felt impelled to pursue such cumulation, are cultural questions that I do not address.[47] My effort here has been to provide another, perhaps little recognized example of how technologists "have developed their own bodies of more or less systematic knowledge to meet the needs of practice."[48] It demonstrates, I believe, that knowledge in this context must be taken to include ways of thinking.

CHAPTER 5

Data for Design:
The Air-Propeller Tests of W. F. Durand
and E. P. Lesley, 1916–1926

From 1916 to 1926, William F. Durand and Everett P. Lesley, professors of mechanical engineering at Stanford University, made a study of aircraft propellers that was described by a distinguished contemporary as "the most perfect and complete . . . ever published."[1] This primarily experimental study supplied much of the data used by airplane designers in the 1920s to select the best propeller for their new designs. It also helped set the nature of American aeronautical research in the decades before World War II. It is therefore important for the history of aeronautics. More important, the Durand-Lesley work involved in a comprehensive way a methodology widely used by engineers but little studied by historians. Examination of the work in terms of this methodology provides a wealth of evidence on a number of fundamental issues concerning engineering knowledge. This chapter will first describe the background, events, and procedures of the Durand-Lesley study in the detail needed to exhibit the methodology. It will then analyze the methodology in light of the evidence and with an eye toward the fundamental issues.

Prominent among the latter are the role of design data and the requirement for methods to acquire such data. To make their calculations, design engineers need numerical information as well as theoretical tools. In the end, dimensioned drawings must be supplied to the shop, and arriving at these concrete results requires a wealth of data of many kinds. Frequently these data concern the performance of some device, process, or material in situations where theory is inadequate. When this happens, the experimental form of the methodology described here becomes imperative. As pointed out in chapter 1, "knowing how" to acquire knowledge is itself a form of knowledge.

These along with other aspects of design appear clearly in the Durand-Lesley work. Specific requirements of design obviously conditioned the knowledge Durand and Lesley were after—they measured quantities airplane designers needed to know and presented them in appropriately useful form. The activity the designers engaged in and for which the knowledge was intended was clearly normal design— choice of propeller was routine in design of all airplanes until the appearance of the jet engine in the 1940s. The activity occurred, more-over, within a typical hierarchy of problems—the propeller was a major component in the larger system of the airplane—and the nature and use of the data were influenced by that fact. In all these ways, the aims and form of the data were closely bound up with design.

As with the preceding study, the topic of this chapter arose when my concern was primarily for the relationship of science and technology. It was, therefore, logical to inquire into the peculiarities of research methodologies in engineering and the differences between research in engineering and the physical sciences. These themes remain relevant here. We need to find out if peculiarly engineering methodologies exist, if so what their features are, and what objectives they have been created to satisfy. The limited attention given to these matters has been mostly theoretical and abstract;[2] the Durand-Lesley study provides an opportunity for analysis in terms of specific events. We find again that the engineering methods in evidence vary in both form and object from those in the physical sciences.

A number of related secondary themes also appear. They include the convergence of knowledge and methodology in different fields of engineering (specifically, marine engineering and aeronautics) and the role of incremental and inconspicuous change in engineering advance (as opposed to discontinuous and dramatic invention).[3]

Background and Content of the Methodology

The engineering methodology appears clearly—the first prominent example—in the work of John Smeaton early in the Industrial Revolution in Britain. His influential study of the performance of waterwheels and windmills, presented before the Royal Society in 1759, contained the two main methodological components: a systematic method of ex-periment and the use of working scale models.[4] A full history of these matters has not been written and is hardly possible here; a brief survey, however, will help establish ideas needed later. Since Smeaton's time the two components, usually (but not always) together, have formed the basis of an autonomous tradition of engineering research.

Smeaton's method of experiment, exemplified in his tests of model waterwheels, was to alter separately the conditions of operation of the wheel (speed and quantity of flow of water and speed of rotation of wheel) and measure the output of power. He thus followed what today may be called the method of *parameter variation*,[5] which can be defined as the procedure of repeatedly determining the performance of some material, process, or device while systematically varying the parameters that define the object of interest or its conditions of operation. Apart from possible earlier use in science, the method goes back in engineering at least to the ancient Greek catapult designers, who established the best proportions for their full-sized devices "by systematically altering the sizes of the various parts of the catapult and testing the results."[6] In addition to his tests of waterwheels and windmills, Smeaton employed parameter variation to arrive at the best composition of cement for his Eddystone Lighthouse and to improve the performance of atmospheric steam engines. Following in the Smeaton tradition, notable engineers have used the method to advantage: William Fairbairn and Eaton Hodgkinson to study the buckling of thin-walled tubes in the 1840s, Frederick Taylor to revolutionize metal cutting from 1880 to 1906, Wilbur and Orville Wright to obtain essential airfoil data in 1901—the list is easily extended.[7] Even more significant is the continuing use of parameter variation by thousands of anonymous engineers to whom the method is now so familiar as to seem only common sense.

Smeaton's innovative use of working scale models as a convenient means for parameter variation helped with one set of difficulties but raised another.[8] As discussed also in chapter 2, a central problem with scale models—a problem that is methodologically distinct from the use of parameter variation—is that data obtained from a model cannot be applied accurately to the full-scale prototype without adjustment. The need to adjust for "scale effects" is obvious in any fluid-flow device, such as a waterwheel, windmill, or propeller, if only because the forces the fluid exerts on a solid surface increase as the surface area affected by the fluid increases. Simple area effects, however, are only the beginning of the problem. Subtler influences have to be considered, such as differences in paths of motion between model and prototype and differences in the effects of various fluid properties (density, viscosity, compressibility) at different scales. Given the complexity of the problem and the inadequate theoretical knowledge of his time, Smeaton was not able to deal with these second-order difficulties. Although his results were invaluable to a designer who could apply them with judgment, Smeaton insisted that "the best structure of machines cannot be fully ascertained, but by making trials with them, when made of their proper size."[9]

Rational application of model data did not come until over a century later with William Froude's experiments on the resistance of ship hulls for the British Admiralty (1868 to 1874). Such application requires a theoretical law of similitude, and this knowledge was slow to develop. A law of similitude states typically that, provided the conditions of operation of a device are varied with its size in such-and-such a way, then the performance varies with size in such-and-such a way. The mathematical details of the statement are different for different devices and different circumstances of use. With such a statement, the designer, knowing the conditions of operation of a model and the resulting measured performance, can calculate so-called corresponding values for the conditions of operation and performance of the prototype.[10] Froude demonstrated the validity of this procedure by making calculations for the resistance of HMS *Greyhound*, using model measurements and the appropriate law of similitude. Agreement between the resistance so calculated and that measured with the ship left little doubt the method was sound.[11]

Knowledge and use of similitude evolved slowly over the next forty years. Shortly after Froude's demonstration, engineers began to understand that the simplest way to express a law of similitude is in terms of dimensionless groups.[12] A dimensionless group is a mathematical product of two or more quantities arranged such that their dimensions (length, mass, and time, or combinations thereof) cancel, leaving a "pure number," that is, a number without dimensions.[13] In applications to a mechanical device, at least one dimensionless group typically involves the size of the device and several of the parameters that specify its conditions of operation. This group then affords a simple rule for calculating corresponding conditions between model and prototype: conditions correspond when the numerical value of the group calculated for the prototype equals that calculated for the model. Corresponding performance is found in a similar way from other dimensionless groups. Specific examples for propellers will appear later. The first application of dimensionless groups to experimental work was apparently by Osborne Reynolds in his classic experimental studies of transition from laminar to turbulent flow in pipes (1883).[14]

Dimensionless groups, and through them laws of similitude, are derived by one form or another of dimensional analysis. This theory enables an investigator, from study of the dimensions of the quantities in any physical system, to deduce limitations on the form of possible relationships between those quantities. The principles of the subject were formulated originally by Jean Baptiste Fourier in his volumes on the theory of heat (1807 to 1822). Dimensional analysis began to attract more than casual attention, however, only in the last quarter of the 1800s, when it was extended and its applications explored by Lord Rayleigh in England and a number of research workers in France.

Dimensional ideas were still catching on among engineers when Durand and Lesley began their work in 1916.[15]

Knowledge of Propellers in 1916

To understand the nature and contribution of the Durand-Lesley work, we must know something about propellers and about the state of knowledge regarding them when the work began.

The propeller can be looked upon as a power-conversion device that converts the rotative power delivered by the engine into propulsive power to pull the airplane forward. The propeller designer naturally strives to produce a propeller that will do this task as efficiently as possible. But the propeller problem does not end there. The propeller operates in combination with both engine and airframe (airplane less engine and propeller), and it must be compatible with the power-output characteristics of the former and the flight requirements of the latter. (Note the existence and implications of hierarchy.) This compatibility requires that the airplane designer be able to select a propeller of suitable size and shape from among a collection of already efficient designs, a selection usually made in 1916 for a single condition of flight.[16] The selection accomplished, the designer must also be able to calculate the performance of the propeller and thence of the airplane over the range of conditions in which the vehicle will operate.[17] In assessing the historical situation we must therefore distinguish between *propeller* design and the selection and incorporation of the propeller into *airplane* design.[18] The former requires detailed knowledge and experience of propeller operation sufficient to arrive at an efficient shape. The latter calls for systematic overall data on the performance of families of geometrically related propellers, all presumably of already efficient form. Such comprehensive data, helpful but not essential for the propeller designer, are crucial for the airplane designer. Obtaining such data presupposes use of parameter variation, that is, systematic variation of the parameters that define the shape of the propeller and its conditions of operation.

By 1916 knowledge required for propeller design was reasonably well in hand, but systematic propeller data for the airplane designer were almost nonexistent. In the United States, aside from the excellent beginnings of the Wright brothers in 1902–3, little of note had been accomplished in either regard.[19] In Europe, Stefan Drzewiecki, beginning in Russia in 1885 and continuing in France, had developed a theory for the calculation of propeller performance with the aid of measured airfoil data. Experimental work on air propellers began in earnest in England, France, and Germany around 1910, and by 1913

comparisons between theoretical and experimental results in England
showed that the theory, though quantitatively unreliable, provided a
useful qualitative guide to the design of efficient propellers.[20] Experiments in England and France, mostly on individual, unrelated propellers, in fact gave respectable maximum efficiencies of 70 to 80 percent.[21] Systematic performance data, however, were limited to the
small amount obtained by Gustave Eiffel, the French structural engineer, at his private laboratory in Paris. Eiffel, pursuing an increasing
interest in aeronautics, had developed a new type of wind tunnel in
which he made extensive aerodynamic studies, including model tests
reported in 1914 of three families of four related propellers each. To
represent his results he had already in 1911 adopted the law of similitude for propellers in terms of dimensionless groups.[22] How widely the
European knowledge was disseminated in the United States by 1916 is
difficult to say, but certainly most of it was known.[23]

Marine propellers are much like air propellers, but the state of
knowledge in the marine field was very different. William Froude,
working on problems of ship propulsion, had put forth the idea underlying Drzewiecki's theory in crude form in 1878.[24] Because of the low
regard for theory among naval architects, however, Froude's lead was
not pursued.[25] Systematic experimental data on propeller models for
use by the ship designer, on the other hand, were plentiful, and Durand had been one of the leaders in producing them. His towing-basin
tests of forty-nine model marine propellers, begun in 1877 while he
was a professor at Cornell and published in 1905, were the most complete of their kind at the time. They were followed by similar tests of
thirty-six propellers by Robert Froude (William Froude's son) in England, reported in 1908, and of 120 propellers by Capt. (later Adm.)
David Taylor in the United States, reported in 1910. This accumulation
of design data represented an impressive application of parameter
variation and model testing.[26] Durand's marine-propeller tests and the
marine tradition generally had obvious influence on the way he and
Lesley approached the study of air propellers.

The Durand-Lesley Propeller Tests

The Stanford propeller study was one of the first projects supported
by the newly established National Advisory Committee for Aeronautics. Durand, who was among President Woodrow Wilson's appointees
to the original committee, proposed the study at the first meeting of the
NACA on April 23, 1915, and his suggestion was favorably received.
Though it had been set up by Congress "to supervise and direct the

scientific study of the problems of flight, with a view to their practical solution," the committee's immediate reason for being was to catch up with Europe in the development of aeronautics. Its members saw clearly that knowledge of aircraft propellers, in addition to being basic among the problems of flight, was one of the elements of European superiority. They also saw that the United States had an asset in "a number of competent authorities on propellers for water craft who are thoroughly equipped to place the design of aeronautical propellers on a satisfactory basis."[27]

William Frederick Durand (figure 5-1) was one of these authorities by virtue of his work at Cornell. He was also one of the outstanding mechanical, marine, and hydraulic engineers of his time, with broad experience in teaching and research, government service, professional leadership, and practical consulting.[28] Most important for present concerns, he was a man of well-disciplined temperament who felt at home in careful, systematic research. Although original thinking is evident in his work, his style was that of the painstaking investigator rather than the brilliant innovator. He used theory capably where necessary, but his approach to research was basically from a physical rather than a mathematical point of view.[29]

Following Durand's proposals to the committee, and after some delay for budgetary reasons, an initial contract for research on air propellers was awarded to Stanford in October 1916. The contract was "entered into . . . with a view to establishing engineering data for design,"[30] and it is clear from the course of the Stanford work that airplane, not propeller, design was what was meant. To assist in the work Durand recruited his colleague at Stanford, Everett Parker Lesley (figure 5-2). Lesley also had experience in naval architecture, with a master's degree in the subject from Cornell in 1905 followed by two years of work with Captain Taylor at the U.S. Naval Experimental Towing Tank.[31] His most important asset for the propeller experiments, however, was that he was a versatile engineer with an outstanding ability to make things work in the laboratory.

In November Durand outlined his plans in detail in a preliminary report to Naval Constructor Holden Richardson, secretary to the NACA.[32] He explained in particular how the study would proceed by tests of model propellers in which the design parameters would be varied systematically over the current range of practical concern. Except for use of a wind tunnel in place of a towing basin, the philosophy and methodology were the same as in the studies of marine propellers at Cornell. The experiments on air propellers, in fact, provide a classic example of the transfer of method, knowledge, and expertise from one

Fig. 5-1. William Frederick Durand (1859–1958). From Stanford University Archives.

Fig. 5-2. Everett Parker Lesley (1874–1945). From Stanford University Archives.

area of engineering to another. This transfer was possible because, while the detailed problems of propellers in air are different from those in water, the basic principles are the same.[33]

The first task confronting Durand and Lesley was design and construction of the wind tunnel and associated equipment, another area in which Europe was ahead of the United States. After correspondence regarding the best choice of tunnel, Durand settled on the type originated by Eiffel as the most convenient for propeller tests. This type had a free-jet test stream inside a closed room, tapering intake and exit channels, and return flow within the surrounding building (see figure 5-3). The circular test stream of the Stanford tunnel was five and one-half feet in diameter and had a maximum speed of fifty-five miles per hour. A variable-speed electric motor drove the model propeller, and a dynamometer measured the aerodynamic thrust and torque on the model. Instrumentation was provided to measure the revolutions per minute of the propeller and the speed of the wind-tunnel stream. All equipment was designed and built in the autumn and winter of 1916–17.[34]

The initial set of model propellers was tested the following spring and summer. In April Durand, now chairman of the NACA, moved to Washington to be involved in the newly begun war effort. During an absence of almost two years, he directed the research through correspondence with Lesley, who sent him detailed weekly reports on the progress of the tests. This correspondence, in the NACA collection in the National Archives, allows us to follow the work with an immediacy not possible from published reports.

To understand the Stanford tests, we must be specific about parameter variation in relation to propellers. As suggested by the definition of parameter variation, a propeller's performance is a function of two rather different sets of parameters: the conditions of operation and the geometrical properties of the propeller. The former includes as primary quantities the forward speed[35] V and the revolutions per unit time n; the latter consists of the diameter D and a number of ratios or numbers r_1, r_2, . . . etc., that specify the details of the propeller's shape.[36] With these symbols, the foregoing statement about performance can be written mathematically as

$$\text{propeller performance} = F(V, n, D, r_1, r_2, \ldots), \qquad (1)$$

which says simply that any measure of the propeller's performance—propulsive power produced, shaft power absorbed, efficiency, etc.—is some function F of the parameters within the parentheses. This relation is approximate in that it leaves out the complicating secondary effects of viscosity and compressibility of the air and elastic bending of

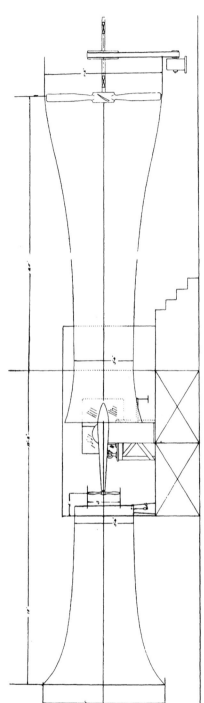

Fig. 5-3. The first Stanford wind tunnel. Flow in tunnel is from left to right. From W. F. Durand, "Experimental Research on Air Propellers," *Report No. 14*, NACA (Washington, D.C., 1917), pl. IV.

the propeller; it is accurate to the extent that these effects are in fact negligible. To this accuracy, experimental parameter variation for the propeller then consists of systematically varying the values of the parameters within the parentheses and measuring the resulting variation of propeller performance.[37]

The experimenters had first to settle on values for the parameters. After preliminary tests to find how large they could make their models without encountering aerodynamic influence from the limited size of the wind-tunnel stream, they chose a diameter of three feet for all models.[38] They also adopted the two-bladed propeller, the normal configuration at the time, as standard. For specifying the complicated shape of the blades, Durand employed five parameters, $r_1, \ldots, r_5$. The primary of these was the mean pitch ratio, a measure of the angular orientation, relative to the plane of propeller rotation, of the blade section at some standard representative radius. Roughly speaking, the larger the mean pitch ratio the higher the angular orientation of all blade sections. The other parameters specified the detailed distribution of pitch ratio along the blade and the type of blade section.[39] Durand then chose three equally spaced values of mean pitch ratio and two values of each of the other four parameters. Using all possible combinations of values, he arrived at forty-eight propellers distributed in a representative way over the then-current field of design.[40] A model of each propeller, made of Pacific Coast sugar pine (see figure 5-4), was then tested through a series of values of rotational speed n at several values of forward speed V.[41]

From Lesley's weekly reports we can follow the trials and triumphs that characterize engineering research but seldom appear in formal accounts. As model construction was beginning, Lesley suggested an altered scheme for varying the shape parameters; Durand's reply is not in the record, but it was presumably negative—the scheme was not changed. Early in the tests Lesley encountered a problem of wood pitch being drawn out of the models by centrifugal force and roughening their surface; two weeks later, after an intermediate statement of success, he reported regretfully, "I am still having trouble with throwing out pitch. . . . I am gradually eliminating this difficulty, however, by revarnishing the propellers and do not think that it will amount to anything in the end." Throughout his reports Lesley recorded the usual struggle to attain acceptable accuracy with new equipment, writing dejectedly at one point, "Although we have tested but 36 propellers the number of tests is now 91."[42]

The tests at last completed, Lesley went to Washington, D.C., in the early autumn of 1917 to assist Durand in preparing a report, published by the NACA in November under the latter's authorship.[43] This report

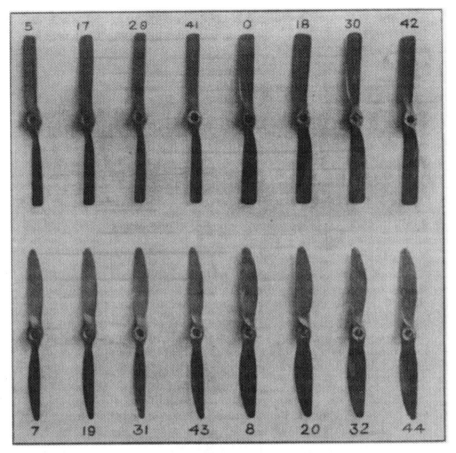

Fig. 5-4. One-third of the forty-eight model propellers used in the initial set of tests. The top row contains propellers of straight blade outline; the bottom, of curved outline. From W. F. Durand, "Experimental Research on Air Propellers," *Report No. 14*, NACA (Washington, D.C., 1917), fig. 27.

of fifty-six pages and thirty-seven plates of drawings and graphs gives a meticulous account—characteristic of Durand—of the equipment and tests, the data and their representation, and the underlying laws of similitude. The entire project, from design of wind tunnel to publication of report, had been accomplished in the remarkably short time of thirteen months.

To make the model results useful for the selection of full-sized propellers, Durand needed the appropriate law of similitude. In the report he derived this law by applying dimensional analysis to relations similar to equation (1), which leads to (among other things) the combination of

V, n, and D into the dimensionless group V/nD. This group, which had been employed by Eiffel, is a dimensionless measure of the forward motion of the propeller.[44] The statement about performance preceding equation (1) can then be transformed as follows: a propeller's performance (expressed in the appropriate dimensionless groups) is a function of the conditions of operation and the geometrical size of the propeller combined into V/nD plus the geometrical shape properties. If we take, for example, the propeller efficiency η (defined as the propulsive power produced divided by the shaft power absorbed, and thus itself a dimensionless group), the mathematical representation of this transformed statement is

$$\eta = F(V/nD, r_1, r_2, \dots). \qquad (2)$$

This simplified relation was among the results Durand obtained by applying dimensional analysis to the relations for propeller performance in the form of equation (1).

The foregoing result has two related advantages. First, instead of having to represent the measured efficiency of a propeller of given shape (r_1, r_2, etc., fixed) by a cumbersome set of curves showing, say, η versus V for a series of values of both n and D, we can supply the same information by a single curve of η versus V/nD. Second and more important, since η is a function only of V/nD (and not of V, n, or D alone), this single curve can be used to find the efficiency of a prototype propeller even though the curve was plotted through test points obtained from a model with a much reduced value of D. The designer need only read off η for the value of V/nD calculated for the prototype. For example, suppose the designer needs to know the efficiency of a 9-foot-diameter propeller operating at $V = 240$ feet per second (fps) and $n = 40$ revolutions per second (rps) ($V/nD = 2/3$); he can read it off immediately from a test point obtained with a 3-foot-diameter model at $V = 60$ fps and $n = 30$ rps (also $V/nD = 2/3$). The law of similitude—specifically the rule for corresponding operating conditions—is thus built into the dimensionless representation of the data.[45] If the designer is interested in power produced or absorbed, the details are more complicated, but the principle is the same. The application of laws of similitude in this manner is general in branches of engineering where scale models are used.

The foregoing ideas were known only in part by Durand and Lesley at the outset. The increase of their knowledge in the course of the work is an instructive example of the complexities that occur in the diffusion and growth of engineering ideas. As Durand knew, laws of similitude had been employed in marine work since the time of William Froude but in limited and specialized ways.[46] In his studies of marine propellers, Durand had presented his results as functions of a customary quantity called "slip," which served much the same role as the still unknown V/nD.[47] When he came to outline his plans to Naval Constructor Richardson in November 1916, it was natural therefore that he should discuss at some length the problem of the law of similitude. He left the matter open, however, and nowhere mentioned the choice of the basic parameter. Shortly thereafter Durand appears to have learned of the use of V/nD by Eiffel, for in a letter to Lesley in May 1917 he writes, "Regarding the abscissa for plotting and studying the results, I would suggest that you start in with the function used by Eiffel, namely, V/nD."[48] At this same time dimensional analysis was finally catching on among engineers as a method for obtaining laws of similitude. This trend stemmed in the United States mainly from a series of papers by Edgar Buckingham, who in 1914 put forth a powerful fundamental theorem for application of the method.[49] Durand adopted Buckingham's ideas as well, and in his final report he gave a detailed application to propellers, including the secondary effects of viscosity, compressibility, and elastic bending omitted in equations (1) and (2).[50] His analysis had become thoroughly modern in approach. Here, as elsewhere in the evidence, we see engineering knowledge growing and consolidating in response to the demands of research and design.

In his report Durand presented the propeller data in graphs as functions of V/nD. By suitable choice of V and n in the tests, he and Lesley had been able to cover the range of V/nD then of concern to the airplane designer, despite the much reduced value of D used for the models.[51] They thus made available to the designer an empirical representation, or graphical mapping, of the performance functions F over the prevailing range of practical interest. The report also contained a brief discussion, of interest to the propeller designer, of the detailed influence of propeller shape on propeller performance. The main focus, however, was on use of propeller data in airplane design.

The tests reported by Durand were the beginning of a ten-year program of propeller research at Stanford. As often happens in research, one thing led to another: growth of understanding broadened the view of the program, new knowledge suggested where further knowledge would be useful, and mastery of technique produced increased confidence. At the same time the requirements of airplane design were continually expanding.

The most pressing need was to extend the range of parameters to explore additional possibilities and to cover new areas of practical interest. Durand and Lesley therefore broadened the range of mean pitch ratio by three additional values. They also added three more types of pitch distribution, six of blade contour, one of blade width, and nine of blade section. Since an impractical total of 7,872 additional propellers would be required to cover all combinations of the parameters (old and new), judicious sampling became necessary. Guided by experience from their first tests, the investigators added fifty-four propellers to their study.[52] Curiously, two of these departed entirely from the systematic series, having highly unlikely cycloidal blade outlines. The report says merely—and tantalizingly—that these were "to test . . . ideas submitted to the committee by Brig. Gen. H. H. C. Dunwoody (retired)."[53]

Tests of the additional propellers were made in installments and reported by Durand and Lesley in three reports in the years 1919 to 1921.[54] For part of the work, from January 1918 to February 1919, Durand was on war duty for the National Research Council in Paris, and responsibility for the propeller work devolved almost completely upon Lesley. This work included construction of a larger and improved wind tunnel, put in use in 1919, with a test stream of seven-and-one-half-foot diameter and a top speed of eighty miles per hour.[55] The reports from this period exhibit the gradual refinements of technique typical of experimental research: improved accuracy of measurement, more convenient presentation of data (the authors experimented with various forms), and growing sophistication of methods and ideas. Here, as in the application of dimensional analysis, a typical learning process was taking place.

By 1922 the individual reports, made at different stages of experimental advancement, were clearly inadequate. Understanding and techniques had improved, and a summary was needed. The two men therefore issued a comprehensive report in which the results for all propellers were painstakingly reexamined, doubtful data checked and improved by repeated tests, and a consistent set of results presented for eighty-eight of the models, representing the practically significant part of the tests. This report, which virtually completed the parameter-variation study, focused entirely on the utility of the data for airplane design and included graphical aids (nomographs) for solution of design problems.[56]

From the wealth of data the effect of pitch ratio deserves special attention. The large effect of this parameter on efficiency is typified in figure 5-5, based on the first of the Stanford reports. This shows measured curves of η versus V/nD for three propellers differing only in the value of their mean pitch ratio.[57] The situation confronting the airplane designer is clear. To optimize propeller performance at a fixed flight condition, one value of pitch ratio will suffice. The designer need only calculate the value of V/nD for that condition and select from the data the pitch ratio giving maximum efficiency at that value (inter-

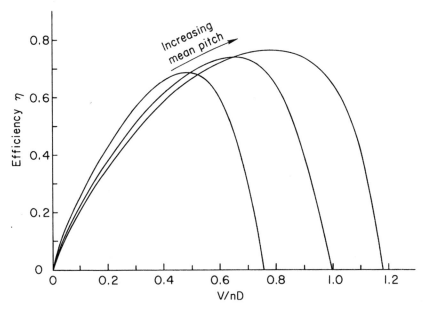

Fig. 5-5. Typical efficiency curves for propellers differing only in mean pitch ratio. Redrawn from W. F. Durand, "Experimental Research on Air Propellers," *Report No. 14*, NACA (Washington, D.C., 1917), pl. VI.

polating between curves if necessary). An airplane does not fly at a fixed condition, however, but at varying values of *V/nD*. To maintain maximum efficiency, therefore, the designer wants a propeller for which the pitch ratio can take on different values; that is, optimizing the propeller over a range of flight conditions requires varying its geometry. To test the aerodynamic feasibility of achieving such variation in a simple way, Durand and Lesley included in their second report (1918) a model whose mean pitch could be set at different values by angular adjustment of wooden blades in a bronze hub.[58] Measurements with the model confirmed that results essentially the same as in figure 5-5 could be obtained by a single mechanical change rather than by reshaping the entire blade. Notwithstanding this success, variable-pitch propellers did not become practical until the 1930s, when the mechanical problems of changing pitch in flight were reliably solved.

Although the parameter-variation study was complete, the experience had expanded Durand's views of what was needed and what could be achieved. A statement of his thinking appears in the sixth Wilbur Wright Memorial Lecture of the Royal Aeronautical Society, which he gave before an audience of two thousand in London on June 25, 1918.[59] Speaking on "Some Outstanding Problems in Aeronautics,"

Durand discussed in detail the relative merits of three methods for obtaining data on propellers: (1) theoretical calculations by the Drzewiecki theory, (2) wind-tunnel tests of model propellers plus a law of similitude to transfer the model results to full scale, and (3) tests of full-scale propellers in flight. Full-scale testing in flight he considered too expensive and time consuming for the number of propellers required to cover the range of design parameters. Consistent with his marine background, he was skeptical of theory on grounds that it contained so many doubtful assumptions it would require a great deal of experimental checking in any event. All things considered, he felt that parameter-variation tests of models in the wind tunnel afforded the best opportunity for success. He thus provided a broader rationale than before for the tests under way at Stanford. Even with these tests, however, correlation among the three methods would still be lacking, and the ultimately desirable application of theory to full-scale propellers would be impeded. To overcome this, Durand extended his earlier thinking:

> It seems likely that this final correlation of computation with ultimate result may best be made in two stages. The first should comprise a careful study of the relation between the results derived by the computations of method 1 and the model tests of method 2. Such a correlation would then permit us to pass readily from the results by computation to the probable results by model. . . . The second correlation should then comprise a series of comparative tests to determine with sufficient generality of application the character and amount of correction to be applied to the results of model test in order to satisfactorily reproduce the results to be expected from full-sized forms. . . . This would by no means require the testing of a full-sized form corresponding to each model. . . . It appears reasonable to expect . . . that a well-selected and not too numerous series of tests, properly distributed among the various characteristics of form and of operation, would serve adequately to give the correlation desired.[60]

This methodology, which was becoming typical in aeronautics, went beyond the practices of marine engineering.[61] From 1922 to 1926 Durand and Lesley attempted to carry out such a program.

They began by examining the correlation between theory and model experiments in a report published in 1924.[62] The Drzewiecki theory (today called the blade-element theory) proceeds roughly as follows: (1) The blade of the propeller is thought of as divided into a large number of elements at different radii. Each element is then imagined as a small airfoil moving in a straight line (instead of its actual helical path) and at a velocity determined by the forward speed of the propeller, the tangential speed of the rotating element, and a secondary speed of flow induced by the aerodynamic action of the propeller itself. (2) The forces on each element at its appropriate velocity are estimated from experimental airfoil data. (3) These forces are summed over all elements to find the performance of the propeller.[63]

The principal difficulty is in calculating the secondary flow induced by the action of the propeller. Neglecting this induced flow (and other complicating effects) as apparently impossible to assess, Durand and Lesley used the theory to calculate performance as a function of V/nD for eighty of their propellers.[64] They thus carried out parameter variation by theory rather than experiment; indeed, nothing in the basic idea of parameter variation requires that experiment be used. The result was a graphical mapping of the performance functions F by theoretical means, to be compared with the earlier one by experiment. When such comparison was made, theory and experiment were found to give the same general trends. Quantitative disagreement was substantial, however, and too erratic to provide the consistent correlation for which Durand had hoped. Durand and Lesley were therefore unable to find an empirical correction "to pass readily from the results by computation to the probable results by model."

Concurrent developments in Britain are instructive. There, at the National Physical Laboratory beginning in 1913, effort was concentrated on improving the theory, aided by flow-field measurements near a propeller and by comparisons of calculated and measured performance for a few representative models.[65] The apparent aim was to develop an experimentally confirmed theory with which to make reliable calculations of the values of the functions F as occasion demanded. There is no British evidence before 1922 of parameter-variation studies, whether experimental or theoretical, to provide a collection of data for the airplane designer. This difference from the American approach was accompanied by an apparent absence of people with marine experience. The British investigators may thus have been less impressed than the Americans with the problems of matching the propeller to the engine and vehicle.[66] On the other hand, a lack of the marine prejudice against theory was undoubtedly an asset. By 1922 to 1924 theoretical work in Britain had reached the point where the induced flow, which Durand and Lesley at that time considered impossible to assess, could in fact be estimated fairly well. The result was considerable improvement in the agreement between theory and experiment.[67] Notwithstanding, in 1922 the British made their own parameter-variation measurements.[68] In fact, because of inability to deal adequately with the secondary effects of viscosity and compressibility, the theory never did become sufficiently accurate for precision design. Even in the 1940s airplane designers were still instructed, whenever suitable information was available, to select propellers on the basis of experimental parametric data.[69]

The comparison with theory completed, Durand and Lesley proceeded to the correlation between model and full-scale results. At that time full-scale testing could be done only in flight. To obtain the needed full-scale measurements, Lesley therefore spent the summer of 1924 at the NACA's new laboratory at Langley Field, Virginia, directing flight tests of five navy-type propellers on a VE-7 biplane. Whereas the earlier model tests in the wind tunnel could be carried out with negligible obstruction behind them (figure 5-3), such was not possible in flight, where the bulky engine and airframe were unavoidable. Durand therefore made comparable tests in the wind tunnel at Stanford with a simulated portion of the airplane (see figure 5-6). These parallel flight and wind-tunnel tests were the first such tests ever made on a family of propellers.[70] In their detailed report, published in 1926, Durand and Lesley found fairly consistent differences of from 6 to 10 percent between the flight and model data.[71] These differences they conjectured to be due to a combination of experimental uncertainties and an interplay between propeller size and the secondary effects of viscosity and compressibility. Later knowledge shows this judgment to have been correct. The observed differences provided guidance to the designer in applying the model results to full-scale design. They were not consistent enough, however, to supply the specific corrections that Durand had anticipated in his Wilbur Wright Lecture.

The correlation study did, however, produce an important idea, derived from the necessity to evaluate the performance of a propeller operating ahead of the obstruction from engine and airframe. One consequence of such obstruction is a local slowing of the airflow at the position of the propeller, causing the propeller to operate at a lower effective forward speed than it would without obstruction and thereby increasing the propeller *thrust*. At the same time, however, the slipstream behind the propeller augments the velocities of flow over much of the airframe, increasing the *drag* of the vehicle from what it would otherwise be. A question arises therefore as to how best to assess the useful work done by the propeller, not in essentially unobstructed operation as in the original wind-tunnel tests, but in its realistic location on an airplane. In confronting this question, Durand and Lesley came to realize that the increment ΔD of airframe drag as a consequence of propeller action serves to decrease the useful propulsive work and should therefore be charged against the propeller. Thus, the effective thrust that should be credited to the propeller in the presence of the airframe is the actual shaft pull of the propeller minus the increment of

Fig. 5-6. Installation in second Stanford wind tunnel for tests of navy-type propellers with simulated portion of VE-7 airframe and engine. From Stanford University Archives.

drag ΔD. The decreased efficiency calculated on this basis they termed the *propulsive efficiency*. This concept, straightforward in retrospect, was, like most such things, not at all obvious at the time. It is instructive to see it develop and its explanation become clearer in successive Stanford reports, an excellent example of the kind of intellectual activity underlying engineering knowledge.[72] The idea proved a lasting one, and propulsive efficiency became a standard theoretical tool in the design of propeller-driven airplanes.[73]

In formulating the concept of propulsive efficiency, Durand and Lesley were learning how to think about the use of propeller data in airplane design. This development of ways of thinking is evident throughout the Stanford work; for example, in the improvement of data presentation to facilitate the work of the designer and in the discussion of the solution of design problems.[74] Though less tangible than design data, such understanding of how to think about a problem also constitutes engineering knowledge.[75] This knowledge was communicated both explicitly and implicitly by the Durand-Lesley reports.

In 1926 Durand returned briefly to parametric studies of model propellers, publishing a report on thirteen variations of a standard blade form adopted by the U.S. Navy.[76] This report was Durand's last on air propellers and brought to a close the investigation he had initiated with his presentation to the NACA in 1915. Lesley and others continued work on propellers at Stanford for twelve more years, but on problems not essential to our concern with methodology.

The achievement of the Stanford study, as Durand and the NACA had intended, was the production of a large quantity of high-grade parametric propeller data for use by the airplane designer. The work had no marked effect, nor could it have, on the design of propellers as such. Wooden propellers of 1920 to 1925 were not visibly different from those of 1914 to 1915,[77] and the peak efficiencies of the best Durand-Lesley models were little improved over the best prior propellers. Even the most advanced metal propellers of later years were only slightly better in this regard.[78] It gradually became apparent that the maximum potential efficiencies had been approached fairly early and that little gain was possible. The true significance of the Durand-Lesley model data was that they greatly improved the airplane designer's ability to match the propeller to the engine and airframe. This improvement was reinforced by the guidance available from the comparison between model and full-scale results and by the ways of thinking that had been developed. The designer now had at hand much needed tools for the improvement of airplane performance.[79]

We may assume that designers quickly put the Stanford data to use. Direct evidence of such use is difficult to find (design calculations rarely survive for long), but there is no lack of testimony from contemporaries.[80] In addition, the results are visible frequently in the secondary literature. They are cited extensively, for example, in Fred Weick's textbook on propellers, the bible of its field in the United States in the 1930s, and were the source, directly or indirectly, for a number of charts in Walter Diehl's *Engineering Aerodynamics*. They were also the basis for an analytical study by Max Munk and for correlation studies by Diehl and (now) Admiral Taylor. These publications helped diffuse the Durand-Lesley data through the engineering literature and make them available for use by airplane designers.[81]

The Stanford work had other, less direct effects. The demonstrated inadequacies of theory, plus the failure to obtain a consistent correlation between model and full-scale results, made it clear that propeller data must ultimately come from tests at full scale. It was equally clear, partly from Lesley's experience at Langley Field, that full-scale testing in flight was slow and expensive. The only alternative was to build a wind tunnel large enough to take a full-sized propeller. Acting on this radical idea, the NACA from 1925 to 1927 built the first wind tunnel capable of full-scale testing, the Propeller Research Tunnel at Langley Field, a project for which the Stanford work provided not only motive but also technical base. From the famous PRT came much of the propeller data used by designers in the 1930s and 1940s.[82]

Another effect is more conjectural. From the mid-1920s, work in NACA laboratories was noted for the comprehensive empirical study of aerodynamic problems by the method of parameter variation. Perhaps the most important (but far from only) example of this was the study of the performance of airfoils as it depends on the parameters that define the airfoil's shape (see chapter 2), a study that continues in the laboratories of the NACA's successor, NASA. The work of Durand and Lesley was the first aerodynamic study under NACA auspices to employ this method. Because of the work's comprehensiveness and quality, and because of Durand's prestige and membership on the committee, we may suppose that the Durand-Lesley effort was an important influence in establishing the empirical pattern of NACA research. To the extent this is true, the methodological effects of the Stanford propeller work went far beyond propeller technology.[83]

Parameter Variation, Scale Models, and Technological Methodology

The Durand-Lesley propeller study suggests a number of observations about the methodology on which it was based. In working toward the improvement of a technical device (the airplane), Durand and

Lesley were clearly functioning as engineers. The data and ways of thinking the two men produced were knowledge useful to this end; they therefore clearly qualify as engineering knowledge. But what of the methods whereby the knowledge was obtained? What general statements can we make about them, and in what ways, if any, are they peculiar to engineering? Most of the observations that follow are illustrated by the Durand-Lesley work, though they can hardly be taken as proven by the single case. Others are prompted rather than supported by the evidence, and all invite further research and discussion. They seem worth making, however, since they bear on engineering methodology generally and most do not appear to have been made elsewhere.[84]

The first group deals with parameter variation:

(1) Parameter variation can be carried out by either experimental or theoretical *means*. Additionally, it can be used for at least two *ends*: to provide a useful collection of systematic design data and to aid the development or refinement of a theory or theoretical model. The work of Durand and Lesley illustrates the four possible combinations of means and ends. Experimental parameter variation provided both design data and one ingredient in the attempt to establish an empirical correction for the blade-element theory. Theoretical parameter variation supplied the second ingredient in this attempt (both means obviously must be employed for a correlation between theory and experiment); if the attempt had succeeded, theoretical parameter variation could then have been used to provide design data.[85]

(2) *Experimental* parameter variation—our sole concern from this point—is not per se peculiar to engineering. It appears also in activities that are clearly scientific. An example is the famous experiment by Hans Geiger and Ernst Marsden in 1913 on the scattering of a stream of α particles (doubly ionized helium atoms) by encounters with atomic nuclei.[86] These investigators systematically varied the angle of deflection at which the number of scattered particles was counted, the thickness of the metal foil that provided the scattering target, the atomic weight of the foil, and the speed of the incident particles, these being the parameters of the problem. Their object was, in their words, "to test a theory of the atom proposed by Prof. Rutherford, the main feature of which is that there exists at the centre of the atom an intense highly concentrated electrical charge."[87] A more common present-day example is the chemist's measurement of the rate of a chemical reaction as it depends on temperature, concentration of reactants, and presence or absence of a catalyst. The object here is to infer a theoretical chemical model (that is, a sequence of individual reaction steps, called a chemical

mechanism) that will account for the observed overall result. Other examples could easily be cited.

(3) As the foregoing examples illustrate, physical scientists use experimental parameter variation for a different purpose than do engineers. Although their work may in the end provide data useful in engineering (e.g., the chemists' rate data may be used by chemical engineers), scientists tend to have only a secondary interest, if any at all, in supplying design data. Mostly they use parameter variation as an aid in establishing a theory or model. Analysis would be simpler if the converse were true of engineers, but the facts are not so obliging. As we have seen in the propeller work, engineers often have not only an immediate interest in design data but a longer-range interest in establishing a theory. In their common use of parameter variation in the quest for theory, however, engineers and scientists have different priorities. Engineers are after a theory they can use for practical calculations, perhaps eventually including the provision of parametric design data. To obtain such a theory they are willing, when necessary, to forgo generality and precision (up to whatever limits of accuracy are set by the requirements of the application) and to tolerate a considerable phenomenological component (as in the use of experimental airfoil data in the blade-element theory). Scientists are more likely to be out to test a theoretical hypothesis (Geiger and Marsden) or infer a theoretical model (the physical chemist). In either case their primary goal is to explain known phenomena or predict new ones. They value generality and precision for their own sake and consider it important that the theory or model be as close to first principles as possible. Essentially, scientists are interested in understanding, in some sense, how the physical world operates. These statements are, of course, an oversimplification in that they ignore those troublesome examples where both theoretical and design ends are pursued together and with seemingly equal regard.[88] The distinctions are observable ones, however, and are consistent with the view of scientists and engineers as communities with different values, the one valuing "knowing" and the other valuing "doing."[89]

(4) The foregoing remarks suggest an answer, insofar as experimental parameter variation is concerned, to our question of how the methods we have observed are special to engineering. As seen above, we cannot contend that experimental parameter variation is a peculiarly engineering method per se. We can contend on the basis of the Durand-Lesley evidence, however, that the method, in its role of supplying design data, is peculiarly important in engineering for producing peculiarly engineering knowledge. Experimental parameter varia-

tion is used in engineering (and only in engineering) to produce the data needed *to bypass the absence of a useful quantitative theory,* that is, to get on with the engineering job when no accurate or convenient theoretical knowledge is available. This is perhaps the most important statement about the role of parameter variation in engineering.[90] Such bypassing of theory was the function of experimental parameter variation in the work of Durand and others on marine propellers. Durand brought this function along implicitly when he transferred the method to aeronautics; he made it explicit in his Wilbur Wright Lecture when he rejected theory as at least temporarily inadequate. Despite the efforts of Durand and Lesley and, especially, the British, propeller theory remained inadequate for airplane design, and experimental methods continued to be used to bypass this shortcoming. Historical examples in other areas are numerous, and the method is much in use in the same way today.[91] A recent instance is the experiments of Fox and Kline on divergent passages (diffusers) used to decelerate fluid flow, a common engineering problem for which no satisfactory theory existed when the experiments were done.[92]

(5) Because of its utility in the foregoing way and because usable design theories are still difficult to establish in many areas, experimental parameter variation tends to be applied in engineering more for provision of design data than for development of theory. Since designs in theoretically intractable areas often have great practical urgency, this predominant use of the method is indispensable in modern engineering. No such indispensability exists in science, whose paramount interest in theory can be served in other ways. For this reason, the method is doubtless more widespread in engineering than in scientific activity.

(6) The absence of a theory usable for engineering application may be due to various causes; a decision whether or not to circumvent the absence by experimental parameter variation involves a number of considerations. The absence may be due to lack of basic scientific understanding, want of a crucial phenomenological component when a theory from first principles is not feasible, or inadequate refinement of the theoretical structure. In the case of the propeller, the scientific principles were well understood, and the phenomenological component needed in the blade-element theory was available from experimental airfoil data. The stumbling block was that the theory could not be refined to the point where it was sufficiently accurate to displace full-scale experimental data when such data could be made available.

In the absence of adequate theory, the decision whether to try to develop one or resort to experiment involves a number of trade-offs. The theoretical approach, which is more problematical and often takes longer to bring to success, has the virtue of providing both theoretical understanding and an economical method of design applicable to any set of conditions. The experimental approach provides design data relatively quickly, is comparatively free of the assumptions and simplifications required by theory, and can help bring to light unforeseen problems; it has the disadvantage of requiring considerable experimental effort and therefore usually greater cost and of being limited to the range of dimensionless groups over which measurements are made. The Americans and the British made, for the most part implicitly, a different assessment of these trade-offs and followed different paths in their propeller work.

Even when an adequate theory of some sort is available, experimental parameter variation may still be employed because of lack of numerical data on the physical properties of the substances involved or insurmountable difficulties of one kind or another in carrying out the theoretical calculations.[93] Whether theory is or is not available, the case for the use of experimental parameter variation often boils down in the end to the very basic one that it provides usable results in an acceptable time, whereas waiting for theoretical understanding or guidance may involve indefinite delay.

(7) A pair of final points on parameter variation. The application of parameter variation requires the assumption of a functional relation such as equation (1); this requires in turn that a proper measure of performance is known and that the parameters on which the performance depends have been identified. In this sense, parameter variation is a second-stage process in research. This first stage—identification of the relevant quantities—is a heuristic, catch-as-catch-can process that depends on insight, intuition, and sometimes luck. In the Durand-Lesley work the requirements for parameter variation were satisfied from prior knowledge in naval architecture and aeronautics. An example of first-stage work, in which the parameters were *not* known, is the experiments by Harry Ricardo on the influence of type of fuel on gasoline-engine knock. Ricardo's finally unsuccessful attempt to identify the fuel properties (parameters) that influenced knock, though done for a clearly engineering purpose, was much akin to a scientist's search for a pattern that may help explain some new physical phenomenon. It could not possibly constitute parameter variation because the relevant parameters were not yet known.[94]

Parameter variation likewise should not be confused with cut-and-try methods, in which one thing after another is tried in an attempt to find something that works or works best. An instance of this approach is Thomas Edison's ad hoc trials of "6,000 vegetable growths" in his celebrated search for a filament suitable for his incandescent lamp.[95] Here there was no question of an attempt even to identify parameters; Edison could not have done parameter variation, since he apparently never knew what the parameters of the problem were in any significant sense. Unfortunately, people often misunderstand parameter variation when they do encounter it and so put down all experimental work by engineers as cut-and-try. In so doing they miss much of the essential richness and complexity of engineering methodology.[96]

The following observations stem from other aspects of the Durand-Lesley work.

(8) Experimental parameter variation does not of itself require the use of working scale models as in the Stanford tests. It can in principle be carried out just as well at full scale. The parametric tests of the ancient Greek catapult designers mentioned earlier were done in this way, as were also the tests in the Propeller Research Tunnel at Langley Field. For reasons of economy or technical limitation, however, scale models are used more often than not. Derivation of the required law of similitude then necessitates again that the parameters of the problem have been identified. Rational model testing thus requires, in effect, the ability to do parameter variation, and this method is in fact usually used when models are employed. In principle a scale model could be tested without recourse to parameter variation; there would be little point in it, however, since the object of testing the model is to obtain data for design. In brief, one sometimes does parameter variation without scale modeling but seldom scale modeling without parameter variation.

(9) As with experimental parameter variation, use of working scale models is not peculiar to engineering.[97] Scientists aim at understanding the nature of things, which take place at some natural scale and are preferably studied at that scale. In some situations, however, they also use working models. This happens especially in fields, like the earth sciences and meteorology, where the size of phenomena places them beyond direct experiment, and complexity has (until the recent advent of large-capacity electronic computers) prevented effective mathematical modeling. Earth scientists, for example, have studied such things as mountain building and glacier flow using small-scale working models in the laboratory. Meteorologists do the same for cyclones, tornadoes, and atmospheric circulation.[98] Their very size and complexity, however, make the phenomena difficult to model. This is particularly true with respect to boundary conditions (the physical and geometrical constraints and actuating agents simulated in the model), which tend to be complicated or far removed or both. For this reason, scientific use concentrates mostly on approximate modeling of simplified or partial phenomena and on results aimed at theoretical understanding (often qualitative) of the mechanisms involved. Because of the approxima-

tions and simplifications, the methods are viewed skeptically by some investigators. Use by scientists is also relatively rare and must be searched for.[99]

Modeling of working devices in engineering differs in all these respects. As with airplanes and ship hulls, many of the crucial boundary conditions are provided by the device under study; they are known and close at hand and can be faithfully simulated both physically and geometrically. Modeling can therefore be done realistically and with detail. It is also possible, with the aid of laws of similitude, to use the models to obtain relatively reliable quantitative design data (the most essential difference from scientific use). Modeling methods have become correspondingly valued, and use by engineers is widespread and relatively easy to find. In all these respects, working scale modeling by engineers has a different character from that by scientists. As before, the method is peculiarly important to engineers for producing peculiarly engineering knowledge.

Because of the need for them in connection with model testing, dimensional analysis and laws of similitude are also widely used in engineering. Accounts of them appear regularly in elementary engineering texts. Scientists seldom employ laws of similitude, and, while they do use dimensional analysis, it is for purposes other than deriving such laws.[100] Similitude accordingly is rarely discussed in textbooks in science. The situation is thus similar to that observed for control-volume analysis in the preceding chapter.

(10) The Durand-Lesley experiments on the variable-pitch propeller illustrate how engineers seek to optimize a device by choice or adjustment of the design parameters. Since optimization procedures are pervasive in technical activity, it is tempting to claim them as peculiar to engineering. In fact, scientists often have to optimize in the design of their experiments to obtain the best possible data or data that could not be gotten otherwise.[101] Chemists try also to maximize the yield of an overall chemical reaction by optimizing the mechanism of the underlying multistep process. (Their work may lead to technologically useful results, but their effort usually derives from a scientific desire to understand the chemical mechanism as fully as possible.) Any attempt to claim optimization solely for engineering would thus be unjustified. It seems fair to say, however, that optimization is a more fundamental and widespread activity in engineering than it is in science. The engineer must in principle design and build products to work as effectively as possible, and under economic and social constraints that are usually of secondary concern to the scientist. Optimization therefore becomes a constant element, implicitly or explicitly, in engineering thinking. For the engineer optimization has the nature of an ethos, a way of life; for

the scientist it plays only an occasional and incidental role. This difference between science and engineering is fundamental, even though it is one of degree and attitude rather than method. It is another aspect of the difference between "knowing" and "doing."

Concluding Remarks

To the casual observer the work of Durand and Lesley might appear to be simply data gathering, though at a high level of technique for the time. Detailed examination, however, has revealed a sophisticated and complex methodology, involving experimental and theoretical parameter variation, scale modeling and laws of similitude, and comparison of both theory with experiment and model results with full-scale results. These elements interacted intimately, and all were conditioned by the need of the designer to optimize the integration of a technical device into a larger system of devices. The methodology is also more general, in whole and in part, than the present example. But to say that work like that of Durand and Lesley goes beyond empirical data gathering does not mean that it should be subsumed under applied science. As we have observed, it includes elements peculiarly important in engineering, and it produces knowledge of a peculiarly engineering character and intent. Some of the elements of the methodology appear in scientific activity, but the methodology as a whole does not.[102]

The engineering utility of the methodology rests primarily on the fact, as our case study makes clear, that there is no *essential* relation between experimental parameter variation and physical theory. Indeed, the strength of experimental parameter variation is precisely its ability to provide solid results where no useful quantitative theory exists. It is of course true that engineers use theory whenever they find it feasible and advantageous to do so. The independence of experimental parameter variation from physical theory, however, makes use of theory often a matter of choice, not of necessity. (The widespread use of quantitative theory by engineers can make engineering appear to an outsider to be no more than applied science, even when the theories are highly phenomenological and have little real scientific standing. The independence of parameter variation from physical theory refutes this superficial view of engineering as simply applied theoretical science.)

When used in conjunction with scale models, experimental parameter variation does involve theory to the extent that dimensional analysis and laws of similitude are used in transferring experimental results from model to full scale. Such secondary use, however, is very different and much easier than devising a theory to predict performance ab initio from more general physical considerations. Durand and Lesley and the British all tried to establish such a theory—the blade-element theory—for propellers. Engineers would have been happy to use the theory if the attempt had succeeded. When it failed to produce a theory of sufficient accuracy, however, they could still fall back on experimental parameter variation and laws of similitude to get on with their job of designing airplanes. And if theoretical laws of similitude had not been available, they could still have run their experiments at full scale (as in fact they ultimately did, though for other reasons).

Insofar as the lack of useful theory is due to lack of scientific understanding rather than other causes, parameter variation provides a way to circumvent even a lack of science. Most of our ability, for example, to design devices involving the ubiquitous but still scientifically intractable problems of turbulent fluid flow derives from this capacity.[103] The use of parameter variation to circumvent science, however, is rarely remarked on in the historical literature. The few discussions that inquire into the method do so in a limited way in assessing Smeaton's pioneering work on waterwheels and windmills. They are therefore understandably preoccupied with the *technique* of the method, which they view as newly introduced into engineering from experimental science. They consequently see parameter variation as essentially scientific in character.[104] The paradoxical fact that the *function* of experimental parameter variation may be to free engineering from limitations of science is perhaps easier to see in the modern context, where the newness and origins of the method are not in question.

From a wider view, the methodology treated here can be seen as part of the broad activity in engineering to develop the analytical capability needed for design. Design engineers must be able to analyze their designs, and research engineers, using whatever methods they can, work to develop the tools the designers need. Durand and Lesley supplied propeller data for performance analysis and attempted to develop the blade-element theory for analytical design purposes. Neither of these activities would have held much interest for a scientist—even the theory, difficult as it was, was only a quantitative elaboration of a principle of operation already well understood. The Stanford investigators also developed the important conceptual tool of propulsive efficiency, but this was to answer a practical design need that would never have arisen in scientific work. All these activities illustrate how "technologists have developed their own bodies of more or less systematic knowledge to meet the needs of practice."[105] This sort of development (that is, of the technological base) is a very different thing from the development of a specific product or device that is usually meant when the word *development* is used.[106]

The Stanford work also illustrates how technology advances by slow and undramatic accumulation, in this case of knowledge (design data) and analytical capability. The methodology of parameter variation and model testing, in fact, is sometimes pointed to as the methodology par excellence for incremental advance.[107] In the Stanford work it did have such result: it aided greatly in optimization of propeller-driven aircraft, but it did not cause radical change in airplane design (as did later development of wing flaps, retractable landing gear, swept wings, and jet engines). The conclusion does not follow, however, that the methodology never leads to revolutionary development. Parametric study of metal cutting was basic to Frederick Taylor's invention of high-speed tool steel,[108] and parametric wind-tunnel tests of model airfoils played a critical role in the Wright brothers' achievement of flight. The Stanford work no doubt typifies the usual result of parameter variation and model testing (hence, in part, its value as a case study). Incremental advance, however, is not the sole or inevitable outcome of the methodology.[109]

Whatever the result, parameter variation, used experimentally and often with scale models, provides a systematized procedure whereby engineers trained in its use can approach the optimum solution of complex problems with minimal dependence on individual genius or insight. It thus contributes to the continual effort in engineering to replace "acts of insight" (unteachable) by "acts of skill" (capable of being taught). The success of this effort is fundamental to the ability of modern engineering to advance on so wide a front and with such sureness—persons of genius are, after all, always in short supply.[110] From its easy communicability from one generation of engineers to another, as much perhaps as anything else, comes the long-term importance of the methodology developed by Smeaton and employed by Durand and Lesley.

Finally, some questions and conjectures. Living closely for several years with the Durand-Lesley work (along with the Britannia Bridge design)[111] makes me wonder if other characteristically engineering methodologies exist besides the one examined here. The iterative techniques associated with design synthesis, for example, may be of such nature, as may also the procedures for engineering optimization. My direct experience in such matters, regrettably, is limited to mechanical technology. Various questions therefore arise: How characteristic and widespread are the present methodology and its component elements in other fields? Are there, as seems likely, other methodologies in mechanical technology and other fields, and if so how do they operate? To what extent if at all do other methodologies serve as subtle and systematized ways for engineering to proceed despite gaps in scientific knowledge? More broadly, I also wonder if it might be profitable to look for something justifiably called "engineering method," analogous to but different from scientific method, which has been one of the centrally productive problems for the history and philosophy of science.[112] (I carry this conjecture a bit further at the end of chapter 8.) Answers to questions of this kind can only deepen our understanding of knowing how in engineering.

Design and Production: The Innovation of Flush Riveting in American Airplanes, 1930–1950

In the early 1930s most metal airplanes of American origin were held together by rivets with dome-shaped heads protruding beyond the external surface of the aircraft. A decade later almost all such airplanes had rivets flush with the surface. To observers outside the airplane industry this change, if they noticed it at all, must have seemed straightforward and even trivial. In fact, since flush riveting had never before been used in plates as thin as those in aircraft, a great deal of learning was necessary. We may therefore ask, Why and how did the change to flush riveting take place? And what knowledge did the learning process generate that was not available before?

The previous chapters have explored examples of engineering knowledge related in varying degree to the interface between engineering and science. My purpose here is to move to the other end of the spectrum and take up a case as remote from science as possible, on the assumption that some features of the cognitive structure of engineering may be more visible there. To be most significant, this example should come from a modern high-technology industry that is usually thought of as science based. Flush riveting in airplanes meets these requirements. Although a far from negligible development in an industry that was depending more and more on engineering science,[1] it did not itself involve anything that could remotely be described as science in the modern sense. Moreover, it did not even require a major technological invention or conceptual breakthrough. As airplane performance improved to the point that the aerodynamic drag of protruding rivets was no longer economically and militarily acceptable, anyone could see at once what had to be done: the rivets had to be made flush. What needed to be learned was "simply" how to accomplish the change in routine design and production, and this required the effort of a large number of people over a considerable period of time. Such humble productive activity in the context of science-based industry has received little attention from historians. It is, I suspect, more prevalent than is generally recognized.[2]

The study of flush riveting does in fact prove intellectually rewarding. Detailed examination of the learning process—and the essence of the learning can be found only in the details—reveals different examples of knowledge with different characteristics. Reflection on these differences then suggests a framework for thinking about technological knowledge that may have interest beyond the present example. Except for some discussion by Sidney Winter in relation to economic production theory, the epistemology of industrial production has been largely terra incognita.[3]

Flush riveting also tells us something—an unexpected bonus— about the nature of industrial innovation. The fundamental decision to remove the rivet heads from the airstream was a design decision based on aerodynamic considerations. Once it had been made, however, implementation would have been impossible without successful changes in the methods of production. Considerable knowledge did have to be generated for detail design, but the pivotal developments were in production, and it was there that the greater part of the innovative activity took place. Flush riveting can therefore be called a *production-centered innovation*. (One can no doubt speak similarly of operation-centered innovations.) Flush riveting thus differs from innovation derived from formal research and development, which has dominated the scholarship on technological change in mid-twentieth-century industry.[4] Comparable investigations of *"non-R&D"* innovation of whatever sort, such as Samuel Hollander's economic inquiry into plant-level process innovations in the rayon industry, or some cases in a policy-oriented study by Arthur D. Little, Inc., are relatively scarce and not much concerned with technical detail.[5] When examined at the level of such detail, the innovative activity of flush riveting appears widespread and simultaneous throughout the entire aircraft industry. This pattern is very different from the model of innovation by diffusion from an initiating source that has occupied the attention of economists and historians.

By dealing with innovation and knowledge that are production centered, this chapter moves outside the concentration on design. As we shall see, however, design and production of flush rivets were intimately related. Although production requirements determined what kinds of rivets and riveting methods designers could call for, knowledge also had to be produced specifically for the demands of design.

We thus encounter another instance of both normal design and design hierarchy. Day-to-day choice of rivet shape and size is as normal as design gets and about as far down the design hierarchy as one can go.

Origins and Nature of the Problem

Flush riveting appeared in American aircraft as part of the development of the modern aluminum stressed-skin airplane in the 1930s.[6] Significantly, however, the revolutionary airplanes that led this development in the first half of the decade did not incorporate flush rivets to any appreciable extent. The Northrop Alpha and Boeing Monomail, the single-engine commercial airplanes generally credited with starting the development in 1930, both utilized rivets that protruded into the airstream. So also did succeeding Northrop designs, as well as the Boeing and Martin twin-engine military and commercial planes that helped inspire the definitive Douglas DC-1, DC-2, and DC-3 of 1933 to 1936. These twin-engine Douglas transports, especially the almost legendary DC-3, set the pattern for multiengine commercial land-planes until appearance of the jets in the 1950s. Besides stressed-skin structure, the Douglas designs incorporated all the major new innovations of retractable landing gear, wing flaps, and controllable-pitch propeller. These and other improvements raised the maximum speed of the DC-3 to 212 miles per hour, compared with 170 miles per hour for the Northrop Alpha. Even the Douglas airplanes, however, continued to use protruding-head rivets.[7]

Though at first apart from the mainstream, flush riveting appeared early in the development of stressed-skin airplanes.[8] Anyone with an elementary knowledge of aerodynamics could see that protruding rivets caused undesirable air resistance; that this resistance could be eliminated by making the rivets flush would have occurred to numerous people almost without thinking. As often happens with obvious ideas, some of these people would be expected to employ such rivets on principle, even before data existed to prove them worth the extra trouble and expense.

In these circumstances (and because details such as flush riveting often go unrecorded), assessing priority is difficult, and pinpointing all the early uses in aircraft is impossible. The first proposal I find in the United States was in a patent granted to Charles Ward Hall in 1926 which covered numerous features of metal airplanes, including flush riveting. By 1929 Hall had equipment for such riveting in operation at his Hall-Aluminum Aircraft Corporation in Buffalo, and he probably employed flush rivets on the Hall PH-1 of 1932, a twin-engine patrol flying boat for the navy. That same year the Boeing Airplane Company used flush rivets on its P-26 single-engine fighter for the army air corps.[9] Over the next few years an increasing number of aircraft apparently incorporated such riveting. A considerable fraction of these were flying boats or amphibians, perhaps because of early awareness by seaplane designers that any protuberances on a planing surface have an especially detrimental effect on takeoff from water. Many pioneering applications were undoubtedly partial; according to a news item of 1934, "some manufacturers" (unidentified) were "compromising the gain and expense" by limiting flush rivets to the upper surface of the wing near the leading edge or to other locations where protruding rivets are especially harmful.[10]

Concurrent with these applications, wind-tunnel tests of a specially built wing, reported in 1933 by the Langley Laboratory of the National Advisory Committee for Aeronautics, confirmed the aerodynamic penalty incurred by protruding rivets. For a hypothetical single-engine airplane, for example, eliminating the rivet-head drag was estimated from these tests to increase the top speed from 200 to 205 miles per hour. This evidence was presumably becoming known in the industry.[11]

Still, the fact remains: flush riveting in the first half of the 1930s was peripheral and lagged behind the major innovations. This lag was not due to problems in the riveting itself. Flush riveting did not depend critically on any prior technology, and its problems could have been solved—indeed, already had been in part—whenever people like Hall set their minds to it. The designers of the DC-3 and its predecessors simply had their eyes on bigger game. Only when the performance improvements caused by retractable gear, flaps, and the like had been realized did it become generally attractive to pursue the lesser gains offered by recessed rivet heads. However, as the speeds of commercial airplanes went up as a result of the other innovations, the cost of dragging exposed rivet heads through the air increased, whereas the extra cost of flush riveting stayed constant. And in military aircraft, as other methods of increasing top speed reached the point of diminishing returns, the smaller gain from getting the rivet heads out of the airstream became more inviting. By the mid-1930s the time was ripe for flush riveting throughout the industry. Only after that time do engineering reports and articles of any depth begin to appear. The circumstances here provide a clear example of Thomas Hughes's concept of a "reverse salient in an expanding technological front."[12]

To understand the problems of flush riveting in aircraft, some terminology and prior history are helpful. Riveting is an ancient art, practiced by at least the second century B.C., for joining together overlapping pieces of metal or other material by means of a malleable metal

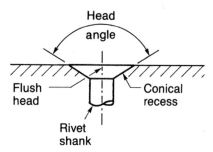

Fig. 6-1. Typical flush rivet head.

pin.[13] In modern times the pin, or *rivet,* as it comes from the manufacturer usually has a cylindrical *shank* with a preformed head, the *manufactured head,* at one end. Riveting is accomplished by putting the shank through matching holes in the pieces to be joined—usually metal plates or sheets of one kind or another—and hammering or otherwise upsetting a second head, the *upset head,* on the end opposite the manufactured head. In structural applications, riveting, besides joining the pieces together, transmits load from one structural element to another.

Both the manufactured and upset heads of rivets may be of various shapes (round, pointed, cylindrical), most of which protrude above the surface of the plates. To meet special needs, however, one (or occasionally both) of the heads may be made flush. This is accomplished by putting the head in a *conical recess* at the entrance to the hole in the plate (figure 6-1). Most often nowadays the manufactured head is the flush head; in this situation the head is preformed with a conical shape the same as that of the recess, as with an ordinary flathead screw. On occasion, however, the upset head is made the flush head; to accomplish this the shank of the rivet is hammered or pressed into the recess, and any excess material is cut off. Either way, the included angle of the conical head is called the *head angle.* Whichever method is used, flush riveting entails extra expense and is employed only when circumstances make it necessary or desirable. Such circumstances led to limited use, beginning at least in the 1830s, in iron and steel ships (to reduce water resistance on the hull or obtain a smooth surface on the deck plating) and in boilers, bridges, and other structural work (in situations where room for the rivet head is restricted by other parts of the equipment or structure). Here the conical recess was made by machine countersinking; that is, by cutting away metal with a rotating conically shaped tool. More often than not in such use, the upset head was made the flush head.[14]

When flush riveting came to be applied to aircraft, a new problem arose. To save weight, airplane skins tend to be much thinner than the plates in ships, boilers, and bridges. In the small airplanes of the 1930s, the skins were often so thin relative to the size of rivet needed for strength that the machine countersink required by the head would have had to penetrate through one sheet of skin and into the next. This situation could be detrimental to the strength and rigidity of the riveted joint, and some other way to form the conical recess had to be found. Again, the solution was easy to perceive in principle: deform each sheet by pressing a conical dimple around the rivet hole and accommodate the flush head in the outermost of the nested dimples. This method was, in fact, employed by Hall and other early users; their experience showed that it worked in practice. With dimpling, all the basic ideas required for flush riveting in aircraft were in hand.

The task that remained for the aircraft industry at the middle of the 1930s was to learn to implement these ideas in various structural situations in volume production. The history of this learning process breaks down conveniently into two major parts, one concerned with production and the other with design. The first has to do primarily with the question of how to fabricate flush-riveted joints economically and reliably; the second deals with the proportioning of such joints to carry the structural loads. In dividing the material in this way, however, we must not lose sight of the fact that the two activities were in reality intimately related. The requirements of aerodynamic and structural design placed restrictions on what the production process had to produce. Conversely, experience about what was feasible in production generated information needed by the designer to decide which type of flush riveting—dimpled, machine countersunk, or some combination thereof—to use in a given situation. Producibility also defined the types to be tested for strength by the structural engineer. Thus, while the innovation was production centered in the sense explained earlier, generation of knowledge for production and for design went forward together.[15] Knowledge of flush riveting from ships, boilers, and buildings was of little use for this purpose, both because aluminum alloys were so different from iron and steel and because dimpled riveting had not been used before. Nowhere in the aeronautical literature do I find appeal to the prior experience.

As we shall see, the aircraft industry solved its flush-riveting problems more or less simultaneously throughout the industry. Only a relatively small portion of this activity, however, was reported in detail in technical publications or even company reports—shopworkers and design and production engineers have more urgent things to do than record their experiences. Even though a fair amount of material *can* be found, one soon develops the uncomfortable feeling of seeing only the tip of a very large iceberg. Some help can be had, of course, by interviewing participants who are still living. Additional insight comes from observing how continuing developments in riveting go on today, on the reasonable assumption that attitudes and activities in the riveting community have not changed greatly since our period. I also remember from my own experience how things took place in related areas of aeronautics in the 1940s and 1950s. I have drawn from all these sources in piecing together the historical account.

Production of Flush-riveted Joints

Although relatively unnoticed, riveting has played a sizable role in airplane manufacture. The airplanes of World War II had as many as 160,000 rivets in a medium bomber and 400,000 in a large bomber. In 1944, design, fabrication, and installation of riveted joints reportedly constituted 40 percent of the cost of a typical airframe.[16] Because of the large number of rivets, airplane riveting is a mass-production operation, even though airplane production itself is not. As with all mass-production operations, small savings per rivet mount up.

Aircraft riveting is also a subject of great complexity. Besides the ever-important matter of cost, the aircraft engineer must pay attention to the often conflicting requirements of weight, production quality, structural reliability, corrosion resistance, service maintenance, and appearance.[17] Even for only *flush* riveting, to grasp this complexity fully from a historical paper is impossible—one must go through a number of the original technical articles in their entirety. The reader will have to imagine the detailed and painfully complicated descriptions of tools and processes, the contradictory assessments of the same process by different people, and the frequently unresolved arguments about the relative merits of different types of flush rivets or riveted joints. Of such is the fabric of complex technological activity.

The widespread, simultaneous nature of flush-riveting development is unmistakable. By the end of the basic stage of development, which occupied the second half of the 1930s, at least fifteen manufacturers were using the new type of riveting. That they worked largely independently is indicated by the wide range of head angles from one company to another (73 to 130 degrees) and the diversity of head dimensions even among companies using the same angle.[18] The need for flush riveting was apparently felt more or less simultaneously everywhere, and, once it was, work on such riveting welled up from below, so to speak, throughout the industry. This widespread simultaneity is the most striking characteristic of the activity we are examining. I will come

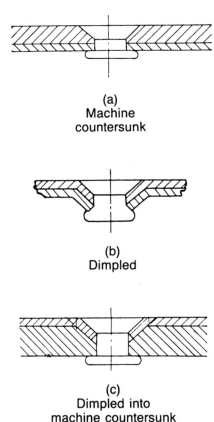

(a)
Machine
countersunk

(b)
Dimpled

(c)
Dimpled into
machine countersunk

Fig. 6-2. Three basic types of flush riveting. From G. Rechton, *Aircraft Riveting Manual. Addendum I. Riveting Methods*, Douglas Aircraft Company (Santa Monica, Calif., 1942), pp. 7, 21, 26, reproduced by permission.

to the reasons for it later, but it needs to be kept in mind throughout.

In the basic stage, manufacturers put their effort into learning to produce the various kinds of flush riveting and finding out which kind was best used where. Attention quickly settled on three types, depending on the thickness of the sheet being riveted (which is dictated by the load-carrying requirements of the structure). Where the thickness of the outer sheet was sufficiently greater than the height of the conical rivet head, old-fashioned machine countersinking proved suitable (figure 6-2a). The limitation here was the desire to avoid the sharp, feather edge that occurs when the countersink extends to or beyond the thickness of the sheet. For the eighth-inch-diameter rivets that predominated in aircraft construction, for example, the thinnest standard sheet that could meet this requirement for most of the various head proportions was 0.051 inches.[19] Some manufacturers tried countersinking thinner sheet by letting the countersink extend into the inner skin, but this was discarded as "poor design practice" when such joints proved weak or the resulting feather edges developed cracks in fabrication or service. For such thinner sheets, recourse was had to dimpling, as described earlier (figure 6-2b). Here production engineers learned from experience that there were minimum and maximum sheet thicknesses beyond which the method could not satisfactorily be used; for example, 0.020 and 0.081 inches, respectively, for eighth-inch rivets. If the thickness exceeded the maximum, the dimple tended to crack during forming; if it was less than the minimum, the sheet between adjacent rivets might buckle under load. These limits, however, were set at considerably different values by different manufacturers. When the inner sheet alone exceeded the maximum, the problem was solved by dimpling the outer sheet and machine countersinking the inner (figure 6-2c). Various limitations also emerged for this combination. For each type of riveting the limits on sheet thickness varied with rivet diameter, and additional considerations entered when more than two overlapping sheets had to be riveted. Learning about these matters was a complicated and protracted business, much more so than a summary description can convey. Knowledge of the various limits, which could come only from manufacturing experience, was essential for the designer.[20]

Since dimpled riveting was new, development concentrated more on this than on countersinking. The problem was not an easy one. In the words of a tool-design engineer for Bell Aircraft, "At first glance, nothing seems simpler than putting a dimple into a sheet of metal, yet it has proved to be no simple problem for the best engineering talent in the industry." Even allowing for self-aggrandizement, this statement is probably not far from the truth. A description of later developments at Northrop Aircraft included the following description of the requirements on a dimple:

> It is essential that the dimple does not crack during forming or after assembly of the sheet. The dimple must be dimensionally accurate to insure fastener flushness and to nest correctly in either machined countersunk recesses or in other dimpled recesses, regardless of sheet thickness. Furthermore, the dimple must provide maximum joint strength under all conditions, at the same time developing the very minimum of warpage or distortion in the sheet. Productionwise, the dimpling method must be rapid and accurate, and at the same time require simple, inexpensive tools and set ups.[21]

Since there existed at the time no theory for calculating the deformation of the sheet in the press-forming process, the problem had to be solved entirely empirically. The nature and diversity of the work on this and other problems can best be described in terms of three aircraft companies for which written accounts exist. Statements in the literature indicate that parallel developments went on elsewhere.

The Douglas Aircraft Company in Santa Monica took up flush riveting for the DC-4E, the same commercial transport described in connection with flying-quality specifications in chapter 3. The DC-4E, which first flew in June 1938, had a maximum speed of 245 miles per hour, 33 miles per hour greater than the DC-3. Douglas aerodynamicists had insisted at the inception of design in 1936 that the time had come to get the rivet heads out of the airstream, and Arthur Raymond, director of engineering, agreed. Development of a suitable method of flush riveting was assigned to Vladimir Pavlecka, a Czech-born engineer in charge of structural research. Working under Pavlecka's direction between 1936 and 1938, a small group of engineers and production workers devised what they called the "Douglas system of flush riveting."[22]

The Douglas system had two distinguishing features. First, the manufactured head of the rivet itself provided the tool for press forming the dimples. (In our account, the manufactured head is always the flush head unless specified otherwise.) In this scheme the manufactured head impressed the dimples into both sheets in a female die in a single operation (figure 6-3a, b); a second operation then upset the rivet shank to form the upset head. The Douglas method of "rivet dimpling" thus differed from the more usual "predimpling," where the dimples were made in the two sheets, usually separately, by pressing between male and female dies (figure 6-3c); the sheets were then assembled, the rivet inserted, and the upset head formed. Pavlecka argued that rivet

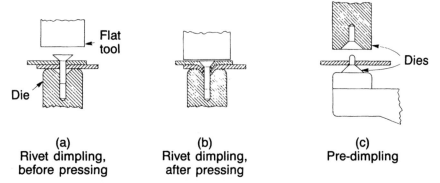

(a)
Rivet dimpling,
before pressing

(b)
Rivet dimpling,
after pressing

(c)
Pre-dimpling

Fig. 6-3. Two methods of dimpling. In all cases the moving die is shown crosshatched. From V. H. Laughner and A. D. Hargan, *Handbook of Fastening and Joining of Metal Parts* (New York, 1956), p. 209, reproduced by permission of McGraw-Hill Book Company.

dimpling would reduce the number of individual operations and hence the time and cost compared with predimpling. He also anticipated a closer fit of the various parts and thus a tighter joint. Rivet dimpling was not new, having been used in Europe before 1934. There is no indication, however, that Pavlecka was influenced by the European work. The idea again was one that could be expected to occur to anyone who thought seriously about dimpled riveting.[23]

The second feature of the Douglas system was use of a rivet with a head angle of 100 degrees. This was a departure from the 78 degrees that had long been used in boilers and pressure vessels and had been taken over by the army and navy as standard for the flush rivets already in limited use in aircraft.[24] Pavlecka made the change after attempts to form dimples by means of 78-degree heads led to cracking of the sheet and to excessive deformation of the rivet head in the dimpling and upsetting operations. The shallower dimple required by the 100-degree head (the outside head diameter and rivet diameter being unchanged) eliminated these difficulties. The Douglas rivet also incorporated a thin constant-diameter addition on the top of the head. This was said to eliminate possible cracking of the edge of the head in dimpling and upsetting as well as damage to the otherwise sharp edge in the shipping of batches of rivets.[25]

At the Curtiss-Wright Corporation in Buffalo developments took a different course. Don Berlin, chief engineer, and Peter Rossman, production design engineer, in a paper read to the Society of Automotive Engineers in March 1939, reported an extensive study begun in 1937 in connection with a large order of military fighter planes. (The SAE concerns itself with airplanes as well as automobiles.) In contrast to procedures at Douglas, Curtiss-Wright opted for predimpling and, for a number of reasons, preferred the 78-degree head angle. Berlin and Rossman, like the Douglas team, encountered cracking problems when they tried to predimple to accommodate the standard 78-degree head. Like Douglas, they solved the problem by going to a shallower dimple. Unlike Douglas, however, they did this by keeping the 78-degree angle but reducing the head height to less than that of the standard head (for a given rivet diameter). They also gave careful attention to the precise shape of the dimpling dies and to details of the predimpling operation. Also in contrast to Douglas, they specifically rejected the constant-diameter addition to the top of the head on the grounds that it led to an undesirable gap between the edge of the flush head and the adjacent sheet. In the discussion following the paper, Pavlecka explained and defended the Douglas system, and B. C. Boulton of the Glenn L. Martin Company told why his company "after exhaustive tests" had adopted a 115-degree rivet. The Curtiss-Wright people, however, had clearly solved the problem to their own satisfaction. As in many technical matters, there was no single road to success.[26]

Bell Aircraft Corporation in Buffalo entered the flush-riveting field in 1939, about the time of the Berlin-Rossman paper. According to an article by Arthur Schwartz, the tool-design engineer quoted earlier, Bell at once "decided that a thorough study was in order before considering the adoption of any of the [diversity of] existing methods." This study concluded that rivet dimpling led to undesirable sunken areas surrounding each rivet head and made excessive demands (as I shall explain later) on the skill of the worker. Also, according to Schwartz, "Exhaustive experiments showed a decided superiority of [a] . . . rivet head angle in the neighborhood of 120 deg." Bell therefore opted for predimpling at this angle. With carefully proportioned dimpling dies, this process was said to avoid the difficulties that Schwartz saw in other methods based on smaller angles. Bell also rejected the constant-diameter addition to the rivet head for the same reason as Curtiss-Wright.[27]

Douglas, Curtiss-Wright, and Bell all struggled with matters besides dimpling method and rivet shape. Both Curtiss-Wright and Bell put a great deal of effort into solving the production problems of machine countersinking as well as dimpling. All three companies found themselves forced to develop special tools for the shop processes required by the different types of riveting. Since obtaining the desired smoothness in a flush-riveted surface in routine production is difficult, Bell gave careful attention to the dimensional tolerances required on rivets, countersinks, and dimples; Bell also developed special gauges to check that the tolerances were attained in the shop. Curtiss-Wright, faced with a large production order, made detailed study of the time required to do each step in the various types of flush riveting; on this basis they arrived at estimates on production costs and how these might be expected to decrease with increased shop experience. And to satisfy themselves that flush joints could provide the necessary strength, Douglas and Curtiss-Wright (and presumably Bell, though this was not stated) made structural tests of a large number of joints under both static and vibrational loads.

Learning to solve these kinds of problems required minute attention to detail. To exhibit the detail, one problem may serve as typical: the shaping of dies for predimpling. Although Curtiss-Wright, Bell, and Douglas all addressed this crucial problem (Douglas went over to predimpling when rivet dimpling proved unsuitable for large sheet thicknesses), the most detailed account came from R. L. Templin and J. W. Fogwell of the Aluminum Company of America. Alcoa, the main supplier of aluminum to the aircraft industry, dealt with such problems to assist and encourage its customers in the use of its product. Templin and Fogwell, working with 100-degree rivets, took it as their task to produce a set of dimpling dies that would (1) leave the sheet flat around the dimple after the dimpling operation, (2) form dimples that would nest tightly into one another, and (3) provide a cylindrically shaped hole so that no undesirably sharp corners would dig into the rivet shank. These goals they attained by trial-and-error variation of the geometrical parameters of the dies (compare chapter 5, note 96). They satisfied the flatness requirement by having the dies extend outside the dimple in such a way as to strain the surrounding sheet slightly in an angular direction opposite to that of the dimple; with the removal of die pressure, elastic springback of the metal then caused the sheet to lie flat. The precise amount of strain proved important, and a difference of half a degree could be crucial. They solved the nesting problem by choice of a suitable die pressure and by careful selection of different maximum diameters for the dimpling cones on the male and female dies. The cylindrical hole was obtained by drilling the initial hole in the sheet to a diameter slightly smaller than that of the mandrel on the male die (which served to center the die in the hole, as in figure 6-3c); the suitably tapered mandrel then forged the hole to the desired shape in the dimpling operation. The initial hole could not be made too small, however, lest radial cracks appear around its edge. To guide their work, the Alcoa engineers used microphotographs of sectioned specimens, as in figures 6-4a and b, which show dimples with sharp-cornered and nearly cylindrical holes, respectively. Other developers of flush riveting used trial-and-error methods and microphotographs in much the same way. People in different companies, however, did not always agree on the best solution to a given problem.[28]

Other aspects of both dimpling and upsetting required attention throughout the industry. Forming the upset head on the shank of flush rivets presented no great problem—the methods already in use for protruding rivets could be taken over with only minor change. In the most usual method a first worker, using a pneumatic hammer with a concave fitting, applied a rapid succession of blows to the dome-shaped protruding head; at the same time a second worker held a heavy metal "bucking bar" against the end of the rivet's shank. The impact of the hammer working against the inertia of the bucking bar thus deformed the shank to produce the upset head. On sufficiently small subassemblies, a single worker upset the rivet by squeezing it between the ends of a C-shaped tool that (depending on the depth of its jaw) could fit around the sheets and reach both ends of the rivet. The only special problem in upsetting *flush* rivets was that of applying the necessary force to the flush manufactured head without marring the adjacent

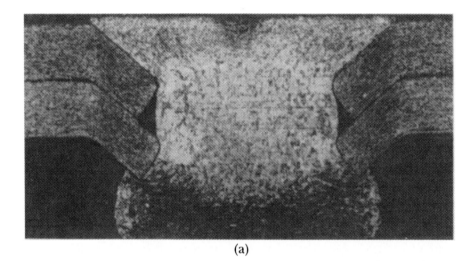

(a)

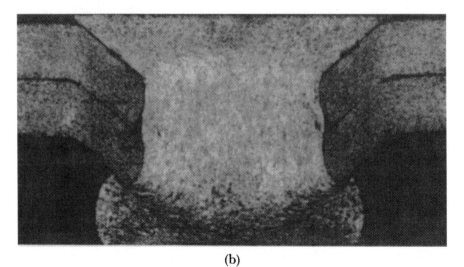

(b)

Fig. 6-4. Microphotographs of sections through dimpled holes, actual rivet diameter one-eighth inch. (a) Drilled hole same diameter as mandrel of male die; note sharp corners digging into rivet. (b) Drilled hole 15 percent smaller in diameter than mandrel; note more nearly cylindrical hole without sharp corners. From R. L. Templin and J. W. Fogwell, "Design of Tools for Press-countersinking and Dimpling of 0.040-Inch-Thick 24S-T Sheet," *Technical Note No. 854*, NACA (Washington, D.C., August 1942), figs. 2, 4.

sheet. For this purpose workers used slightly convex fittings on the hammer or squeezer, sometimes in conjunction with a swiveling device that could adjust itself to the surface of the sheet even if the tool was not oriented precisely.[29]

Where dimpling was needed, manufacturers found they could supply the required die pressure with the same hammers and squeezers used for upsetting by simply replacing the upsetting fittings with appropriate dimpling dies. In rivet dimpling by a two-worker team, a problem arose in that the female die (figure 6-3a, b) had to be hand held carefully at right angles to the sheet for the head to seat properly. This was the difficulty that helped discourage Bell from using rivet dimpling. Douglas, however, claimed to have overcome the problem by using a special swivel-mounted die. Douglas also found that rivet dimpling was done more satisfactorily by one or two heavy blows than by a rapid succession of weaker ones; they developed and patented a special pneumatic "one-shot" hammer for this purpose. (Considerable argument went on about the relative merits of rapid hammers, one-shot hammers, and squeezers for different operations and different dimensions of sheet and rivet.) Douglas also patented an air-operated two-position bucking tool useful for both the rivet-dimpling and upsetting operations. Such specialized tool development took place throughout the industry for both dimpling and machine countersinking.[30]

In such fashion—widespread, detailed, and entirely empirical—the aircraft industry had learned by the end of the 1930s how to produce flush-riveted joints. At the same time, additional aerodynamic tests plus articles about them in technical magazines had emphasized the gains from such riveting.[31] With this knowledge of how and why, no designer would any longer consider using protruding rivets in a high-performance airplane. The learning process, however, did not stop there. Three additional developments between the early 1940s and the first years of the 1950s require mention.

(1) Although flush riveting had come a long way, it still left something to be desired regarding tightness of joint and smoothness of riveted surface. If the manufactured head did not seat firmly in the recess, the joint would yield under load; if the top of the head were too high or too low, owing to the recess being made too shallow or too deep, the surface would not be aerodynamically smooth. About 1941 researchers at the NACA's Langley Laboratory had the problem of building a wing with the extreme degree of smoothness required by the new laminar-flow airfoils (see chapter 2). After much consideration, they realized they could attain their goal by making the top of the manufactured head deliberately too high and carefully milling away the excess after the upsetting operation. At the same time, tightness was improved because

the upsetting hammer, with its impact now concentrated entirely on the top of the head, forced the head firmly into the recess. They also learned with obvious pleasure that tightness proved even better if a round-heat rivet was inserted from the back side and the upset head formed by pressing the shank into the recess (after which it was made smooth by milling). The last arrangement, reported in 1942, came to be known as the "NACA rivet." To perform the milling operation the Langley people developed and patented a power-driven, hand-held milling tool that they said would not mar the surrounding surface of the skin. The basic idea of the NACA rivet—of having the upset head be the flush head—had long been used, of course, in ships and boilers. Whether the NACA people got the idea from the earlier work is impossible to say, but it seems unlikely. Again, the concept is one that could easily have been rediscovered.[32]

Despite its increased difficulty and cost, the NACA rivet found application where extreme smoothness and tightness were paramount. More important, the different but related practice of installing the top of a *manufactured* head deliberately too high and milling away the excess—so-called flush-to-high installation—became widespread. Two of the NACA workers obtained a patent on this procedure, which did not exist in earlier practice. Problems of milling the head without marring the skin, however, continued to trouble the industry in routine production.[33]

(2) The dimpling methods developed up to the early 1940s worked satisfactorily with the aluminum-plus-copper alloys then in use. The new higher-strength aluminum-plus-zinc alloys of the mid-1940s, however, were less ductile, and the problem of cracking of the sheet reappeared.[34] Various aircraft companies tried different approaches with differing degrees of success. In the long run, the old solution of modifying the rivet head or the dimpling dies did not work. Heating the sheet around the hole and thus causing a local and temporary increase in ductility finally proved successful. Engineers tried various methods of heating, including resistance heating by passing an electric current through the sheet, induction heating by means of a high-frequency magnetic field, and conduction heating from electrically heated dies. Conduction won out, and aircraft tooling companies developed integral heaters and associated automatic controls for both squeezer and hammer-plus-bucking-bar applications. As in earlier aspects of dimpling, extensive trial-and-error and parametric tests were necessary to establish die temperature, pressure, and contact time as they depended on sheet thickness and rivet size.[35]

"Spin dimpling" also achieved some success with the higher-strength alloys. In this method a special male die, shaped so that only two small radial sectors rested on the (cold) sheet, was spun rapidly around its longitudinal axis during dimpling. Compared with hot dimpling, spin dimpling found less favor in the United States than in England, perhaps because the 100-degree rivet adopted as standard in this country in the early 1940s (as explained later) required greater deformation of the sheet than did the British standard of 120 degrees. The experience with the improved alloys illustrates the close relationship that exists among design, material properties, and manufacturing processes.[36]

(3) In the first half of the 1940s, the need for increased production for World War II, coupled with a lack of skilled workers, caused a trend toward riveting by machine. The highly skilled two-person teams of the 1930s were replaced increasingly by larger and larger squeeze riveters operated by one worker. Stationary pedestal-type versions provided jaws as deep as five feet, and some were capable of upsetting a number of rivets simultaneously. At the same time structural engineers produced designs incorporating more and larger subassemblies to take advantage of the new machines.

In the second half of the decade, in response to the ever-present desire to reduce costs, the flush-riveting process became increasingly automated. The Engineering and Research Corporation, in cooperation with various aircraft companies, developed the Erco Automatic Riveter, which could punch and dimple the rivet hole on subassemblies and insert and upset the rivet, all in rapid automatic sequence. The General Drivmatic Riveter provided the equivalent for machine-countersunk rivets. The latter machine allowed North American Aviation to reduce the riveting time for the complete wing panel of an F-86 fighter from one day to two hours. At the same time, only 1 percent of the rivets had to be rejected as imperfect and redone, compared with 12 percent by previous hammer or squeeze riveting. Moreover, the machine was so precise in operation that rivets could be installed to the desired flushness without head milling. Thus, what had been largely a hand process in the early 1940s became a highly mechanized procedure ten years later, with improvements in precision as well as economy.[37]

In our concern for specifics, we must not lose sight of the fact that the entire development from the second half of the 1930s to the early 1950s was a continuous and integral process. Experience in the shop and in service was a stern teacher that required numerous changes, mostly in close relation. From observation of what did and did not work in practice, for example, the limits on the different types of flush riveting became more sharply defined; as a result, the choice of which type to use in a given design situation became more apparent. When rivet dimpling was found lacking in tightness (contrary to original Douglas expectations), such dimpling tended to be avoided in structurally critical locations. As time went on, shopworkers and production engineers learned, frequently to their dismay, the idiosyncrasies and intractabilities of different materials in different situations. At the same time, agreement grew, though it was never complete, on which kind of tool best suited a particular operation. Most important, increasingly sophisticated dimpling methods made stricter tolerances feasible on the size and shape of dimples. Such improvement was prompted by growing realization that poor nesting of dimples could bring about sizable decreases in strength and that small gaps around rivet heads or a consistent departure of the heads from exact flushness by as little as 0.005 inch could cause significant increases in aerodynamic drag. North American made an important improvement in this regard in the late 1940s with so-called coin dimpling. In this method, the female die for predimpling was made in two concentric parts that exerted different pressures on the sheet, thus forging a more sharply defined and precise dimple. All this affords an example of how increasing experience leads to increased knowledge for both production and design.[38]

As usual, generation of knowledge depended in part on its interchange. Although the various companies operated essentially independently in the beginning (and maintained their individual programs later), communication grew as time went on. As with the Berlin-Rossman paper to the SAE, presentations and discussion at technical meetings provided a useful if somewhat delayed channel. So also did technical journals and trade magazines of the sort cited in the endnotes. More immediate exchange took place by word of mouth; according to an engineer at North American, "When I got in a jam, I'd call my college friends at other companies." As production activities increased and became internally specialized, an identifiable fastener community appeared. (Rivets are only one of many kinds of fasteners needed in airplanes.) By 1952 and probably earlier, a Panel on Fastening and Mechanical Joining was holding sessions at the annual meeting of the SAE. Here experts from four or five companies (Convair, Douglas, Northrop, Lockheed, and Ryan in 1952, for example) presented short papers on many matters, including flush riveting, and answered questions from the audience and each other. News also spread via engineering representatives of Alcoa and of tool suppliers such as Erco, who traveled about the industry promoting their company's products. Employees at one aircraft firm also resigned and moved to another, taking their growing knowledge with them. Numerous means thus existed to cross-fertilize the learning process.[39]

Means also existed to pass the knowledge on to routine users. One of these was the "process specifications" that aircraft manufacturers issue for guidance of their detail designers and production engineers. These proprietary documents, jealously guarded by the companies, contain (among many things) specification of the company's standards for flush riveting. These appear as tables of the limits the company considers necessary in the various types of such riveting and statements about the fabrication processes and precautions to be specified on the drawings and followed in the shop. The requirements, like others in the documents, differ somewhat from company to company. In such process specifications, which are continually revised on the basis of experience, the knowledge of flush riveting was packed down for everyday use.

From 1941 to 1943, because of the need to train numbers of detail designers and shopworkers for wartime production, a spate of books on aircraft riveting also appeared.[40] These included information on the various types of flush riveting and detailed explanations of the tools and processes for their manufacture. Since much of the knowledge in any shop process is difficult or impossible to put into words, these books included many pictures and drawings. An especially impressive effort was the *Aircraft Riveting Manual*, compiled in 1941–42 by George Rechton, a process engineer at Douglas, for internal use in that company. This manual and its addendum contained 226 pages of detailed photographs and drawings of tools and processes, many of them on flush riveting. Only a small fraction of the material was written text.[41]

But even a pictorial work cannot convey all the knowledge involved in a demanding industrial process. Some of it can be imparted only by personal demonstration. In 1942 newly hired riveters, at the Douglas plant built hurriedly in Chicago to increase wartime production of C-54 transports, were having trouble getting their flush riveting right. John Buckwalter, head of the plant, made an appeal to Santa Monica for Rechton, who came and was able to show the Chicago workers how to do it.[42] In the words of the North American engineer quoted earlier, "You can't build an airplane from the drawings." By the nature of things, some of what had been learned about flush riveting could only remain in the neuromuscular skills of the workers and the intuitive judgment of the engineers.[43]

Design of Flush-riveted Joints

Before a riveted joint can be manufactured, it must of course be designed. For a flush-riveted joint, this calls first for a choice (perhaps tentative) of rivet diameter and a decision on the type of riveting—dimpled, machine countersunk, NACA. The rivet diameter is conditioned by the general level of loading, and the type of riveting depends

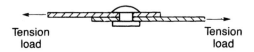

Fig. 6-5. Protruding rivet with tension loading in plane of skin. (In a flush-riveted joint, measures are also taken to eliminate the step from one sheet to the next, but that does not concern us here.)

on skin thickness, smoothness requirements, and production considerations, as we have seen. Both decisions usually involve enginering judgment based on knowledge of current and past practice; the choice of type of riveting must also conform with the tables of limits in the company's process specifications. Completion of the design then requires calculation of the number of rivets needed per lineal foot of joint. On the assumption that the load on the joint is shared equally by the rivets, this calculation comes down simply to dividing the load per lineal foot (known from the overall structural design) by the load that can be carried by a single rivet (taken from empirical tables obtained as described below).[44] If the number of rivets per foot turns out to be unsuitable for some practical reason, the design may have to be repeated with a different choice of rivet diameter. (Finally, the layout pattern of the rivets must be decided, but that need not concern us here.) The key piece of knowledge in the design of a riveted joint is thus the load that can be carried by a single rivet, specified as a function of rivet diameter. Engineers call this the rivet's *allowable strength.*

The important loading on a riveted joint in an airplane is a tension (or compression) in the plane of the skin (figure 6-5). With a protruding rivet the loading is transmitted from one sheet to the other by bearing of the sheet on the rivet shank and by resistance of the shank to shearing. Engineers calculate the maximum sustainable value of these forces, and hence the rivet's allowable strength, by simple theoretical considerations involving the dimensions and material properties of the rivet and sheet. The theory neglects added forces resulting from friction between the overlapping sheets and work hardening of rivet and sheet materials caused by upsetting the rivets. These oversimplifications are clearly conservative, however, and the theory provides a useful approximation. For flush rivets, even this crude theory fails. In the machine-countersunk joint (figure 6-2a), wedging action between sheet and rivet head seriously alters the mechanics of the situation; in the dimpled joint (figure 6-2b), bearing between the nested dimples complicates things further. As a result, when protruding rivets had to be discarded for use on aircraft surfaces, so also did the theory. Knowledge of the strength of flush riveting came entirely from experiment.

Such knowledge grew in parallel to that of production. In 1933–34, when flush riveting was in its infancy, tests by the air corps's Matériel Division showed that machine-countersunk rivets were somewhat weaker than protruding-head rivets and should therefore be used with care. The tests were rough and fragmentary, however, and far from conclusive. After the mid-1930s, the many companies working on production problems also conducted strength tests to support their designs. Company test departments established allowable strengths by gradually increasing the tension load on specially constructed joints having a small number of rivets and noting the load per rivet at which the joint failed. Each company used the rivet-head angle it favored and covered a limited range of rivet diameter, sheet thickness, and material pertinent to its own design needs. In 1940 the National Bureau of Standards, working at the request of the NACA, reported tests of 865 specimens supplied by seven manufacturers, the navy, and itself. Considerable diversity existed in type and details of riveting, and even for a given type the load-carrying ability varied considerably from one manufacturer to another. Microphotographs revealed that "in the stronger [joints] the sheets conform more closely to each other and to the rivet, with little or no void between them and a greater portion of the manufactured head in contact with the sheets." The industry was learning of the close relation between strength and manufacturing technique.[45]

The variation in strength among different manufacturers caused problems for the government agencies responsible for certifying airplane designs. In early 1942 the navy's Bureau of Aeronautics therefore issued a specification of allowable strengths "for use in the design of naval aircraft," reasonable values being arrived at by analysis of test results from Boeing, Martin, the army air corps, and the Bureau of Standards plus "other pertinent test data." A great deal of interpolation and what the engineer calls "educated guessing" was undoubtedly involved. Experience had also shown by this time that loads less than those for failure could cause flush-riveted joints to become unfit for service as the result of excessive permanent yielding. This possibility was taken account of in the values by a method too complicated to go into here. The final result was a tabulation of allowable strength as a function of rivet diameter, sheet thickness, and rivet and sheet material for each of three rivet-head angles (78, 100, and 115 degrees). Values were given separately for machine-countersunk and dimpled rivets, with the former turning out generally weaker and the latter stronger than the corresponding protruding rivets. Later in the year the same tables appeared in *ANC-5*, the handbook of structural requirements issued jointly by the army air corps, the navy Bureau of Aeronautics, and the Civil Aeronautics Administration. Companies had to see to it that their fabricated joints came up at least to the specified value.[46]

Knowledge of allowable strengths, however, was still far from complete. More refined testing by aircraft companies, the Bureau of Standards, the NACA, Alcoa, and several universities had by 1944 thrown doubt on the values in *ANC-5*. In July the Engineering Committee of the Aircraft War Production Council formed a special subcommittee with members from seven aircraft companies to study the problem, and an extensive experimental program ensued. Over the next eighteen months, nine firms cooperated in testing 3,500 specimens of machine-countersunk joints and 3,600 of dimpled joints. The tests covered three rivet materials and six sheet materials (all aluminum alloy), in addition to the necessary range of rivet diameter and sheet thickness. Most specimens used only the 100-degree rivet head that was by then standard. A successor to the original subcommittee, under the chairmanship of Milton Miner of Douglas, analyzed the mountain of data. Its initial findings appeared in a first report in April 1946, with final reanalyzed values in an appendix in May 1948. The latter were incorporated in a revised edition of *ANC-5* dated May 1949.[47]

The *ANC-5* of 1949 showed clearly the growth and refinement of knowledge since the edition of 1942. This was especially evident in the allowable strengths for dimpled rivets, which now exhibited a complex interplay of rivet diameter and rivet and sheet material. The values for failure and for permanent yielding were now listed separately, giving the designer a basis for more sophisticated judgments. In general, the allowable strengths for 1949 (now for standard 100-degree rivets only) were somewhat higher than for 1942; this increase may have come from more thorough and precise testing or from better manufacturing techniques—the record does not say. The tables, however, carried a cautionary note that any deviations from "'good' manufacturing" practice could lead to significant reductions in strength. The subcommittee's report also contained the admonition that "individual manufacturers should ascertain that their methods are adequate to substantiate the values of [the tables] before using the data for design purposes."[48] The lesson about the relation between strength and manufacturing technique had been well learned.

Thus, as with production, the aircraft industry was well on top of the strength problems of flush-riveted joints by the early 1950s. How this was accomplished is an impressive, if undramatic, example of successful industrial response to a mundane but crucial general problem. Passengers who glance out at the upper surface of an airplane's wing probably rarely notice the flush riveting. Accurate knowledge of allowable strengths, however, is essential for their safety.

Standardization of Flush Rivets

As mentioned, the development of flush riveting included adoption of a standard head angle of 100 degrees. How this happened deserves a closer look, since standardization is characteristic of items used throughout an industry.

The early diversity of head angles—from 78 to 130 degrees—is understandable in view of the complexity of considerations in production and design. It could not be tolerated for long, however. The combinations of rivet length and diameter that had to be stocked by the military for field replacement of protruding-head rivets were already burdensome and expensive; to compound the problem by a variety of head proportions of flush rivets was unacceptable, especially in wartime. Within industry, the variety also presented costly difficulties for subcontractors, who were compelled to maintain different sets of countersinking and dimpling tools to do business with different prime contractors employing different head angles. To alleviate these problems, the aircraft companies formed intercompany standards committees at their own initiative in the eastern and western parts of the United States, and by the end of 1939 the western committee had settled on a standard head with the 100-degree angle. (I find no record of action by the eastern committee.) Apparently, however, different companies did not eagerly give up their separate ways (see note 18). Finally, to coordinate activities throughout the country, the industry-wide Aeronautical Chamber of Commerce on November 7, 1940, established the National Aircraft Standards Committee. As part of a broad program of standardization, the committee in April 1941 also approved a 100-degree rivet head. In December the Aeronautical Board of the army and navy made the same head mandatory in all aircraft supplied to the military services. The 100-degree angle then spread rapidly throughout the industry.[49]

The 100-degree head adopted by the Standards Committee was essentially the one developed by Pavlecka's team at Douglas. I have not been able to unearth the reasons for the committee's choice, but a reasonable conjecture can be made based on what had been learned. Experience in tests had indicated that too large a head angle produced a wedging action by virtue of the sheet's pushing on the head under load; this caused prying up of the head and roughening of the surface. Large angles were also known, according to the navy specification of 1942, to have lower allowable strengths, perhaps from the same wedging action. Too small an angle, on the other hand, caused the cracking problems we have witnessed in the dimpling process. These findings would have pushed the committee toward some intermediate angle, and, since 100 degrees was already in use by Douglas and Lockheed, two of the largest manufacturers, that angle would have seemed a reasonable and expeditious compromise. With the angle settled, the head height and other dimensions of the Douglas rivet may have seemed desirable because of the large number of airplanes that company was already putting into the field.[50]

Observations and Reflections

Getting the rivet heads out of the airstream thus took a good deal of doing—more, I dare say, than people outside the aircraft industry would suspect. And all of it despite the fact that the essential ideas were already readily available. Moreover, the development did not stop with the early 1950s but has continued to the present. Already in the 1940s structural engineers were concerned about the failure of flush-riveted aircraft joints from fatigue; that is, from a reduction in strength caused by frequently repeated (as compared with static) loading. Despite more than forty years of testing and study, the question of the effect of fatigue on allowable strengths still remains a matter of design judgment. Over the years, increasing size and wing loading of aircraft and consequent increases in skin thickness have resulted in less emphasis on dimpling and more on machine countersinking. This shift was accelerated by flight experience in the 1950s showing that the bend in a dimple was particularly subject to fatigue cracks in the higher-strength aluminum-plus-zinc alloys. The detailed shape and tolerances of rivet heads have also undergone refinement, with an eye especially toward obtaining smooth installation without head milling. And, finally, increased replacement of aluminum alloy in the 1970s by stainless steel, Monel Metal (nickel-and-copper alloy), and titanium has required changes in design considerations and riveting techniques. Flush riveting today is still not a closed book. But our story must conclude somewhere, and by the early 1950s the original innovation of flush riveting was completed in the sense that design and production had been reduced to routine, relatively trouble-free procedures. Both design engineers and shopworkers now had a wealth of knowledge with which to carry out their tasks.[51]

Science, it should be evident, played no role in our account. No scientific knowledge was involved (unless one counts such commonplace things as the concept of force and the awareness of electric-resistance heating) and no enabling scientific discovery was necessary. The development of production techniques and the determination of allowable strengths took place entirely empirically by trial and error or parameter variation of a commonsense engineering kind (see preceding chapter). No scientific theories were called for, and few mathematical equations appeared in the articles and reports, and then only for elementary engineering calculations. A good deal of analytical thinking *was* evident, but such thinking is not solely a province of science. One looks in vain in the story of flush riveting for anything that could seriously be identified as scientific activity. This is neither to commend nor disparage the methods the industry found necessary; it is simply a statement of fact.

A second set of observations has to do with the widespread and simultaneous character of the learning process. At the beginning— and, curiously enough, even after communication became established in the fastener community—each major aircraft company pursued its own innovation program. Reasons are not difficult to find. The first has to do with necessity. Although much could be learned about forming dimples or upsetting flush rivets from books or word of mouth, complete mastery required hands-on experience. All the acquired knowledge in matters of this kind is not susceptible of codification or communication. Each company knew it had to obtain its own firsthand knowledge no matter how much it might pick up from others. A great deal of duplication resulted, of course, but it was unavoidable. Situations of this kind appear frequently in production technology. A second reason is what some engineers call "the not-invented-here syndrome." Capable engineers, like any skilled professionals, take pride in their ability and consider that they—or at least their company—can do things better than the competition. They also have their share of human obstinacy. Desperation might rule in a jam, but otherwise "why should we go to *them?*" Necessity and pride are both reflected in a passage from "the typical day of [the] fastener engineer" as described in the report of an SAE panel: "He is sometimes influenced by innovations reportedly in use by competitors, but hard headedness obtained through experience prevails, enabling him to uphold his company's reputation and satisfy the customer."[52]

But reasons are one thing, the ability to act on them another. That such widespread, simultaneous innovation could occur when aerodynamic requirements prompted it depended on two circumstances. (1) As I have emphasized, all the fundamental ideas about flush riveting were either already known or conceptually obvious. (2) The aptitudes and training needed for their implementation were available throughout the industry. The latter circumstance was connected in part with the nonscientific nature of the activity. Just as no new or nonobvious ideas were required, so also no training in engineering science was needed to carry them out. Neither the creative insight of a James Watt nor the scientific training of a Rudolf Diesel was essential. A great deal of knowledge and resourcefulness about production and design *was* required, but it was of a kind that could be obtained from experience on the factory floor or in the engineering office. Such knowledge and resourcefulness were already available in every aircraft company. Every company thus could and did get into flush riveting as soon as the aerodynamic need was felt throughout the industry.

The pattern here—what I described earlier as a welling up from below simultaneously throughout the industry—is very different from the familiar one of diffusion downward and outward from one or more initiating sources. This latter, well-studied pattern has appeared clearly in this book in chapters 4 and 5, where the activities were close to science. In both cases the basic ideas were not obvious, and the protagonists, who necessarily had to be versed in engineering science, were academics at one or two universities. The circumstances, as well as the pattern, were thus very different from those seen here.[53] The diffusion pattern has also appeared in countless instances *far* from science, an obvious case being Eli Whitney's cotton gin. Although the basic notion of the device was simple and no great training was required, the idea again was new and far from obvious before Whitney's creative insight brought it to light. To the extent that such examples are representative, the circumstances of the diffusion pattern, whether close to or far from science, appear to be different from those of the simultaneous pattern. The differences appear to lie in the nature of the required ideas and the range of availability of the necessary aptitudes and expertise. It would be interesting to know whether the simultaneous pattern exists elsewhere and whether it is more prevalent in production-centered than in other types of innovation.[54]

Flush riveting also exhibits interesting differences regarding a number of aspects of engineering knowledge. Since science was not involved, these differences pertain unambiguously to such knowledge alone. Table 6-1 lists the differences without elaboration for various aspects of knowledge, not mutually exclusive, according to whether the purpose was production or design. Knowledge for design must be further subdivided depending on whether it derived from requirements of airframe production or structural strength (specific examples appear in row 6). The table concerns only knowledge that can be set down explicitly in words, tables, diagrams, and pictures. The implicit knowledge involved in skill and judgment is temporarily ignored.

The table shows clearly that knowledge generated for production and knowledge generated for design but derived from the requirements of production have a great deal in common. As we saw, the latter also fits naturally into the historical section on production methods. The common elements are not surprising; they reflect the fact that

Table 6-1. Differences in Knowledge According to Purpose

Aspect of Knowledge	Knowledge for Production	Knowledge for Design	
		From Production Requirements	From Strength Requirements
1. Use	To translate design drawings into reality	To supply design drawings of riveted joints	
2. Character	Qualitative and quantitative	Mostly quantitative	Almost entirely quantitative
3a. Generation: actors	Production engineers and shopworkers		Structural and design engineers
3b. Generation: locale	Almost entirely in industry; government laboratory (NACA) in one instance		Industry, government laboratories, and universities; final resolution in industry
3c. Generation: organization	Teams within companies; no formal industry-wide organization		Formal industry-wide committees
3d. Generation: communication	Descriptive articles in technical journals and trade magazines; interchange via personal acquaintance and technical meetings		Evaluative technical reports; committee meetings and correspondence
4. Codification	Process and other specifications, differing from company to company		Military-commercial requirements, binding throughout industry
5. Published literature	Technical journals, trade magazines, books, descriptive manuals		Government bulletins
6. Example	Shapes and dimensions of dimpling dies	Limits on sheet thickness for different types of riveting	Allowable strengths

both sorts of knowledge, despite their difference in purpose, were grounded in productive activity. They cause one to wonder, moreover, whether the first two examples of knowledge in row 6 may not differ epistemologically in some shared way from the third. Reflection shows that this is indeed the case.

The difference lies in the distinction between *descriptive* and *prescriptive* knowledge. Descriptive knowledge, as the term suggests, describes things as they are. Prescriptive knowledge, by contrast, prescribes how things should be to attain a desired end. Descriptive knowledge is thus knowledge of fact or actuality; it is judged in terms of veracity or correctness. Prescriptive knowledge is knowledge of procedure or practice; it is judged in terms of effectiveness, of degree of success or failure. It follows that descriptive knowledge, while it may be more or less precise, is not subject to willful adjustment by engineers to serve their needs. Prescriptive knowledge can be altered at will to be more or less effective.[55]

Of the examples of row 6, the allowable strengths obviously qualify as descriptive. Although they serve a practical purpose in design, they are statements of fact. They say that a rivet of a certain type and diameter combined with sheets of a certain material and thickness will fail at a certain load. The accepted values of the failing loads may become more precise as experimental techniques improve, but they cannot be altered purposively by the research engineer or designer. As we have seen, they are the same for all companies.[56] The limits on sheet thickness for certain types of riveting, on the other hand, are clearly prescriptive. They tell, for a given material, type of riveting and rivet diameter, what thicknesses should be avoided to make a producible and effective joint. The prescribed limits differ from company to company depending on the judgment of the production engineers. The designer in a given company must follow them, but the production engineer may modify them on the basis of changing judgment or increasing experience. The shapes and dimensions of the dimpling dies for a given rivet and sheet are likewise prescriptive knowledge. They differ from company to company and can be modified by the production engineer to make a more or less smooth (and hence aerodynamically more or less effective) riveted joint. In both the limits on sheet thickness and the shapes and dimensions of dimpling dies, engineers are free within bounds to make a tradeoff between technical effectiveness and other considerations such as cost. They have no such freedom with regard to allowable strengths. The limits on thickness and the contours of dies are thus both examples of prescriptive as against descriptive knowledge.[57] The common elements between the two center columns in table 6-1 are thus paralleled by a commonality in the associated knowledge (though this does not imply that *all* knowledge used for or derived from production is necessarily prescriptive; see below). These observations suggest that, for epistemological discussion, classification of engineering knowledge according to its nature as descriptive or prescriptive may be more fundamental than according to its purpose for production or design.

"Descriptive" and "prescriptive" denote varieties of explicit knowledge. To these we must add the implicit, wordless, pictureless knowledge essential to engineering judgment and workers' skills. Such *tacit knowledge* was evident here, for example, in the ability of workers to form dimples and upset rivets by hand. It appeared also in the attempt of production engineers to forecast the decrease in riveting cost with time and in the "educated guessing" by which structural engineers arrived at allowable strengths in the face of early incomplete data. Words, diagrams, and pictures can help suggest and promote tacit knowledge (as in the photographs of Rechton's *Riveting Manual*). The knowledge itself can come in the end, however, only from individual practice and experience. The fact that tacit knowledge is inexpressible does not mean that it is any the less knowledge, a fact we recognize when we say such things as "Mary knows how to ride a bicycle" or "those workers know how to drive a good rivet."[58]

Tacit knowledge and prescriptive knowledge are closely related in practice in that they have to do with procedures. They can thus both be described as procedural knowledge, or what I referred to earlier as "knowing how." (Tacit knowledge thus does not constitute the entirety of knowing how, as I use the latter term.) With this addition, our schema of engineering knowledge can be summarized thus:

explicit knowledge

descriptive knowledge	prescriptive knowledge	tacit knowledge

procedural knowledge

The distinctions here are not meant to be sharp and discontinuous. In general, we can expect gradations of knowledge and cases that do not fit neatly into just one category. What we have is more in the nature of a framework for thinking about the substantive structure of technological knowledge. The extent to which it is general and useful must remain to be seen.[59]

Learning to *produce* flush-riveted joints required generation of a great deal of procedural knowledge of both the tacit and prescriptive varieties. It also drew upon descriptive knowledge with regard to material properties, though this was mostly already known or became known for other purposes. Learning to *design* flush-riveted joints required both descriptive and prescriptive knowledge that did not previously exist; tacit knowledge for design was minimal, though it did enter in various ways, such as the "educated guessing" mentioned above. How the proportions of knowledge vary in technology and engineering generally will require detailed examination. Production by its nature is likely to rely heavily on both varieties of procedural knowledge, though the tacit component will tend to decrease as automation increases; descriptive knowledge must also enter, however, especially through the physical properties of materials. Design of the detailed sort exemplified by flush riveting would be expected to depend mostly on the two kinds of explicit knowledge. Design at the higher levels of hierarchy would also be expected to call for a great deal of tacit knowledge related to design feeling and judgment.[60]

Engineering knowledge and the activities that generate it are rich and complex. Such knowledge is motivated and conditioned not only by design but by production and operation as well. Edwin Layton, writing about "Technology as Knowledge," viewed technology "as a spectrum, with ideas at one end and techniques and things at the other, with design as a middle term."[61] He went on to recommend to historians the value of focusing on design, a recommendation I trust this book supports. Since "things," the sine qua non of technology, have to be made and used, however, as well as designed, the problems of production and operation will ultimately need their share of attention.

The Anatomy of Engineering
Design Knowledge

The foregoing cases provide a wealth of evidence for the epistemology of engineering. This and the next chapter offer generalized analysis prompted by that evidence.

The risks of generalizing from limited case studies—even five of them—are obvious. As pointed out in chapter 1, the present evidence, ample as it is, does not represent all of aeronautical engineering, let alone engineering generally. Within these limits, however, each case does cover a characteristically distinct facet of engineering design. The dangers can also be lessened by drawing from ideas and studies by others and from my own experience and observations as an engineer. One needs an empirical foundation of some sort, and historical case studies provide such basis. In something as complex and essentially practical as engineering, theorizing not grounded on factual knowledge is a good deal riskier.

Although the cases all come from aeronautics, the generalizations of this chapter are intended to be more universal. Design in other branches of modern engineering (mechanical, electrical, etc.), though different in detail, proceeds in much the same fashion. It therefore involves the same broad categories of knowledge and activities that generate it. The specifics from my experience and the studies of others supply illustrative evidence for this fact. As stated in chapter 1, I believe that extension of the generalizations to the other branches will call for addition and modification rather than fundamental revision.

Problems, Design, and Knowledge

We can start with the obvious statement that engineering is a problem-solving activity. Engineers spend their time dealing mostly with practical problems, and engineering knowledge both serves and grows out of this occupation. Rachel Laudan, in a valuable article on "Cognitive Change in Technology and Science," puts it this way: "Change and progress in technology is achieved by the selection and solution of technological problems, followed by choice between rival solutions. Viewed this way, cognitive change in technology is a special case of problem solving."

Laudan elaborates the foregoing theme in terms of what I shall call "devices" and "systems." (She calls them "individual technologies" and "technological complexes.") Devices are single, relatively compact entities such as airplanes, electric generators, turret lathes, and so forth—my examples, not Laudan's. Systems are assemblies of devices brought together for a collective purpose; respective examples are airlines, electric-power systems, and automobile factories. (For want of a better word, I shall here use *technologies*—in one of its several restricted senses—to denote devices and systems taken together.) The distinction between devices and systems, while useful for analysis, is to some degree arbitrary; a complicated technology can often be regarded as a device—I shall so regard the airplane for most (but not all) of my discussion—or as a system composed hierarchically of individual devices. For airplanes these might be taken to be power plant, aerodynamic components, structural elements, electronic equipment, etc. Laudan discusses at some length the problems posed for and solved by both devices and systems—how to fly through the air, generate electricity, and machine pistons by the devices mentioned earlier, or transport people, distribute electricity, and manufacture automobiles by the systems. She examines where such problems originate and how they are selected for attention (more of this later). She also discusses how the solutions are assessed as successful or not. She treats in only a general way the "heuristics" by which the solutions are obtained, that is, how they are worked out once the problems for the technologies are settled upon.[1]

The relation of Laudan's ideas to the ones given here can be put thus: Laudan's work concerns mainly the problems for which technologies provide solutions. How these solutions are contrived she lumps together in her generalized treatment of heuristics. The work here examines the design problems that arise in bringing a technology into being and the knowledge such problems require, in effect, with the details of an essential aspect of Laudan's heuristics. The problems leading to technologies are important here as background and underlying causes; they are not central. Put briefly, Laudan's work deals with problems *solved by* (or problems *for*) technologies, the present work with problems *within* such entities.

The problems whose solutions we traced in the historical cases can be stated as follows:

Chapter 2: What shape airfoil to use for the long-range airplanes at Consolidated Aircraft in the late 1930s and how to design (i.e., shape) airfoils for airplanes generally.

Chapter 3: What engineering requirements to design for to obtain flying qualities satisfactory to pilots.

Chapter 4: How to think about and analyze flow situations in mechanical design generally (including airplane design).

Chapter 5: What propeller to select in an airplane design.

Chapter 6: How to design (and produce) flush-riveted joints for aircraft.

These are clearly internal problems *within* the design of airplanes. They arose from requirements of normal, everyday engineering design at the lower levels of hierarchy described in chapter 1. Their solution was needed so designers could arrive at airplanes that would solve the operational problems for which the airplanes were intended, that is, the problems to be solved *by* the device. As we have seen, solution of such internal design problems demanded in turn the generation of new knowledge. The cases were assembled in part to exhibit these epistemological implications of design. As explained in chapter 1, however, such focus emerged after three of the five had been chosen with other things in mind. The fundamental role of knowledge and the determining role of design should perhaps have been obvious a priori; the fact is, they became apparent from the historical evidence.

In examining how problems get posed for devices and systems, Laudan sees a number of circumstances at work. After noting the rare case (at least in modern times) of problems given directly by the environment and not solved by existing technologies, she sees four internally technological sources of problems that lead to cognitive change: (1) functional failure of current technologies; (2) extrapolation from past technological successes; (3) imbalances between related technologies at a given period; (4) potential rather than actual technological failures. These sources all call for new or modified technologies, the design of which calls in turn for new or extended knowledge. From the vast number of problems arising from such sources, the technical community makes its choice. Selection often hinges, as the literature on technology normally stresses, on "the social or economic utility assigned to [the problems] by society at large," that is, on the contextual needs and desires of society. Intratechnological considerations, however, also play a role. Important among the latter are the problem's centrality within some larger technological system and the likelihood of its being solvable. Such purely technological concerns arise within an already "well-structured world" of technological knowledge and influences.[2]

The contextual needs of society, which Laudan treats as mediators of selection, may alternatively be regarded (as I shall) as *sources* of problems in their own right. Such sources—economic, military, social, personal—then underlie or interact intimately with the technological sources given by Laudan. They may even give rise to problems solely on their own, independently of failures, imbalances, or extrapolations. An economically or socially prompted need to span a river, for example, may call for a bridge in an unproblematic situation well within the scope of existing knowledge. Contextual sources show up in the present studies mostly in a peripheral motivating role, as in the military and economic requirement for the airplanes at Consolidated (chapter 2) and Douglas (chapter 3). They appear centrally only in the story of flying qualities (chapter 3), where the needs and reactions of individuals (i.e., pilots) were both the source of the problem and an essential element in the narrative.[3] Such limited consideration is not meant to suggest that contextual sources are in some way secondary; as recognized in chapter 1, they lie at the foundation of engineering problems.[4]

The technological sources that Laudan sees supplying problems *for* devices and systems (see above) also operate *within* such entities, that is, directly at the levels of design and knowledge. Examples from our cases (except for potential failure) are not hard to find: *Functional failure* of conventional airfoils as speeds went up (chapter 2) necessitated search for knowledge of how to design high-speed airfoils. *Extrapolation* from Boeing's success with the B-17, based on knowledge generated subsequent to that design, led Fleet to believe Consolidated could design a still better bomber (also chapter 2). *Imbalance* between increasing flight speeds and the drag of protruding rivets (chapter 6) required that engineers learn to design rivets flush with the surface. The considerations that influence problem selection also appear: The NACA's perception of propeller performance as *central to the system* represented by the airplane (chapter 5) helped inspire Durand and Lesley's measurement of the data needed for propeller selection. If Edward Warner and others in the aeronautical community had not seen flying qualities as a design problem with a *likelihood of solution* (chapter 3), they would not have moved actively in that direction.

Other technological sources exist in addition to those cited by Laudan. One, which can supply problems both for and within a technology, is the *perception of a new technological possibility*, that is, that some technical advance is conceivably in the cards though no one has an idea how to achieve it. This is a different thing—and prior to—Laudan's selection criterion of *likelihood of solution*. Design for desirable flying qualities had been perceived by the aeronautical community as possible well before Warner and others saw that advances in ideas and techniques made it likely and went to work on it in earnest. Aeronautical engineers of the mid-1930s had realized for some time that lower-drag airfoils were possible if extensive laminar flow could somehow be managed. The problem had been perceived some years before Jacobs began work on it; he did so intensively when he saw a likelihood of solution if he could design an airfoil to maintain a falling pressure over a sufficiently long distance. Edward Constant regards such perception of a radically new possibility as joining with the prediction of potential failure (see Laudan's list) to constitute "presumptive anomaly."[5]

Three more sources supply problems solely within a technology:

(1) The first of these, often insufficiently recognized, is the *internal logic of technology*. That is to say, once a device or system and its goal have been decided on, physical laws and practical requirements (including cost) take over and mandate that certain things be done and certain design problems solved. A classic example here is Thomas Edison's invention and development of the incandescent lighting system. Briefly, Edison's decision to strive for a system that could compete with existing gas lighting, where individual lights could be turned on and off separately, demanded that the lamps, distribution network, and dynamo have certain compatible (and unprecedented) electrical characteristics, characteristics that wouldn't be necessary if the components of the system had been unrelated. If the job were to be done at all, physics and economics dictated that it had to be done *this* way. These electrical characteristics then defined the design problems for the lamps, network, and dynamo.[6] (That Edison and his staff had to discover the logic as they went along does not alter the fact of its inherent nature.)

An example of internal logic from the airplane, though not from our studies, is the conventional airplane landing gear. Design of this device has evolved continuously since 1900 purely from necessity to get larger and faster aircraft on and off the ground safely. Decisions at upper levels of the design hierarchy set the weight, landing speed, and general arrangement of the airplane. Once this is done, the logic of the landing requirement takes over and fixes the specifics of the landing-gear problem. Remarks of this kind can be made about numerous devices required at lower levels of design hierarchy.[7]

(2) *Internal needs of design* form another source of problems. If engineers are to design airplanes that are satisfactory to pilots, they need to know what quantitative flying-quality criteria to use. If they are to analyze flow situations in their designs, they need appropriate theoreti-

cal tools like control-volume analysis. If they are to select a propeller for a new airplane, they need a collection of propeller-performance data. In general, if engineers are to design devices for given purposes (however these may be posed for devices and systems), there are certain concrete things they need to know and know how to do in order to arrive at a detailed, dimensioned design that can be constructed. (Engineering is an intensely quantitative activity, even if that fact is not always reflected in the historical and philosophical literature.) Such design necessities define the problems of knowledge generation that must be solved when necessary data and methods do not already exist. A vast amount of research effort arises out of problems from this source.

(3) *Need for decreased uncertainty*, though an internal need of design for reasons discussed in the final section of chapter 2, deserves mention on its own. It poses problems for how engineering research is done and how engineering data are produced. As explained in the earlier chapter, the need to reduce scale-effect and other uncertainties in airfoil data in the 1930s gave rise to problems in wind tunnels and wind-tunnel techniques; solution of these knowledge-production problems was critical for Jacobs's development of the laminar-flow airfoil. At present, the need to reduce uncertainty in fluid-flow calculations with high-speed computers poses problems for engineers who devise algorithms for such calculations. These kinds of problems take on a life of their own and can provide focus for an entire engineering career.[8]

As touched on above, engineering problem solving is permeated by evidence of numerous kinds of hierarchy. The relationship between systems and devices is clearly hierarchical. So also is the relationship between problems *for* and problems *within* such technologies. The design hierarchy based on task, component, and discipline, described in chapter 1, plays a related and particularly important role for present concerns. Here one level of hierarchy poses problems for some level below: Military project definition for the B-24, for example, posed problems of detailed airfoil design for the engineers at Consolidated; combined aerodynamic and structural requirements for wing design in the 1930s posed problems of flush riveting for aircraft companies generally. Knowledge generation itself also contains various kinds of hierarchy. Need for one kind of knowledge poses need for another: Learning how to design for desirable flying qualities required learning how to represent static stability in terms of pilots' actions; providing suitable propeller data for designers required finding how best to represent performance of a propeller operating ahead of an obstructing body. Generating knowledge as free as possible of uncertainty poses problems that produce yet another hierarchy: The need for accurate aerodynamic data posed test problems that spawned a whole wind-tunnel technology with its own internal hierarchy of design and operational problems. And intersecting and interacting with these kinds of hierarchy is a hierarchy of communities, exemplified by the flying-quality community and its array of subcommunities, all included within the larger aerodynamic community. Engineering problem solving as a whole reflects the same hierarchical characteristics we have observed in aeronautics. Spelling out in detail the intersecting roles of the various hierarchies would be a daunting task. As suggested in chapter 1, such scrutiny would demand deeper consideration of contextual sources and influences.

Examination of engineering as problem solving confirms the importance of design emphasized in chapter 1. To implement the devices and systems that solve both societal and technological problems requires that the devices and systems be designed. Such design then raises its own problems. At the working level in engineering, problem solving and design are almost synonymous.

Whatever the source of design problems, their solution depends on knowledge. The knowledge, however, need not be new. Even when resolution of a new problem is achieved, the problem does not disappear—after our stories were finished, propellers and airfoils still had to be selected for particular aircraft and flush-riveted joints proportioned and detailed. Developments and refinements continue, of course, and solutions in individual cases may call for considerable ingenuity. Once understanding and information are established, however, solutions are typically devised without generation of a great deal of additional knowledge. What is needed is available in textbooks or manuals and can be looked up, taught to engineering students, or learned on the job. Today the problems we have discussed (with the exception of flying-quality specifications for new and advanced types of aircraft) are for the most part old hat. They are solved—and this holds for most day-to-day design problems—mainly on the basis of stored-up engineering knowledge. (The unproblematic bridge mentioned earlier is a case in point.) The important role of stored-up knowledge tends to be overlooked when one concentrates, as does Laudan, on the problems of new technologies.

When knowledge is *not* at hand, as in the circumstances treated by Laudan, it has to be generated. Even stored-up knowledge had to be brought into being at some time in the past. Either way, the needs of design impel, condition, and constrain the growth and nature of that knowledge. The new knowledge may be generated very close to design (e.g., development of flush riveting in aircraft companies), rather far from it (formulation of control-volume analysis in universities), or some complicated combination of the two (evolution of flying-quality specifications in airlines, aircraft companies, and government research establishments). Wherever its generation, the knowledge is judged in the end on the basis of whether it helps to achieve a successful design. As we have seen, the knowledge and the circumstances and processes of its generation are rich, diverse, and complex. Despite the widespread utility of stored-up knowledge, how the body of knowledge grows is the compelling question historically and epistemologically.

The fundamental role of engineering design knowledge encourages generalization about its nature and acquisition. The remainder of this chapter therefore attempts a categorization of such knowledge and the activities and agents that generate it, using the historical cases as the main (though not sole) source of inspiration and example. Given the breadth and complexity of engineering knowledge, a realistic anatomy must be basic to any theories about it that may ultimately develop. Chapter 8 will then suggest a general theoretical model for the growth of such knowledge.

The analysis that follows does not attempt to be exhaustive. As stated earlier, engineering has to do not only with design but with production and operation as well. The flush-riveting story was concerned largely with production and the flying-quality study to some extent with operation; such activities will need careful attention in any complete study of engineering knowledge. For reasons explained in chapter 1, however, the historical cases focus on knowledge for normal, everyday design at the lower levels of hierarchy, and the detailed analysis will be limited correspondingly. Although activities that generate the knowledge are sometimes hardly normal and everyday, the knowledge examined is ultimately so used. I will need to discuss knowledge derived from invention and radical design in one respect, but I shall only mention in passing how such knowledge comes into being. Even within these limits, there remains a great deal to sort out and account for.

Any detailed analysis of engineering knowledge runs the risk of seeming to divorce such knowledge from engineering practice. As I hope the historical cases make clear, the inseparability of knowledge and its practical application is in fact a distinguishing characteristic of engineering. I shall return to this point later, but it needs to be kept in mind throughout.

The following two sections are unavoidably detailed and complex. To help distinguish the forest from the trees, a summary table will be provided toward the end of the chapter (table 7-1).

Categories of Knowledge

Categorizing engineering design knowledge is no less complicated than analyzing the relation between problem solving and design. Some items of knowledge are clearly distinguishable, others are not. In the categorization that follows, the divisions are not entirely exclusive—some items of knowledge can embody the characteristics of more than one category. They are also probably not exhaustive—although the major categories are presumably complete, the subspecies within them most likely are not. I shall take up the categories under the following headings:

1. *Fundamental design concepts*
2. *Criteria and specifications*
3. *Theoretical tools*
4. *Quantitative data*
5. *Practical considerations*
6. *Design instrumentalities*

Except for the last two, which I have had to devise, these terms are used by engineers with essentially the same meaning I give them here. Such categorization and its content, however, are not usual in the profession.

1. *Fundamental design concepts.* Designers setting out on any normal design bring with them fundamental concepts about the device in question. These concepts may exist only implicitly in the back of the designer's mind, but they must be there. They are givens for the project, even if unstated. They are absorbed—by osmosis, so to speak—by engineers in the course of growing up, perhaps even before entering formal engineering training. They had to be learned deliberately by the engineering community at some time, however, and form an essential part of design knowledge.

Designers must know first of all what Michael Polanyi calls the *operational principle* of their device. By this one means, in Polanyi's words, "how its characteristic parts . . . fulfil their special function in combining to an overall operation which achieves the purpose" of the device—in brief, how the device works.[9] Every device, whether a mobile machine such as an aircraft or a static structure such as a bridge, embodies a principle of this kind. An example is implicit in Sir George Cayley's famous statement from 1809 of the principle of the fixed-wing aircraft we now call the "airplane":

> to make a surface support a given weight by the application of power to the resistance of air.[10]

Cayley's concept, revolutionary at the time but normal today, separates lift from propulsion. As Cayley meant it, it says in effect that an airplane operates by propelling a rigid surface forward through the resisting air, thus producing the upward force required to balance the airplane's weight. It was fundamental in that it freed designers from the previous impractical notion of flapping wings. All modern airplane designers have this concept in the back of their minds. Engineers dealing with any device must similarly know its operational principle to carry out normal design.

Operational principles also exist for the components within a device. An airfoil, by virtue of its shape, in particular its sharp trailing edge, generates lift when inclined at an angle to the airstream. A conventional tail-aft airplane, by virtue of a certain arrangement of wing, tail, and control surfaces in relation to the airplane's center of gravity, achieves desired stability and control. Analogous statements can be made about propellers and flush rivets in accordance with Polanyi's definition. At this level, as at that of the overall device, the designer must be aware of the operational principle as basis for design.

As elaborated by Polanyi, it is the operational principle that provides the criterion by which success or failure is judged in the purely technical sense. If a device works according to its operational principle, it is counted as being a success; if something breaks or otherwise goes wrong so that the operational principle is not achieved, the device is a failure. The operational principle also, in effect, defines a device. The members of a group of vehicles all qualify, for example, as airplanes— as against, say, helicopters, which obtain their lift from rotors—only if they share the operational principle laid down by Cayley. (A device is thus defined by its technical nature, not its economic or other use.)

Finally, the operational principle provides an important point of difference between technology and science—it originates outside the body of scientific knowledge and comes into being to serve some innately technological purpose. The laws of physics may be used to analyze such things as airfoils, propellers, and rivets once their operational principle has been devised, and they may even help in devising it; they in no way, however, contain or by themselves imply the principle. Polanyi makes essentially the same point a bit differently: "The physico-chemical topography of the object may in some cases serve as a clue to its technical interpretation, but by itself it would leave us completely in the dark [about how it achieves its operational purpose]. . . . *The complete* [i.e., scientific] *knowledge of a machine as an object tells us nothing about it as a machine.*"[11]

A second thing the practitioner of normal design takes for granted is the *normal configuration* for the device. By this I mean the general shape and arrangement that are commonly agreed to best embody the operational principle. Such qualitative restriction, though not definitive in the same manner as the operational principle, plays a strongly determining role. Designers of the highly diverse collection of airplanes that Gilruth's group tested for flying qualities in the late 1930s (see figure 3-7) probably never had it enter their minds that their airplane should be other than a tractor monoplane with tail aft. Automobile designers of today usually (but not invariably) assume without much thinking about it that their vehicle should have four (as against possibly three) wheels and a front-mounted, liquid-cooled engine. They doubtless assume other things as well. Other features may be left open to be decided in the course of the design (whether, for instance, power is to be applied via the front wheels, the rear wheels, or all four). Whatever the details, the preferred configuration for a given device with a given application is knowledge that has to be learned by the engineering community, usually by experience with different configurations in the early stages of a technology. Part of the experience leading to the normal configuration for airplanes is described in chapter 8.

A shared operational principle and normal configuration define the normal technology (or design) of a device in the sense used by Constant as mentioned in chapter 1, that is, as denoting "the improvement of the accepted tradition or its application under 'new or more stringent conditions.'"[12] Durand and Lesley, for example, were clearly pursuing the normal technology of the propeller, a device that had by their time an established operational principle and an accepted configuration. Radical technology (as I used that term in chapter 1) involves a change in normal configuration and possibly also in operational principle. In the latter event, the configuration must in fact be established ab initio, since it obviously cannot be known at the outset. When this happens, the resulting technology may even justifiably be described as revolutionary in its early stages. It was in this sense that Constant titled his book *The Origins of the Turbojet Revolution*. (Constant, however, did not use the term *revolutionary technology* or describe the change from piston engines in terms of a change in operational principle.) Radical and even revolutionary technology may occur at the level of an overall device, as happened in Constant's instance, or at the level of a subsidiary component, as with Jacobs's idea about how pressure distribution can advantageously operate to reduce airfoil drag. Radical developments that fall short of revolutionary may involve a modification of operational principle, in contrast to complete change. The distinctions here are relative and not always easily defined. As explained earlier, the knowledge and knowledge processes considered in the present work are those required in normal technology, that is, in design under an essentially fixed operational principle and normal configuration.

Every device possesses an operational principle, and, once the device has become an object of normal, everyday design, a normal configuration. Engineers doing normal design bring these concepts to their task usually without thinking about them. By virtue of being so little noticed, other concepts may possibly exist that I have overlooked.

The operational principle and normal configuration provide a framework within which normal design takes place. To translate these concepts into a concrete design requires knowledge from the categories that follow. In a broad sense, the rest of the knowledge for normal design serves such purpose.

2. *Criteria and specifications.* To design a device embodying a given operational principle and normal configuration, the designer must have at some point specific requirements in terms of the hardware. That is to say, someone—the designer or somebody else—must translate the general, qualitative goals for the device into specific, quantitative goals couched in concrete technical terms. The economic goals of United and its associated airlines had to be translated into performance specifications for the DC-4E; the necessity for a bridge to carry traffic over a river has to be translated into specific span and loading requirements. To accomplish such tasks, the people responsible must have knowledge of technical criteria appropriate to the device and its use (in earlier terminology the design problem must be "well defined") and must assign some kind of numerical values or limits to those criteria. Without such technical specifications, the designer cannot start to contrive the details and dimensions that must ultimately be supplied to the builder.[13] Assignment of the values or limits is usually (but not always; see below) particular to particular designs and is best looked upon as part of the design process. The criteria themselves—the essential key to engineering specification—constitute an important element of general engineering knowledge.

The importance of engineering criteria appears in the situation concerning flying qualities before establishment of codified requirements circa 1940. As indicated by the intraoffice memorandum quoted from Captain Hatcher of the Bureau of Aeronautics (chapter 3), designers in that period had to guess what their customers wanted; the evidence suggests they were not consistently successful in doing this or in setting valid goals of their own. This situation stemmed in part from the fact that many of the necessary criteria—properties of the short-period oscillation, stick force per g, plus lateral criteria we did not examine— were either unknown or only partially understood. Such a situation is not unusual in the learning phases of a technology, but engineers work to eliminate it as expeditiously as possible. In the instance of flying qualities, establishment of the needed criteria was part of our story; for airfoils, propellers, and riveted joints, it had been worked out before our narratives began.

Design criteria vary widely in perceptibility. Sometimes, as with flush rivets and the performance specifications incidental to the chapter on flying qualities, the necessary quantities (e.g., loads and dimensional tolerances on flushness for rivets; speed, altitude, and horsepower for performance) are simple and obvious; they can be discerned at once and without much effort. In other cases, such as airfoils (e.g., lowest possible drag at cruising lift) and propellers (maximum propulsive power and efficiency in relation to operating conditions), the criteria are not immediately clear; they have to be devised consciously and deliberately over some period. In still others, such as flying qualities (stick force per g, time for the short-period oscillation with elevator free, and so forth), the criteria are obscure and require great effort over a protracted time. Criteria may apply at the level of the overall device, as with performance and flying qualities of an airplane, or at the level of a component, as with the characteristics of airfoils and propellers. In the latter case, they must take account of integration of the component into the overall system. Whatever their nature, they should, insofar as possible, be general over the class of devices and conditions to which they apply and easy for engineers to comprehend and use.

Assignment of values or limits to appropriate criteria is essential for design. For an overall device like the bridge or DC-4E mentioned earlier, such assignment takes place at the project-definition level in the design hierarchy described in chapter 1. People functioning at this level produce the technical specifications required for the concrete design activity that follows. At the levels of detailed design, designers themselves set requirements for the subsidiary components. Engineers working on the B-24 at Consolidated decided what they were looking for in a wing and its airfoil section. Designers of any airplane know its projected gross weight and landing speed and set specifications for the landing gear accordingly. At whatever level, such specifications for individual projects are a transitory step in a design process and not usually thought of as "knowledge."

When circumstances are sufficiently general, specification of values or limits can apply across an entire technology. Such was the case with Gilruth's flying-quality requirements of the early 1940s, which applied to all conventional airplanes. A second example is the boiler code the American Society of Mechanical Engineers promulgates for all designers of such devices. These sorts of specifications tend to arise where general utility or safety is involved; they are often made administratively or legally binding. Such universal specifications, like the criteria on which they are based, become part of the stored-up body of knowledge about how things are done in engineering.

Translation of the utilitarian, usually qualitative, goals of a device into concrete technical terms—and the knowledge required to do it— are crucial for engineering design. They deserve more attention than they have received from historians and philosophers of technology. Determining the essential criteria often draws on the theoretical tools, quantitative data, and pragmatic judgment discussed in the remaining categories. Because of its special purpose, however, the knowledge that results deserves recognition on its own.

Finally, an important point of difference exists here between science and engineering. Scientists, in their search for understanding, do not aim at rigidly specified goals. Engineers, to carry out their task of designing devices, must work to very concrete objectives; this requires that they devise relevant design criteria and specifications. We thus see here another unique feature of engineering knowledge.[14]

3. *Theoretical tools.* To carry out their design function, engineers use a wide range of theoretical tools. These include intellectual concepts for thinking about design as well as mathematical methods and theories for making design calculations. Both the concepts and methods cover a spectrum running all the way from things generally regarded as part of science to items of peculiarly engineering character.

Mathematical methods and theories, which I take up first, may be simple formulas for direct calculation or more complex calculative schemes. The crucial thing is that their mathematical structure makes them useful for quantitative analysis and design. It is this body of mathematically structured theoretical knowledge, excluding items at the far engineering end of the spectrum and supplemented by relevant quantitative data from the next category, that is frequently referred to as "engineering science" (see chapter 4).[15]

At the scientific end of the spectrum, we find purely mathematical tools having no physical content in themselves. Engineers acquire many of these from prior mathematics, either directly or with some modification or extension. An example of direct transfer is engineering's widespread dependence on the analytic geometry of Descartes, or, from the present studies, Joukowsky's use of classical conformal transformation to derive a special class of theoretical airfoil profiles (chapter 2); an instance of modification and extension is Theodorsen's adaptation of conformal transformation to analysis of arbitrarily shaped airfoils (chapter 2). Mathematical tools may also be devised deliberately for engineering use. Such is the case, for example, with the computer algorithms now seeing widespread application to complex engineering problems.

Next along the spectrum is mathematically structured knowledge that is essentially physical (in contrast to purely mathematical) and has scientific interest for its general explanatory power. Such knowledge often comes from prior science, as happened with control-volume analysis, which embodied mechanical and thermodynamic theory already available from physics. As we saw, however, even such scientific knowledge must typically be reformulated to make it applicable to engineering problems. My colleague Charles Kruger and I, for example, performed this task in a book that recasts and extends parts of the physics and chemistry of gases into a form useful to engineers confronted by problems in high-temperature gas dynamics.[16]

Farther in the direction of engineering are theories based on scientific principles but motivated by and limited to a technologically important class of phenomena or even to a specific device. Such theories have the earmarks of the physical sciences—they are mathematically structured and have explanatory power. They make up what Polanyi calls "systematic technology," which "can be cultivated in the same way as pure science." Their essentially engineering character, however, appears in the fact that they would "lose all interest" and be forgotten if the class of phenomena or device to which they apply should, for some technological, economic, or social reason, cease to be useful.[17] Examples of such essentially engineering theories centered on a class of phenomena are those dealing with fluid mechanics (illustrated here by control-volume analysis), heat transfer, and solid-body elasticity; others are easily found.

When limited to specific devices, these engineering theories frequently depend for their mathematical feasibility (a point not much noted outside the engineering literature) on some approximation peculiar to the device in question. Such device-restricted theory (and its associated approximation) appear in the present studies in the induced-drag theory for airplane wings, employed by the engineers at Consolidated (chapter 2; average chord small compared with span), in Munk's thin-airfoil theory (chapter 2; maximum thickness and camber small compared with chord), and in the theory of dynamic stability of aircraft (chapter 3; in the case of longitudinal stability, vertical velocity of tail as a result of oscillation small compared with forward velocity). Examples outside aeronautics include elementary beam theory in strength of materials (deflection and depth of beam small compared with span) and the dynamic modeling of transistors in electrical engineering (variation of input-signal current small compared with average current). With the availability of large-scale electronic computers, such approximations to render the mathematical manipulations feasible are no longer as essential. Device-specific theories, however, still abound.[18]

Still more toward the engineering end of the spectrum we encounter an assortment of theories that, while they may go back to scientific principles in part, involve some central, ad hoc assumption about phenomena crucial to the problem. Such "phenomenological theories," which often are also device specific, have little explanatory power or scientific standing. They serve almost solely for engineering calculation. Engineers devise them because they must get on with their design job and the phenomena in question are too poorly understood or too difficult to handle otherwise. Phenomenological theories are in this sense a theoretical counterpart of the method of experimental parameter variation discussed in chapter 5. An example of phenomenological theory is the blade-element theory of propellers (chapter 5), which assumes that the forces on a chordwise element of a rotating propeller blade can be obtained from experimental data for the same blade profile in appropriate rectilinear flight. A more broadly applicable example is the so-called eddy-viscosity models of turbulent flow, which have been used for the calculation of such technologically crucial flows over most of this century. These theories all involve some assumption about the turbulent mixing process underlying the apparent stresses (turbulent equivalent of viscous forces) within the fluid. Such assumptions, which have been progressively refined, are required because turbulence has been prohibitively difficult to treat from basic principles (a situation that is also changing with the availability of large-capacity electronic computers). Phenomenological theories, on the whole, are seldom rigorous physically and may even be demonstrably faulty in some degree. They are used because they work, however imperfectly, and because no better analytical tools are available.

A last kind of mathematical tool lies at the far engineering end of the spectrum: quantitative assumptions introduced for calculative expedience but too crude even to be dignified as theories. An example here is in the analysis of riveted joints (chapter 6), where the load on the joint is assumed to be shared equally by the various rivets. As in this example, such analytical assumptions are often realized to be wrong. They again are used for practical reasons and because they are known from experience to give conservative or otherwise acceptable results. Without them, a great deal of everyday design would not get done.

The second class of theoretical tools I have termed *intellectual concepts*. To repeat my earlier words, "Although engineering activity produces artifacts, conceiving and analyzing these artifacts requires thoughts in people's minds." Intellectual concepts provide the language for much of such thinking. They are employed, not only in quantitative analysis and design calculation, but also for the qualitative conceptualizing and reasoning engineers must do before such quantitative activity begins (as well as while it is being carried out).

The concepts exhibit wide diversity. They range from highly scientific to intensely practical, from (though not correspondingly) specifically mathematical to explicitly physical. They sometimes derive from the mathematical theories described above, sometimes from physical reasoning, and sometimes from both. Some apply only to specific devices and some have broader application.

The combinations of characteristics being so diverse, I do not attempt to describe the examples completely. Concepts of broadest applicability include basic ideas from science, like force, mass, electric current, and so forth, and things of a more engineering nature, such as efficiency and feedback. Concepts we have encountered that apply broadly within the more restricted realm of fluid mechanics are the control volume (chapter 4) and the boundary layer (chapter 2). Examples that are device specific include the separation of thickness and camber for airfoils (chapter 2) and the distinction between long and short periods in the longitudinal oscillation of aircraft (chapter 3); both of these originated in sophisticated mathematical theories. Device-specific examples obtained at first by physical reasoning include the propulsive efficiency of propellers in the presence of an obstruction (chapter 5) and moment-curve slope, as well as elevator-angle and stick-force gradients, for analysis of static longitudinal stability (chapter 3); the stability criteria were subsequently derived also by mathematical theory. The ease with which examples can be multiplied shows how basic such concepts are to engineering thinking.

4. *Quantitative data.* Even with fundamental concepts and technical specifications at hand, mathematical tools avail little without data for the physical properties or other quantities required in the formulas. Other kinds of data may also be needed to lay out details of the device or to specify manufacturing processes for production. Such data, essential for design, are usually obtained empirically, though in some cases they may be calculated theoretically. They are typically represented in tables or graphs.[19] As discussed with regard to flush riveting in chapter 6, they divide into two kinds of knowledge, descriptive and prescriptive.

Descriptive knowledge is knowledge of how things are. Descriptive data needed by designers include physical constants (acceleration of gravity, for example) as well as properties of substances (failing strength of materials, coefficient of viscosity of fluids, etc.) and of physical processes (rate of chemical reactions and so forth). Occasionally they deal with operational conditions in the physical world (frequency and strength of atmospheric gusts for aircraft fatigue-loading calculations). As we have seen with flying qualities, they also encompass information on human beings (maximum forces exerted by pilots). In situations where performance of an element or component of a device cannot be predicted from theory (and sometimes even when it can), descriptive data will also include overall measurements of performance. Parametric data on the allowable strengths of rivets and the performance of airfoils and propellers are cases in point.

Prescriptive knowledge is knowledge of how things should be to attain a desired end—it says, in effect, "in order to accomplish this, arrange things this way." The (in some respects misnamed) "process specifications" that manufacturers issue for guidance of their detail designers consist largely of such prescriptive data.[20] An example we observed was the limits on sheet thickness for different types of flush riveting. As pointed out, quantitative values of such prescriptive data differ somewhat from one company to another. Other kinds of data, though prescriptive, are applied by fiat uniformly across the industry. These include the safety factors government agencies prescribe to insure that an airplane's structure will safely carry the imposed loads, as well as engineering standards (e.g., dimensions of flush rivets) agreed upon in the interest of maintenance and economy. Other industries exhibit similar diversity of prescriptive data.[21]

This may be a good place to point out that the distinction between descriptive and prescriptive extends to more than quantitative data. Mathematical theories are descriptive in that they allow calculation of how devices and processes will perform under given assumptions and conditions. They are sometimes employed to compile descriptive data to be used by designers.[22] Operational principles, normal configurations, and technical specifications are prescriptive by virtue of prescribing how a device should be to fulfill its intended purpose. Much of the design procedure of the next category is also prescriptive in the sense that it suggests to the designer how to go about achieving a required design.

5. *Practical considerations.* Theoretical tools and quantitative data are, by definition, precise and codifiable; they come mostly from deliberate research. They are not, however, by themselves sufficient. Designers also need for their work an array of less sharply defined considerations derived from experience in practice, considerations that frequently do not lend themselves to theorizing, tabulation, or programming into a computer. Such considerations are mostly learned on the job rather than in school or from books; they tend to be carried around, sometimes more or less unconsciously, in designers' minds. Frequently they are hard to find written down. The practice from which they derive necessarily includes not only design but production and operation as well, though such practice may not be—typically is not—by the designers themselves.

An example, a particularly simple one, of a practical consideration known from production is the depth of jaw of the single-operator, pedestal-type riveting machine available in a company's shop (chapter 6); for economy of manufacture, a designer wants to make structural subassemblies as large as possible, but the size of the machine puts an upper limit on what can be designed in this regard. Designers likewise must know whether a given kind of part designed to be made of a given metal is most feasibly and economically produced by machining, forging, or casting; such knowledge may dictate the details of the part. Another seemingly simple but actually complex example is knowing the clearance that must be allowed for tools and hands in putting together assemblies for which the worker must reach inside. Such knowledge is especially difficult to define, since the requirement depends in part on the spatial relationship of the parts of the assembly and the position required of the worker. Knowledge of this kind defies codification, and a mock-up or prototype must often be built to check the designer's work.

A practical consideration coming from operation is the judgment, the origins of which were narrated in chapter 3, that a piloted airplane should be aerodynamically stable but not too much so. This important piece of knowledge came from flight experience with airplanes over a period of years. Other examples from operation of airplanes would include knowledge of the best position and arrangement of access panels for field maintenance and how best to design baggage-door handles and locks to insure complete latching by baggage handlers at the loading gate. Aircraft designers are also learning from tragic experience about how to increase the likelihood of passengers surviving in a crash, as, for example, by avoiding cabin materials that are found to give off toxic fumes upon burning. In this era of frequent liability suits, designers of devices of all kinds need to know increasingly how to design to make self-injury by even careless operators as difficult as possible. Such considerations from operation can easily be extended over all branches of engineering. Engineering devices are by definition made to be used, and feedback of knowledge from use to design is essential.

Experience in design also produces knowledge useful in further design practice. Such knowledge often takes the form of design rules of thumb, such as the statement, used by aircraft designers, that up to a certain increase in empty weight of a transport airplane is worthwhile if it will make possible a certain decrease in drag. Engineers distill such rules from their detailed experience to allow rapid design assessments; quantitative values again may differ somewhat from company to company. The design community similarly knows from years of experience that for successful jet airplanes the ratio of thrust of the engines to weight of the loaded aircraft always comes out somewhere between about 0.2 and 0.3. This knowledge supplies a rough check as a new design proceeds; if the calculated ratio falls outside this range, the designer suspects a misjudgment or miscalculation. Rules of thumb derived from design experience show up in all branches of engineering.[23]

Occasionally practical considerations become well codified and are more logically put in another category. The limits on sheet thickness for different types of flush riveting started out as rough design rules, but, as production experience increased, they became precise enough for tabulation and inclusion in process specifications. They now seem best characterized as quantitative data. Flying-quality specifications provide an analogous case from operation. In the late 1920s, to the extent that they aimed deliberately for flying qualities at all, designers apparently used only rough, qualitative considerations, such as that the moment-curve slope should be negative but not too much so (chapter 3). In the early 1930s, Stalker, Johnson, and Diehl all tried to convert this requirement into a more quantitative rule of thumb. By 1942–43, thanks to the efforts of Warner, Gilruth, and others, the aeronautical community had refined the requirements to the point where they could be codified into what I earlier categorized as criteria and specifications. In the final move, the idea from the mid-1930s that objective specification of flying qualities might in fact be feasible played the key role. The overall sequence illustrates how engineers work to embody requirements from practice (in this case, operational) into as concrete and definite a technical formulation as possible. It also shows how our categories of knowledge are not mutually exclusive and sometimes grade into one another.

6. *Design instrumentalities.* Besides the analytical tools, quantitative data, and practical considerations required for their tasks, designers need to know how to carry out those tasks. The cases here (of what I have referred to variously as "knowing how" and "procedural knowledge") have concentrated mostly on the generation of new theoretical and quantitative knowledge used in the design process. Only in the adoption of the Davis wing at Consolidated did we follow events of the process itself, and one example scarcely provides adequate basis for generalization. The instrumentalities of the process—the procedures, ways of thinking, and judgmental skills by which it is done—nevertheless must be part of any anatomy of engineering knowledge. They give engineers the power, not only to effect designs where the form of the solution is clear at the outset, but also to seek solutions where some element of novelty is required. A large professional and analytical literature exists on the design process by engineering teachers and others, including specialists in artificial intelligence.[24] To compare and reconcile these writings in relation to historical studies would

be a formidable undertaking; analytical studies of the design process pose an important future task for historians. I limit myself here to some general observations, more for the sake of balance than for any intent at complete coverage.

Designers doing normal design call upon a number of well-recognized, more or less *structured procedures* (my term, not theirs). Of these, the division of an overall system—in the present instance, the airplane—into subsystems is fundamental; it is a basic element in the hierarchy of design described in chapter 1. Such procedure appeared here in the division of the airplane into airframe, engine, and propeller in the Durand-Lesley account and in the subdivision of the airframe to the level of airfoil section implicit in the Davis-wing story. A second procedure—optimization, in this case of a subsystem—was used to the extent possible by the engineers at Consolidated in their aerodynamic wing design; in this they employed a combination of theoretical tools and quantitative data. Though they probably saw themselves as optimizing, in the end, because of the complexities and uncertainties in the problem, they achieved no more than what Herbert Simon calls "satisficing," that is, not the very best solution, but one that was satisfactory.[25] In engineering design, procedures for satisficing (a term engineers rarely use) are less formal than for optimizing and depend more on judgmental skills of the kinds mentioned below. Most engineering design is, in fact, of this kind; given the large number of interacting variables in most of their problems, engineers seldom are able truly to optimize. Additional design procedures, evident explicitly or implicitly in the work at Consolidated, are the carrying out of related activities in parallel and the use of iterative techniques (i.e., successive improvement of a design based on analytical or test experience with earlier versions). How to employ these and other procedures productively constitutes an essential part of design knowledge.[26]

Between more or less structured procedures and the judgmental skills I shall come to later—less tangible than the former, more so than the latter—are what I have referred to as *ways of thinking*. By this I mean the habitual ways in which design engineers formulate the "thoughts in people's minds" that I mentioned earlier.

Many ways of thinking in engineering derive from and depend on the intellectual concepts discussed as theoretical tools. The ways of thinking needed in design, however, involve more than simply the concepts; they have to do with the mental processes the designer follows. They provide shared ways for apprehending the operation of a device and imagining the effect of alterations in its design. In control-volume thinking, for example, the designer sees the operation of a fluid-flow device as consisting of changes taking place inside a suitable volume in relation to transport of quantities through the enclosing surface; alterations in the device are then imagined in terms of that mechanism. In thinking about airfoils in the context of Munk's thin-airfoil theory, designers conceive of alterations in thickness and camber as acting separately and their effects adding up to give the performance of the airfoil. Such thought processes may occur before as well as during quantitative design calculations; they thus have value apart from the direct application of the mathematical theories from which they derive. They can be illustrated and taught to young engineers and are part of the shared body of engineering knowledge. Though the instances I have cited employ concepts from mathematically structured theories, examples based on less formal sources can also be found.[27]

The foregoing ways of thinking originated from specific intellectual concepts. Designers also think in ways that start from a particular mode of thinking and find concepts to fit the situation. One of these, as in all creative thinking, is by analogy. Aeronautical engineers, for example, think about directional stability (the stability of an airplane in angular motion about its vertical axis) as similar to that of a weathercock. That such stability is called "weathercock stability" shows how ingrained this simple analogy has become. Other, more complex analogies are commonplace. Teaching engineering in the classroom takes place frequently in such terms.[28]

Still another mode of thinking has a different nature entirely. As emphasized in an important article by Eugene Ferguson, designers think also in ways that are "not easily reducible to words." Such nonverbal thinking uses for its language, not expressible concepts as above, but "an object or a picture or a visual image in the mind." Thinking of this kind does not figure explicitly in our examples; the layout designers at Consolidated would have had to visualize the B-24 and its Davis wing at least tentatively, however, in order to set pencil to paper. (And one reason for the power of control-volume thinking is that it combines visual with verbal elements.) Aids to visual thinking include sketches and drawings, both the formal kind and the informal ones that engineers make on place mats at the luncheon table and on the backs of envelopes. They also embrace models, graphs of data, and (nowadays) images on viewing screens in computer-aided design. But the thinking itself is a mental process; knowing how to do it is an aspect of the tacit knowledge described near the end of chapter 6. (Persons interested in the nature and history of this little-examined way of thinking should read Ferguson's article.) Outstanding designers are invariably outstanding visual thinkers. Knowing this, engineering schools make efforts to teach this form of knowledge, and courses and books exist with "visual thinking" in their titles. Practical experience, however, is indispensable.[29]

Finally, designers need the pragmatic *judgmental skills* required to seek out design solutions and make design decisions. Such skills, like visual thinking, call for insight, imagination, and intuition, as well as a feeling for elegance and aesthetics in technical design. They too exhibit a wide variety applying to a broad range of tasks. At one limit are highly specialized technical judgments, such as where to place the control volume for analysis of a fluid-flow device. At the other limit, we find broadly based considerations, like those facing Fleet and Laddon in deciding whether to adopt Davis's airfoil in the face of uncertain data and conjectured military preferences. Judgmental skills in such situations must include an ability to weigh technical considerations in relation to the demands and constraints of the social context. Design operates "within the contingent world of funding priorities, time pressures, conflicting biases, personal and institutional politics, and the like," and designers must know how to respond and take such influences into account.[30]

Whatever the situation, knowledge of how to exercise judgmental skills, like knowledge of how to think visually, is mostly tacit. Though the importance of such skills can be pointed out to engineering students in the classroom, they can be learned in the end only through practical experience. This must include experience not only with what works but also with what doesn't. As pointed out by Gary Gutting, "The mere fact that a system fails to perform properly in certain circumstances in itself constitutes a piece of knowledge essential to the technological enterprise."[31] Undergirding judgmental skills, then, must lie a wide range of experience from past and current practice. Though such skills are to some degree shared by the design community, they are, to a greater extent than the other kinds of knowledge (excepting, perhaps, that involved in visual thinking), an individual matter. Knowing how to exercise such skills is, as much as anything else, what separates an outstanding designer from an ordinary one.

As is obvious from the historical cases, the six categories of knowledge set out above interact intimately.[32] The need for technical specification applied to a given operational principle calls for appropriate theoretical tools and quantitative data. Theoretical tools guide and structure the collection of data, and the data in turn suggest and push the development of tools for their presentation and application. And all these things feed into and in some cases provide the basis for the various design instrumentalities. Though engineering knowledge has many threads, it is itself a tightly woven fabric.

Within the categories, knowledge, like the structure of design that conditions it (see chapter 1), is hierarchical. All six categories contain knowledge pertaining to subsidiary components as well as to overall devices. Cayley's operational principle for the complete airplane, for example, is one thing; the operational principle for the airplane's retractable landing gear is yet another. Changes in knowledge may differ correspondingly—a revolutionary change may take place in the operational principle at the level of an overall device, producing a new device that includes subsidiary components operating on traditional principles. (Very little change would occur if new operational principles had always to be devised at all levels.) Similar remarks can be made about types of knowledge from other categories. Although I have pointed to such things in passing, the whole matter of hierarchy deserves more definitive analysis. This will be complicated by existence of the many different kinds of hierarchy pointed out in the discussion of problem solving. In drawing on knowledge to solve their problems, designers move up and down hierarchically within the categories as well as back and forth from one category to another. Epistemologically minded scholars will need to give attention to such interrelated matters of structure; engineers for the most part go instinctively about their business without giving them a second thought.

The implications of social context for the various kinds of knowledge also deserve more attention than my brief mention under judgmental skills and elsewhere. Such implications—an aspect of what John Staudenmaier calls the "tension between technical design and its ambience" (see chapter 1)—are closely tied to the intersecting structure of knowledge categories and design hierarchy; social impact tends on the whole to be largest for fundamental concepts, and criteria and specifications, at the upper levels of the hierarchy described in chapter 1. It is there that the design-ambient tension most asserts itself. Economic and military demands on aircraft, for example, determine the performance criteria and specifications that are essential to project definition and overall design. The needs of pilots determine the nature of the criteria and specifications for flying qualities of the overall airplane. With regard to fundamental concepts, it was the military desire for a high-speed airplane able to take off and land vertically that led to the development of the operational principle of the airplane known as the Harrier, where the jet exhaust can be deflected vertically for ascent and descent. And as shown by Trevor Pinch and Wiebe Bijker, what we now take to be the normal configuration of the bicycle was in large part determined by the interacting requirements of a number of social groups.[33] Though less pronouncedly, contextual influence also exists for other categories and at lower levels of hierarchy. Flying-quality criteria, such as stick force per g, which were mandated by pilots' needs, became at the same time analytical tools useful at the lower levels of design. Passenger and cargo size directly condition the data needed to proportion the cross section of the fuselage on a commercial airliner. In many ways, social context becomes built into the cognitive structure and content of engineering. Like so many things in this study, the relation between context, categories of knowledge, and hierarchy of design is highly complex.

A danger exists that the categories presented here may appear overly static. Engineering knowledge is obviously highly dynamic; one must at least entertain the possibility that categories useful to classify it may not be constant. Design, the needs of which determine the knowledge, changes over time—quite a bit over extended periods. As remarked in chapter 1, the historical cases given here, and hence the anatomy based on them, pertain to twentieth-century engineering, engineering more or less as it exists today. Three hundred years ago, theoretical tools, to the extent they existed at all, obviously played a far lesser role compared with other categories than they do now. And, while devices in ancient times had to have operational principles and even normal configurations, one can wonder about the nature of whatever design criteria and specifications were used to implement them. The next chapter discusses the dynamics of knowledge generation, including something about how the generation process itself evolves over time. The discussion will not come back, however, to the dynamics of the present anatomy in detail. The extent to which the categories put forward here are valid for earlier times, particularly for the craft era of design, will need careful examination.

Engineering knowledge, especially theoretical tools and quantitative data, get packed down and stored for normal design in textbooks, handbooks, professional journals, government publications, proprietary company reports, and the memories of individuals. Design engineers look there for it first. When it is not to be found, it may have to be generated for the problem at hand, as we have seen in detail in our five cases. Moreover, as pointed out earlier, the stored-up knowledge likewise had to be generated at some time in the past. How to carry out such generation itself constitutes a category of engineering knowledge—an aspect of the knowing how mentioned in chapter 1. It is such a basically different category, however, involving as it does the cognitive learning process, that it requires separate treatment. The next chapter puts forward a model for the knowledge-generation process in engineering.

For the remainder of this one, I shall attempt to classify the technical activities and personal and social agents that bring such generation about.

Knowledge-Generating Activities

In all but one of our stories, growth of engineering design knowledge originated primarily out of prior engineering knowledge and was achieved primarily by engineering activities. Both Davis and Jacobs, different as they were, based their ideas on the existing engineering tradition about airfoils and pursued their work by characteristically engineering methods. The flying-quality community that grew up after 1918 founded its work on the experience of earlier designers and fliers and solved its problems largely through engineering means. Similar statements can be made about the Durand-Lesley propeller studies and the innovation of flush riveting. Only for control-volume analysis can it be said that engineers looked indispensably outside engineering, in this case to prior science, and even then it was to science that had already had considerable engineering application. Moreover, the transfer was made and the knowledge refined by engineers as part of their engineering activity. I would argue that the situation observed in these studies is general—that, while ideas for radical design may (or may not) come from elsewhere, knowledge used in normal design originates and develops mainly within engineering. Other observers of technology have reached a similar conclusion.[34]

To argue this way is not to say that science's contribution to normal design can be ignored. Any attempt to take such contribution into account, however, runs into the notoriously troublesome problem of distinguishing between science and engineering and defining the science-technology relationship. Implicit distinction has already been introduced, in fact, in characterizing the spectrum of theoretical tools. When we try to sort out the activities that generate design knowledge, we encounter the problem in earnest. Though it may be unsolvable in a rigorous sense, it can hardly be ignored without giving up the persuasive view achieved by historians of technology (see chapter 1) of engineering knowledge as a distinct epistemological species.[35]

To orient my thinking, I find the diagram of figure 7-1 helpful. It is intended as a framework for discussion more than a set of hard and fast divisions; the following elaboration should be read in that spirit. The fundamental feature, on which the rest of the diagram depends, is the differentiation between knowledge *used* by scientists and engineers (lower boxes) and knowledge *generated* by those communities (center, solid-line band).

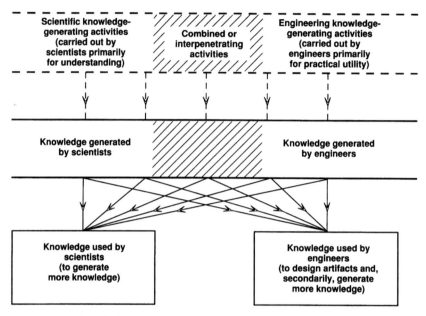

Fig. 7-1. Diagram of knowledge and its generating activities.

At the level of knowledge use in the figure, distinction between engineering and science is relatively straightforward. As stated earlier, engineers use knowledge primarily to design, produce, and operate artifacts, goals that can be taken to define engineering. (We concern ourselves here only with the first of these, design.) Scientists, by contrast, use knowledge primarily to generate more knowledge. As observed and discussed by Hugh Aitken, "Most of the informational output of science—the new knowledge generated—is channeled back into science itself."[36] The reason for this feedback process is that scientists are engaged in an open-ended, cumulative quest to understand observable phenomena. Insofar as understanding is an aspect of knowledge, we can say alternatively that scientists are engaged in an open-ended, cumulative quest for knowledge. The feedback nature of the scientific knowledge process then follows—the use of knowledge in science is to generate more knowledge. Engineers, of course, also use knowledge to generate more knowledge, but such generation is not for them the primary concern. To the extent that engineering knowledge has two uses instead of one, an asymmetry thus appears in the uses of knowledge in science and engineering.

Given the foregoing primary differences, epistemological distinction between science and engineering becomes operationally possible, and representation of collections of knowledge by discrete boxes makes sense. To implement the distinction, one need only examine the use to which knowledge is put by engineers and scientists in the very different institutions in which most of them work—industrial design offices for engineers and research laboratories for scientists.[37] The categorization of the preceding section is an attempt, in effect, to do this for engineering and thus unpack the contents (pertinent to normal design) from inside the right-hand box.

The distinction between engineering and science becomes less objective at the level of knowledge generation and, especially, of the activities that generate it (upper, dashed-line band). This well-known difficulty is exemplified by the career of Irving Langmuir, whose forty years at the General Electric Research Laboratory have been examined in a valuable article by Leonard Reich. Langmuir's studies of (among other things) the physics of incandescent filaments and the conduction of electricity through gases at high voltages led to both fundamental scientific understanding and technical information essential to GE's development and design of new products. For his accomplishments, Langmuir received a Nobel Prize in chemistry as well as the highest awards of American engineering societies.[38] Langmuir's example, though perhaps atypical, is not unique; knowledge generation for science and engineering goes on deliberately in a combined way, though not always by the same individual, in any number of industrial and governmental research laboratories and in applied-science and some engineering departments in universities. In these institutions engineers and scientists work side by side on the same research, and some individuals, like Langmuir, defy classification as one or the other. The knowledge they produce serves both understanding and design.

Instances of clear separation, however, do exist. Examples abound (some appear in earlier chapters) of people, obviously engineers, generating knowledge solely and expressly for design. Numerous scientists likewise pursue their labors purely for the understanding they bring. Both types work intentionally to add knowledge to only one box in our diagram.

All things considered, both knowledge generated and the activities that generate it are best represented as spectra. Toward the right, engineering knowledge-generating activities are pursued *primarily* (and increasingly, the farther to the right) to produce practically useful knowledge; toward the left, scientific activities are carried out *primarily* (and increasingly) for the sake of understanding, or, in light of my earlier remarks, for the sake of the knowledge itself. Engineers and scientists are correspondingly the people who pursue *primarily* those goals. The epistemological distinction is one of priority and degree of purpose rather than method. Though it is deliberately blurred, it is nevertheless real.[39] Even Reich finds such distinction necessary; to establish his point that Langmuir worked simultaneously as both scientist and engineer, he refers throughout his article to the differing goals of scientific understanding and engineering utility.

Knowledge can and does feed into the use boxes from anywhere along the spectrum of generated knowledge. Items flowing from the engineering side of the spectrum into scientific use include, among other things, knowledge of instruments and of scientific problems posed by engineering devices. Examples flowing in the mirror-image direction are well known in the historical literature; some appear in our cases, as I shall come to presently. Location of departure on the spectrum of generated knowledge, in fact, has already provided a criterion for sorting out theoretical tools in the right-hand box. (Whether the frequency of the feeding-in process diminishes uniformly with distance between the point of departure and the box is an open question. I am inclined to guess that roughly it does.) Langmuir, who produced knowledge squarely at the center of the spectrum, made his results useful to both scientists and engineers by publishing in their respective journals. Findings from his work on incandescent filaments, for example, appeared in the *Physical Review*, journal of the American Physical Society, and in the *Transactions of the American Institute of Electrical Engineers*.[40] Intimately related knowledge from the same investigation thus fed into both boxes from the same point along the knowledge spectrum. In many cases, as with the spectroscopic data mentioned below, the selfsame item of knowledge can be useful to both science and engineering and show up identically in both boxes.[41]

An interesting circumstance deserves note: The term *engineering knowledge*, as I understand it to be customarily employed, refers to the knowledge *used* by engineers. *Scientific knowledge*, by contrast, usually means the knowledge *generated* by scientists. This practice probably reflects the fact that scientists are perceived mainly as producers of knowledge and engineers as users. Only recently have scholars begun to look seriously at engineers and engineering activities as knowledge producers. Though I tend to go along with the customary usage, I do not subscribe to the customary perception.

What follows, then, is mainly a spelling out of activities in the engineering half of the upper spectrum. Knowledge originating from the scientific half will be lumped together as "transfer from science," with no attempt to address the activities that generate it. While I thus deal in detail with the engineering side of the diagram (in accord with the topic of the book), I make no attempt to analyze the scientific side. Such an approach is consistent with the historical material, which concentrates on the engineering sources of knowledge. This focus in turn reflects the already mentioned presumption—certain in my estimate—that knowledge used in normal design stems mainly from engineering activity.

The knowledge-generating activities will be described under the following headings:

A. *Transfer from science*
B. *Invention*
C. *Theoretical engineering research*
D. *Experimental engineering research*
E. *Design practice*
F. *Production*
G. *Direct trial*

These activities, like the knowledge categories discussed earlier, frequently overlap and interact.

A. *Transfer from science.* In keeping with the aforementioned presumption, putting this source first carries no implication of first rank in importance; it is a convenience before going on to the intrinsically engineering sources. As indicated under "theoretical tools" in the preceding section, transfer from science appeared in the present studies in Joukowsky's and Theodorsen's use of conformal transformation in airfoil theory and in the entire development of control-volume analysis. It was also evident in Bryan's use of rigid-body dynamics as basis for his mathematical theory of airplane stability. As emphasized earlier (and exemplified particularly by control-volume analysis), such transfer from theoretical science often entails reformulation or adaptation to make the knowledge useful to engineers. Other authors have studied such alteration in a variety of cases.[42]

The examples just mentioned were of theoretical tools. Scientifically generated knowledge, however, can also contribute quantitative data. I recall from personal experience how, when thermally radiated transfer of energy became important in practical gas dynamics in the 1960s, engineers drew on spectroscopic data already obtained by scientists to understand molecular structure. Other examples are not hard to find. Data being what they are, reformulation here is neither as feasible nor as necessary as with theoretical tools, though changes in the parameters used for tabulation or expression may sometimes be required.

All these examples involve transfer from prior, well-established science. Transfer can also take place from current scientific activity. A minor but typical instance from the present studies (chapter 5) is Edgar Buckingham's general theorem from 1914 for application of dimensional analysis to the similitude theory engineers use to guide and interpret tests of scale models. Buckingham, a physicist at the U.S.

Bureau of Standards, apparently received inspiration from engineering problems. His aim, however, was to understand, in the most general way possible, the constraints that dimensions of physical quantities place on possible relationships between those quantities. Almost as soon as his theory appeared in the scientific literature, it was put to use by aeronautical research engineers in the collection of design data.[43]

While engineering design is an art, it is an art that utilizes (increasingly) knowledge from developed and developing science. This is a far thing from saying, however, that science is the sole (or even major) source and that engineering is essentially applied science.

B. *Invention.* Though I have deliberately paid little attention to inventive activity, it requires at least mention as source of the operational principles and normal configurations that underlie normal design. Contriving such fundamental concepts—or coming onto them by serendipity—is by definition an act of invention. Other of our activities may contribute to invention, but it is this elusive, creative enterprise that produces the concepts.

Inventive activity appears only incidentally in our cases—in Jacobs's establishment of the principle of the laminar-flow airfoil and, in a mundane way, in the invention of the dimpled flush rivet. (Achievement of the normal configuration for the airplane will figure in the next chapter.) The historical literature, however, abounds with examples—the act of invention has been a major concern in the study of emerging technology.[44] Though it is apart from normal design, I mention it here because, without the appropriate fundamental concepts for the device in question, the engineer doing such design could not be doing it.

C. *Theoretical engineering research.* A large number of modern-day engineers, mostly in academic institutions and industrial and governmental research laboratories, work at producing knowledge via theoretical activity. (Engineers take "theoretical" as synonymous in this context with "mathematical," as shall I.) Our cases contain numerous examples: Prandtl and Shapiro, in devising control-volume theory, supplied general analytical tools and a way of thinking for a class of engineering devices. Munk and later Jacobs and his group worked out new mathematical tools to design a particular device, the airfoil. Jones and Cohen extended Bryan's theory of dynamic stability to free controls and thus provided both an analytical tool for the designer and important results applicable to the specification of flying qualities. Gates carried out sophisticated theoretical analysis to substantiate stick force per g as a criterion to specify maneuverability. Though absent from our examples, formulas obtained by theoretical research also provide means to calculate parametric design data (parameter variation by computation instead of experiment). Engineers also do theoretical research to provide design procedures, computer programs for optimization calculations being a case in point.

Theoretical research in engineering, which thus contributes to various categories of design knowledge, has much in common with theoretical research in science. Like scientific research, it is systematic, conceptually demanding, and often mathematically difficult. In industrial research laboratories, in the hands of someone like Irving Langmuir, the two types are often impossible to distinguish. The distinction is also arguably difficult in some of the examples mentioned above. Nevertheless, differences in style and emphasis do exist; theoretical research done in schools of engineering, for example, differs in discernible though perhaps subtle ways from that done in departments of physics and chemistry. Engineering research, as pointed out earlier, has as its ultimate goal the production of knowledge useful for design (as well as production and operation); scientific research aims basically at explanation and understanding. As a result, research in engineering is pursued with different priorities and attitudes and remains more intimately tied to specific devices. It emphasizes application rather than illumination, and decisions about alternatives, methods, and significant results are influenced accordingly. Theoretical research as an activity in engineering differs from that in science probably less than does the resulting knowledge, which is noticeably more slanted toward the requirements of design (as discussed in detail in chapter 4). The differences in process, however, are not insignificant.

D. *Experimental engineering research.* An even larger number of engineers engage in experimental research. This vast activity is indispensable to engineering as the major source of the kinds of quantitative data described earlier. Except in connection with flush riveting, our cases did not include acquisition of essential data on materials and processes. They did, however, illustrate how data are obtained on performance of devices, such as airfoils, aircraft, propellers, and flush rivets. Such research requires special test facilities, experimental techniques, and measuring devices—in our cases, such things as wind tunnels, flight-test methods, strength-testing machines, and the instruments needed in conjunction with all of these. The situation differs in detail in fields of engineering other than aeronautics, but not in principle. Since quantitative data of some kind are essential to design in any field, so also is the experimental research from which they come.

Experimental research provides more than design data, however. As with Durand and Lesley's idea of propulsive efficiency of a propeller in front of an obstruction, such research also produces analytical concepts and ways of thinking. The same was true with Warner and Norton's arrival at the elevator-angle and stick-force criteria for static longitudinal stability. These examples, as with many others one could find, arose out of the theoretical thinking that is part of any first-rate experimental program. (One might be tempted to classify this part of the work as *theoretical* research. When, as here, the thinking is predominantly physical rather than mathematical, however, engineers look on it as an integral aspect of the experimental activity.) Flight research on a cross section of airplanes by Gilruth's group also produced the special kind of information needed for technical specifications, in this instance of flying qualities. Such examples illustrate how experimental research contributes to various of the categories of engineering knowledge.

In the hands of a person like Langmuir, who was adept both in the laboratory and with mathematics, experimental research in engineering is again impossible to separate entirely from experimental research in science. Approach, techniques, and instrumentation are basically similar. Overall, however, differences are greater than with theoretical research, and engineering utilizes certain methods in its own characteristic ways. As we saw with the Durand-Lesley propeller tests, engineers make wide use of experimental parameter variation to supply design data when no effective theoretical knowledge is available. They are similarly special in the way they employ working scale models, plus similitude theory to take account of the accompanying scale effects. And procedures of destructive testing—like simply pulling things apart and noting the failing load, as was necessary to find the allowable strengths of flush-riveted joints—would have no standing at all in modern science. A great deal of engineering experiment has a character very much its own.

Experimental *and* theoretical research in both engineering and science (and this also is illustrated by Langmuir) are often most fruitful when done together—or at least in interactive proximity. Development of the laminar-flow airfoil, if not its original idea, could hardly have occurred without coordinated theoretical advances and wind-tunnel tests by Jacobs's group. Sorting out the relationship between dynamic response and flying qualities came from close interplay between theoretical research by Jones and Cohen and the flight tests of Gilruth and Gough and their co-workers. Experiment and theory spark and depend on each other. Though they need to be distinguished for epistemological analysis, they are best pursued in actuality when minimum distinction is made between them.

E. *Design practice*. Day-to-day design practice not only uses engineering knowledge, it also contributes to it. With regard to fundamental concepts, theoretical tools, and quantitative data, such contribution is indirect—practice reveals problems and needs that call for research to generate such knowledge. We saw such influence at work, for example, in the way design of the DC-4E helped motivate flying-quality research. As to criteria and specifications, practical considerations, and design instrumentalities, the contribution is more direct. Though I do not have an example, I can imagine a design criterion of general applicability turning up in the course of practice. Among instrumentalities, visual thinking and judgmental skills, as noted earlier, are best (perhaps only) learned through doing; the same holds true for many practical considerations. Engineering educators recognize this fact when they set out to teach such matters by imitating design situations in the classroom. Though again I do not have examples, I imagine that designers have also been active in producing the structured procedures and verbal ways of thinking they employ. Frequently, ad hoc methods devised in design are later codified by academics and research engineers.

F. *Production.* As we saw in the chapter on flush riveting, production provides another source of design knowledge. The shop floor informed the designer that a certain size of pedestal-type riveting machine limited the size of subassembly that could be designed for riveting by that means (a practical consideration). Production also revealed that head angles that were too small or sheet thicknesses that were too large led to cracking in the forming of dimples (again practical considerations). Production experience similarly contributed to formulation of tables of thicknesses of sheet suitable with different size rivets in different types of flush riveting (quantitative data). This kind of experience is typical. To the extent that production is universal to all branches of engineering, so also is its contribution to design knowledge.

G. *Direct trial.* Engineers deliberately test the devices they design and build; customers who buy them also put them to use in everyday operation. Both kinds of direct trial provide essential design knowledge.

When feasible, engineers conduct intentional *proof tests* to find if their designs perform as intended. They want to find out how well a device achieves its goals; that is, lives up to its technical specifications. Such findings serve to satisfy both designer and customer that the device will do what it is meant to—or, if it falls unacceptably short, to suggest how it might be redesigned or corrected. (Testing finished devices to obtain knowledge of *general* applicability, as in the flight research of Warner, Norton, and Gilruth and others at Langley Field, is best classified as experimental engineering research. It feeds the same categories of knowledge noted there.) Proof testing appears only incidentally in our cases, parenthetically in reference to the Douglas DC-4E and more substantively with regard to traditions of testability (chapter 3); it is, nevertheless, an essential part of engineering. Though aimed at design-specific information, the check it provides between predictions and attainments contributes to the growth of a designer's judgmental skills. If a series of such checks gives consistent differences, it can also supply empirical correction factors (i.e., quantitative data) useful in future designs. Proof tests are also essential for establishing an operational principle; and, as will be illustrated in chapter 8 by early airplane trials in France, they can help in arriving at a normal configuration. They can also reveal that a theoretical tool used in design is inadequate. Knowledge that this or other things don't work in practice is among the important results of such testing.

Similar remarks apply about *everyday operation*. (What I earlier called simply operation—by which I meant everyday operation—is thus included here as one form of direct trial. The reason for this grouping and terminology will appear in chapter 8). In some situations, as with large bridges, proof tests are not feasible; even when they are, problems often do not show up. Everyday use may then supply the critical trial. That manufacturing irregularities and other real-life circumstances would prevent the laminar-flow airfoil from attaining its goal was settled by operational experience. That the Douglas system of flush riveting produced joints that did not remain tight appeared only in service. Other instances where essential knowledge came from routine operation are the disastrous explosions of the pressure cabin of the de Havilland Comet from metal fatigue and, more recently, explosive destruction of the space shuttle *Challenger* from O-ring failure in a joint of a booster rocket.[45] Such learning is not limited to failures, however; day-to-day use can also produce positive knowledge about what is desirable. The flying-quality criteria that Pan American pilot Harold Gray proposed came from his company's operating experience with the Martin M-130. A great deal of knowledge about what customers of various kinds want in the devices they buy comes by feedback from use.

If this and the preceding section seem intimidating in their detail, the summary in table 7-1 (and the ensuing explanation) may help.

The Xs that appear in the table indicate which knowledge-generating activities (on the left) contribute to which categories of knowledge (across the top). The table is probably best comprehended vertically; theoretical tools, for example, come by transfer from science, as the product of theoretical and experimental engineering research, and (insofar as it provides evaluation and shows what doesn't work) from direct trial. Only immediate contributions are indicated; for example, theoretical research provides an immediate source of theoretical tools. Indirect influences—like the role of design practice in revealing problems whose

Table 7-1. Summary of Knowledge Categories and Generating Activities

Activities \ Categories	Fundamental design concepts	Criteria and specifications	Theoretical tools	Quantitative data	Practical considerations	Design instrumentalities
Transfer from science			X	X		
Invention	X					
Theoretical engineering research	X	X	X	X		X
Experimental engineering research	X	X	X	X		X
Design practice		X			X	X
Production				X	X	X
Direct trial (including operation)	X	X	X	X	X	X

theoretical study leads to theoretical tools—are not included.

Neither the categories nor the activities should be thought of as mutually exclusive. As pointed out earlier, an item of knowledge can belong to more than one category; a theoretical tool or an item of quantitative data, for example, can be at the same time part of a technical specification. Among activities, both research and invention can similarly take place in the context of design practice. Such a situation is illustrated by the design of the Britannia Bridge, which Nathan Rosenberg and I have examined elsewhere. For this unprecedented tubular bridge, built for railway use in northern Wales in the 1840s, detail design had to be performed under pressure of a stringent deadline. At the same time, the engineers experimented, urgently and successfully, to devise a structural operational principle to avoid local buckling of the thin plates of the tubular beam. Such buckling was a new and puzzling phenomenon they had encountered in experimental research aimed originally at other design data. (The use of a gigantic tubular beam as the entire structure was itself also a new operational principle.)[46] Such simultaneous activities and overlapping categories of knowledge are not unusual. Like the earlier division of knowledge into descriptive, prescriptive, and tacit elements in chapter 6, the table here provides more a framework for orientation and analysis than a rigid set of distinctions. (The division in chapter 6 can be thought of, if one wishes, as applying orthogonally to the present table.)

The orientational function of the table is illustrated by the following: Once the operational principle (the essential fundamental concept) of a new technology has been invented, the engineering community goes about establishing the necessary "technological base," that is, the body of knowledge needed for design application. This is usually considered to comprise the theoretical analytical tools and quantitative data. In some cases, practical considerations may also be involved. The learning process the community follows will then typically entail theoretical and experimental engineering research, plus transfer from science as well as direct trial (proof testing) of at least the earliest instances of completed designs. To the extent that practical considerations are involved, tentative exploration of production and operation may also be required.

As stated at the beginning of the chapter, I believe the table and the ideas behind it apply to design in all branches (aeronautical, mechanical, electrical, etc.) of modern engineering. I believe, in addition, though I haven't thought about the matter in depth, that they they can also be adapted without major difficulty to the engineering that occurs in production and operation. This, I imagine, will call again for addition and modification, and perhaps some change in terminology, rather than basic revision. If this proves true, the categories and activities that hold for design (and are shown in the table) may very well hold, though with different relative intensity, for these other areas of engineering. Should both my beliefs be correct, a table much like that given here would then apply over all branches and areas of modern engineering.[47]

The table and the material supporting it add emphasis, if such is still needed, to the view, by historians of technology, of engineering knowledge as a distinct epistemological species. Although such knowledge shares elements with science as discussed earlier, other features are peculiar to engineering. As explained by Polanyi, operational principles lie outside the realm of science. The same is true of normal configurations, the other subcategory in fundamental design concepts. Criteria and specifications, practical considerations, and design instrumentalities are, almost by definition, the province of engineers. By the same token, any special activities needed to expand these categories of knowledge fall outside science. (In making these statements, I count the engineering knowledge and activity used to provide equipment for large-scale scientific experiment as logically part of engineering.)

The foregoing distinction in knowledge reflects a broader distinction noted by Herbert Simon. As Simon observes, the natural sciences deal with how thing *are*. Engineering design, like all design, deals with how things *ought to be*. (The same can be said of engineering in production and operation.) The practical consideration that an airplane should be made stable but not too much so, incorporated into the flying-quality criteria and specifications described in chapter 3, provides a case in point. It is in no sense knowledge about how an airplane innately is; rather, it is knowledge about how an airplane ought to be to enable the pilot to fly the machine with ease, confidence, and precision. Such knowledge had (and has) no interest or importance for scientists; it was discerned and almost entirely generated by engineers and pilots working together for essentially engineering purposes. Similar remarks can be made about knowledge we have examined in other categories. In general, all knowledge for engineering design (as well as for the engineering aspects of production and operation) can be seen as contributing in one way or another to implementation of how things ought to be. That, in fact, is the criterion for its usefulness and validity. Such implementation obviously requires procedural knowledge (that is, "knowing how") of both the prescriptive and tacit varieties (see chapter 6). As we have seen, however, it also requires a great deal of descriptive knowledge, which is synonymous with "knowing that" or knowledge of how things are. Part of this knowledge comes from science, but much of it (e.g., the allowable strengths for flush rivets) arises within engineering itself. (Though the sciences deal with how things are, they are not the sole source of such knowledge.) Clearly, if engineering knowledge is to be understood fully, it must be addressed on its own terms.[48]

Personal and Social Agents

Cutting across and embodying the knowledge-generating activities is a host of personal and social agents. How these function in relation to each other and to the knowledge they generate will need careful analysis by the growing community of historians and sociologists of technology.[49] Given my absence of background in sociology, I can little more than point them out. I make no attempt to theorize about the social process. It is in evidence, however, throughout the historical cases.

Generation of engineering knowledge involves an astonishing range of individuals. The cases of this book display people from the following (not mutually exclusive) groups: aerodynamicists, design engineers, research engineers, applied mathematicians, instrumentation engineers, production and process engineers, aircraft-manufacturing executives, inventors, pilots (airline, military, and research), academics, military-service monitors, and airline executives. Cases from other branches of engineering would lengthen the list. This remarkable diversity of skills and training is perhaps greater than for any other major branch of knowledge. It reflects the fact that engineering is an extraordinarily broad and ramified area of human endeavor. Engineers do a multitude of things in modern society, and representatives from a large number of areas contribute to the vast knowledge this complex activity requires.

Such contribution also calls for a wide range of aptitudes. Throughout their careers, Munk and Jacobs (chapter 2) both exhibited keen insight and originality, though Munk was more mathematical and Jacobs more physical in their modes of thinking. Bryan (chapter 3) was an accomplished and rigorous mathematical analyst along more or less traditional lines. Prandtl (chapter 4) combined remarkable physical insight, more than average mathematical competence, and an unusually broad and synthetic view of both theoretical and applied aerodynamics. Warner (chapter 3), though very different from Prandtl in other respects, had a similarly broad view of aeronautics generally and unusual feeling and animating ability for the knowledge required (and possible) at a given time. Gilruth (chapter 3) understood how to organize and carry through a large-scale flight-test program involving numerous people, including Mel Gough, whose enthusiasm and piloting skills helped provide the necessary flying expertise. Durand's thorough-going temperament (chapter 5) sustained the parametric study of propellers over ten years, and Lesley supplemented Durand's experimental skills with his own keen ability to make things work in the laboratory. And the many production and process engineers who learned how to do flush riveting (chapter 6) found satisfaction in detailed tasks that would surely have exhausted the patience of most of the aforementioned. These are only a sample of the people who have appeared in our stories. It is part of the genius of modern society that it is able to orchestrate the activities of such a diverse group for the production of knowledge aimed at a single device (or system of devices) such as the airplane.

The means of such orchestration—the social agents for the production of knowledge—can be divided into formal institutions and more or less informal communities. The essential role played by communities of practitioners, which I take up first, was clearly and crucially evident in the story of flying qualities. A fastener community also appeared explicitly in the development of flush riveting, and an international fluid-mechanics community was obviously at work in the formulation and dissemination of control-volume analysis. A wing-and-airfoil subgroup to the general aerodynamics community can be seen struggling with the airfoil problem, and Durand and Lesley were in the vanguard of what would become a similarly identifiable subgroup devoted to propellers. These communities had more than an incidental function; they provided the central agency for the long-term generation and accumulation of knowledge in their areas. While individuals are the immediate producers of engineering knowledge, there can be no doubt that Constant is correct in pointing to communities committed to a given practical problem or problem area as "the central locus of technological cognition."[50]

Such communities supply a number of functions essential to the learning process. Competition between members helps supply motivation to tackle difficult practical problems; at the same time, cooperation provides mutual support and aid in overcoming difficulties. These contrary yet reinforcing processes—and the tension between them—appeared explicitly in the fastener community in chapter 6 and implicitly elsewhere. The interchange of knowledge and experience that goes with cooperation also fosters the generation of further knowledge; such activity was particularly evident in both the fastener and flying-quality communities. Once generated, knowledge that proves useful is disseminated by word of mouth, publication, and teaching; see, in particular, the story of control-volume analysis. It is also preserved by incorporation, through various means, into the community's tradition of practice, as, for example, in the case of flying-quality specifications. And, though we haven't had occasion to examine the matter directly, it is the community at some level that provides the recognition and rewards (honorific as well as monetary) that help fuel the cognitive enterprise. Engineering knowledge is thus the product of communities "committed to 'doing' and having a sense of collective identity fostered by complex interaction based in part on a shared problem."[51]

Formal institutions provide the structure and support within which knowledge-producing communities function. The major institutional roles in the present studies were supplied by government research organizations (mainly the NACA), university engineering departments (e.g., at Caltech, MIT, and Stanford), aircraft manufacturers (Consolidated, Douglas, Curtiss-Wright, etc.), and military services (the navy and army air corps). Institutions performing important but less prominent functions included airlines (particularly United), professional societies (Institute of Aeronautical Sciences and Society of Automotive Engineers), government regulatory agencies (Civil Aeronautics Board), and equipment and component suppliers (Alcoa, Erco, etc.). In fields other than aviation, the mix of institutions would be similar, with the addition in some cases of university science departments and (especially) industrial research laboratories. (The preponderant role of government research organizations—as against industrial laboratories—is an atypical characteristic of aviation.) Of these varied institutions, government and industrial research organizations and university engineering and science departments exist mainly for the generation of knowledge, as well as its dissemination. Aircraft manufacturers and equipment and component suppliers, by contrast, have to do with design and production of devices, and military services, airlines, and regulatory agencies with their operation; these institutions, though they play a lesser role in generating knowledge, have, as we have seen, determining influence on its purpose and nature. Engineering professional societies cut across all the aforementioned activities and condition and contribute to how they are carried out.

The specifics about how these engineering institutions foster and shape the activities of communities and individuals are too myriad to lay out completely. Airlines and the military provide information about what is needed (e.g., for design of the DC-4E and B-24) and about what works and, especially, doesn't work in everyday use (the laminar-flow airfoil and the Douglas riveting system). Research laboratories bring together a community of people of different skills (the Langley Field flight-research group), and from the interaction of these people new knowledge emerges. Government research organizations through their advisory committees (the NACA Committee on Aerodynamics and its chairman Edward Warner) help motivate and focus research in a particular field (flying qualities). Universities, by requiring organization of knowledge for teaching, foster its growth and dissemination (as with control-volume analysis). Professional societies organize meetings of practitioners in a given community and encourage interchange of knowledge and experience (regarding flying qualities and riveting procedures). And groups of manufacturers combine with government regulatory agencies to bring about both standardization of devices and knowledge of their performance (witness flush rivets). Other institutional roles can easily be cited. As with individual engineers, formal institutions do a complex multitude of things that promote and channel the generation of engineering knowledge. They do not, however, constitute the locus for that generation in the crucial way that informal communities do. Their role, in the limited sphere of engineering cognition we address here, is to supply support and resources for such communities.[52]

A Variation-Selection Model for
the Growth of Engineering Knowledge

In the historical cases, I have inferred the nature of engineering knowledge more by examining knowledge production and the motives for it than by looking directly at its use in design. Except for the account in chapter 2 of the design process at Consolidated Aircraft, I have focused on the generation of new knowledge when knowledge required by designers was lacking. Such focus is natural in historical writing—change has more attraction than customary activity. It is also perhaps more important epistemologically. To repeat the words of Karl Popper quoted at the start of chapter 2, "The central problem of epistemology has always been and still is the problem of the growth of knowledge."

This final chapter explores further the idea, raised in the concluding section of chapter 2, that growth of knowledge in engineering can be described in terms of the blind-variation-and-selective-retention model put forth by Donald Campbell. It seemed to me important that a theoretical model for such growth be attempted by someone and that Campbell's concepts would be a useful place to start. Though the work takes Campbell's model as a springboard and depends critically on his concepts, it also incorporates ideas of my own, some of which do not appear in preceding chapters. The result provides an exemplar of the variation-selection approach that is more explicit than usually appears in epistemological or historical studies. The model should in no way be taken, however, as the be-all and end-all of the book. I did not aim originally for such an outcome; the thought that it might be desirable emerged from the case studies and the analysis of the preceding chapter. If the reader finds the model mistaken, the material of those chapters can still stand on its own.

As stated in chapter 2, Campbell contends that some version of the variation-selection model is fundamental to all genuine increases in knowledge, from that embodied in genetic codes arrived at by biological adaptation to the theoretical structures of modern science.[1] I cannot hope to do justice here to his extensive overall argument. Suffice it to say that his model contains, in Campbell's words, three essential elements: "(a) Mechanisms for introducing variation; (b) Consistent selection processes; and (c) Mechanisms for preserving and/or propagating the selected variations."[2] The discussion here will concentrate on the mechanisms of variation and the processes of selection; the methods of preservation and propagation of engineering knowledge (journals, handbooks, textbooks, engineering-school teaching, design traditions, word of mouth, and so forth) are obvious in our cases and do not require elaboration. I shall also bypass criticisms that have been directed at Campbell's model, particularly with reference to scientific advance. Such concerns, which might also be raised here, have been painstakingly examined by Thomas Gamble. Though the model continues under discussion and attack, Gamble's arguments convince me that it "survives such criticism."[3] The pattern that emerges in the present study reflects both the process by which engineering knowledge grows in the short term as well as a longer-term shift in the method by which the process is carried out. The model encompasses, though not always explicitly, the categories of knowledge and knowledge-generating activities of chapter 7. As we shall see, features of the model are also visible throughout the five case studies.

The notion of "blindness" reflected in the name of Campbell's model (which has been the focus of much of the criticism with regard to scientific advance) enters via the mechanisms of variation. For Campbell, any variation that leads to truly new knowledge—knowledge that has *not* been attained before—must be blind in the sense of going "beyond the limits of foresight or prescience." It is important to be clear here, since the point tends to be misunderstood, that "blind" (in this sense) *does not mean "random" or "unpremeditated" or "unconstrained."* It simply denotes, in accord with Campbell's characterization, that the outcome of the variation cannot be foreseen or predicted insofar as the matter in question is concerned—if it could, the knowledge obtained would not be new. As has been argued by Gamble, blindness in this sense is not inconsistent with variations constrained by physical requirements or previously attained knowledge. The situation, in my view, is somewhat like that of a blind person setting out unaided down an unfamiliar alleyway. The person may have familiarity with the general disposition of the neighborhood and a cane to provide tactile information and make apparent the constraint, provided by side walls, to move in a particular direction. Even so, such a person cannot see ahead to know whether the passage leads somewhere or is a blind alley. That can be settled only by proceeding ahead "blindly" (though, note, *not* randomly). Consistent with Campbell's characterization, knowledge grows (that is, blindness is reduced) through extension of the limits of what can be foreseen or predicted. Whatever the disagreement in other fields, such a statement seems to me valid in engineering design.[4]

Blindness, however, is far from absolute. As also stated by Popper, "To the degree that past knowledge enters, . . . blindness is only relative: it begins where the past knowledge ends."[5] For much of normal design, the degree of blindness involved in the generation of new knowledge may very well be small. The important idea is that when the outcome is not completely foreseeable, the variation must in some degree be blind.

The variation-selection process in engineering appears in its elementary and direct form in attempts of the French to design flying machines between 1901–2 (when Ferdinand Ferber learned of the Wright brothers' gliding experiments) and 1908 (when Wilbur Wright demonstrated controlled powered flight in France). As emphasized by Frederick Culick, the French proceeded mostly by direct trial and error, that is, by building various types of machines and trying them by proof test in flight. They did little systematic research and development (in contrast to the Wrights) and gave little emphasis to analysis or attempts to arrive at basic principles. An essential element in the knowledge they were after was how best to configure an airplane to embody Cayley's operational principle (chapter 7). In pursuit of such knowledge, designers tried various arrangements, the most diverse being those of Louis Blériot. After two unsuccessful biplanes in 1906, Blériot in 1907 built in rapid order—or, better, disorder—a monoplane with a canard ("tail-first") horizontal surface and pusher propeller, a tandem-wing monoplane with tractor propeller (similar to the machine of Samuel P. Langley in the United States), and a tractor monoplane with aft tail. The last put him on the road finally to a practical airplane. No other method being apparent to them, Blériot and the others tested their ideas by attempting to fly, sometimes with disastrous results (but, fortunately, no fatalities till 1909). These unsystematic efforts led to only modest achievement by 1908, but the body of experience and community of designers and fliers that resulted put the French in position to forge ahead rapidly once Wilbur Wright had shown them how truly successful flight could be accomplished. From the French work (mainly) came the engine-forward, tail-aft biplane (and occasionally monoplane) configuration that became normal in airplane design by the beginning of World War I.[6]

The variations in the French aircraft, like all design variations, originated in people's minds. As with all creative processes, the cognitive mechanisms involved are difficult to assess. They obviously included appeals to what little was known of the work of the Wrights and Langley, mental imaginings of what might or might not succeed in flight, and guidance from flight experience as that experience grew. The variations the French tried (even Blériot's) were thus hardly random or unpremeditated. Since information and experience were meager, however, the level of blindness, at least at first, was well nigh total. The designers had a clear goal—to fly—but no way at all to foresee whether a given variation would be successful, and most of them were not. Since the French were not inclined toward theoretical analysis, variations could be selected for retention and refinement only by trials in flight. The criterion for selection from these trials (as well as in the mental processes leading to the variations to be tried) was, "Does it (or is it likely to) work?" That a process of blind variation and selective retention was going on—in the sense of Campbell and Popper—seems clear. (The example is not from normal design, but it serves to illustrate the ideas.)

A point that appears here is important: As noted, final selection took place by visible, direct trial of a number of variations. To arrive at these variations, however, the designers must have gone through some kind of mental preselection process to winnow the much larger number that were certainly conceivable. In subsequent discussion I shall regard such thought trials as part of the mechanisms of variation and take "variations" to mean only those that are in some way examined *overtly*. The division is, of course, to some extent arbitrary, since engineers do a great deal of visible sketching and doodling as they think. It is reasonably realistic, however, and provides a useful division for discussion. (The idea of "overt" as distinguished from hidden "shortcut" processes derives from analogous considerations by Campbell.)[7]

As exemplified by events in France, design itself constitutes a variation-selection process of knowledge generation. This statement holds true not only in the early stage of a new technology when the knowledge sought is that of a workable general configuration; it applies also after the configuration is settled and the object of design is a particular instance of it. In this more normal situation, the desired knowledge is how to arrange and proportion the particular device so as to accomplish its task given the constraints of the normal configuration. The design of most airplanes after the early 1910s was more or less of this kind. In this situation, the designer usually lays out a number of plausible variations on some basis and selects the final design by some sort of analysis or experimental test or combination of both. More often than not, the process takes place iteratively, with the results from one variation suggesting the properties and proportions of the next.[8] It is the mechanisms of variation and processes of selection in connection with such normal design, in contrast to the exploratory activities of such as the French, that I intend to examine. I shall take up that task after a few additional remarks.

Normal design requires detailed knowledge of the kinds categorized in chapter 7. This knowledge, too, to the extent it is (or was at some time) truly new, has to come from a subsidiary variation-selection process of some sort (subsidiary from the point of view of design, that is). The cases of this book offer ample instances of such process, as I shall indicate later. These processes in turn may call upon knowledge derived from still subsidiary variation-selection processes, as will also appear. The overall scheme, then, is one of a *nested hierarchy* of blind-variation-and-selective-retention processes in which the knowledge produced at one level is used in the process at the next outer level. All interior levels contribute finally to the knowledge required in the primary process of design. The notion of nested hierarchy is an essential feature of Campbell's model, though he employs it in a different context.[9]

The details of how the variation-selection process works in engineering are not static; as implied above in relation to the early history of the airplane, they evolve over time. Broadly put, the cumulative growth of engineering knowledge as the result of individual variation-selection processes acts to change the nature of how those processes are carried out. This long-term methodological shift complicates the attempt to generalize about the process. The character of the shift itself, however, can be described fairly simply: At all levels of hierarchy, growth of knowledge acts to increase the complexity and power of the variation-selection process by (1) modifying the mechanisms for variation, with resulting effects on degree of blindness and size of the field of overt variation (that is, the number of variations from which visible selection is made); and (2) expanding the processes of selection by trying out overt variations *vicariously* through analysis and experiment in place of direct trial in the environment. The idea of selection mediated vicariously instead of by direct trial also derives from Campbell.[10] It is an essential part of his general variation-selection model. Since the first decade of the 1900s, long-term evolution of the variation-selection process in normal aeronautical design has followed essentially the foregoing pattern. This fact is also reflected in our cases.

With the above overview, we can look more closely at the two elements of the variation-selection process in relation to normal design and how these elements evolve over time. Details will vary considerably from case to case, but some generalization seems warranted.

The mechanisms for producing overt variations, whether at the level of design or at knowledge-producing levels within the nested hierarchy, include at least three hidden, mental activities:

1. Search of past experience with similar situations to find knowledge that has proved useful. Such search also reviews knowledge about variations that *haven't* worked; unless the situation has an unusual degree of novelty, such variations are likely to be discarded a priori.
2. Conceptual incorporation of whatever novel features come to mind as called for by new circumstances and that might have some chance of working. To the extent these features depart from what has worked in the past, the resulting variation can only be in some degree blind. This remains true even though the originator may, for some reason, place high odds on their working. As concluded by Gamble, "In any situation in which the advances in knowledge are not *wholly* explicable in terms of previously attained knowledge, a [blind-variation-and-selective-retention] process must be at work" (emphasis added).[11]
3. Mental winnowing of the conceived variations to pick out those most likely to work. The criterion in this hidden preselection process is, "If it were to be tried in some way would it be likely to work (or would it be likely to help in the design of something that would work)?"

The ordering here is not meant to suggest that these activities take place sequentially. Rather, they go on concurrently and interactively in a more or less disordered way in the mind of the designer or research engineer. Much of the process takes place unconsciously, as anyone who has participated in it knows.[12] It is likewise obviously fallible. A priori discard and mistaken winnowing may narrow the area for overt search unduly, with the result that useful variations may be missed. The process may produce variations that when overtly tried do not in fact work—the blindness leads down a wrong path. From outside or in retrospect, the entire process tends to seem more ordered and intentional—less blind—than it usually is. It is difficult to learn what goes on in even the conscious minds of others, and we all prefer to remember our rational achievements and forget the fumblings and ideas that didn't work out. Luck can also play a role.

Modification of these mechanisms of variation as knowledge cumulates in a technology takes place in several ways. First, the body of experience about what has and hasn't worked in the past increases, making a priori judgments easier. Second, experience within an established technology will for a time enhance the ability to conceive of novel features that have a chance of working; ultimately, however, the degree of novelty that is possible tends to be exhausted (in the absence of some radical input from outside, in which case the technology is super-seded, in effect, by a new technology).[13] Third, expanded processes of vicarious overt trial (to be detailed later) enlarge the framework within which engineers conceive what is likely to work. They consequently develop more accurate feeling for not only how a device or item of knowledge might work in direct trial, but also how it might fare under theoretical or experimental test.

Since they work at cross purposes, the net effect of these influences, over time, on the degree of blindness and field of overt variation is difficult to assess. With regard to the latter, experiential enhancement of the ability to conceive of novel features tends, for a while at least, to widen the field of overt variation; a priori discard on the basis of experi-ence and more accurate feeling for what is likely to work both act to narrow it. The enlarged framework that engenders the latter feeling, however, produces at the same time a contrary effect—to the extent that the expanding processes of trial become cheaper and more rapid, they too widen the field of variation that is open to overt examination. (Electronic computers have augmented this capacity enormously in recent years.) Which influence predominates in the long run is a ques-tion that most likely has no answer; the field that is overtly searched probably depends very much on the individual case. What happens in the separate but related matter of blindness I shall return to later when I take up the question of uncertainty in the entire variation-selection process.

The processes of selection, the second element in the variation-selection model, all entail overt trial of one kind or another. As stated earlier, growth of knowledge in a technology characteristically acts to expand the power of vicarious in place of direct trial. Such expansion is achieved by two means:

(1) Substitution of partial experiments or complete simulation tests for proof test or everyday use. Such vicarious trial is especially visible in aeronautics in the widespread dependence on wind tunnels. It appears, however, in any technology where working scale models are employed, in use of "breadboard" devices in electrical engineering, and in trials with pilot plants in chemical engineering, to mention only a few exam-ples. Trial of this kind may aim for knowledge required to design a specific device or some component thereof or for some item of knowl-edge needed in design generally. Facilities built for specific simulation tests (e.g., wind tunnels that accommodate a diversity of airplane mod-els) serve as well for research aimed at general knowledge.

(2) Conduct of analytical "tests" in place of actual physical trials. This too constitutes a form of vicarious trial—every performance or strength calculation made in the course of design is, in effect, a test run

on paper. Henry Petroski, in his instructive book *To Engineer Is Human: The Role of Failure in Successful Design*, says much the same thing in discussing iterative structural design: "As each hypothetical arrangement of parts is sketched either literally or figuratively on the calculation pad or computer screen, the candidate structure must be checked by analysis. The analysis consists of a series of questions about the behavior of the parts *under the imagined conditions of use after construction*" (emphasis added).[14] Analysis, that is, can be seen as a means for vicariously trying out different variations. As with experiments and simulation tests, analytical tests sometimes produce items of general design knowledge (e.g., general analytical techniques) as well as specific designs. Use of vicarious trial, both experimental and analytical, was a strength of the Wright brothers in comparison with their French contemporaries.

Evolution of both of the foregoing means involves development of increasingly sophisticated experimental and analytical techniques, which are themselves products of variation-selection processes in the nested hierarchy. As this evolution continues, the hierarchy becomes ever larger and more complicated. And, as indicated earlier, the intellectual framework afforded by these means of vicarious overt trial gets incorporated into the hidden mental winnowing that goes into the choice of variations to be tried. (The winnowing itself can be seen, in fact, as a kind of hidden vicarious trial.)

Although vicarious trial forms an essential part of modern technology, in the end all designs and design knowledge must prove out in operation. As discussed in chapter 7, this direct trial may be supplied by proof test of a completed device. It may also come from the everyday use that is any device's ultimate purpose; as explained earlier, devices or ideas that appear satisfactory in vicarious trial or proof testing may fail or otherwise prove inadequate when routinely employed.

In direct and vicarious trial of both specific designs and general design knowledge, the criterion for selection of a variation for retention is, Does it work? or, more precisely, Does it help in design of something that works? This question, perhaps unexpressed, exists in the mind of anyone laboring to add to engineering knowledge, even in the most abstract reaches of engineering research.

I have deferred consideration of the long-term change in blindness of variation till now for the following reason: The entire variation-selection process—variation and selection together—is filled with uncertainty, and one can ask how, if at all, the level of this overall uncertainty changes with time. One contribution to uncertainty comes from the degree of blindness in the variations. A second stems from what I shall call, for want of a better term, "unsureness" in the process of selection. Changes in blindness and unsureness, and their effect on overall uncertainty, can advantageously be discussed together.

Uncertainty in the growth of knowledge in a given technology (that is, in the overall variation-selection process by which knowledge grows in that technology) must surely, in some sense, diminish as the technology becomes older. Aircraft design engineers of today clearly operate more surefootedly than did the French designers in the early 1900s; present-day research engineers probe for aerodynamic knowledge with more assurance than did my colleagues and I at the NACA in the 1940s. From what does this decrease in uncertainty stem, and why does it seem so evident?

Temptation exists to attribute the decrease in part to a decrease in blindness in the necessary variations. As a technology matures, increments of novelty become on average smaller, and so also (one might assume) does the degree of blindness involved in their pursuit. Against this, however, is the fact that advances are more difficult to come by and more sophisticated; the degree of blindness might thus be increased. Since blindness is a subjective attribute that resists measurement, the point is impossible to settle. Perhaps the temptation to see a net decrease stems from an illusion. Primary problems in a technology (e.g., airplane configuration or basic airfoil shape) necessarily get solved early on, and the subsidiary problems that follow do not appear so critical or so dramatic. Such problems also move to lower nested levels of hierarchy, where they are less visible to outsiders. Whether blindness in variation really diminishes is thus difficult to say. Talented engineers who struggle to advance a mature technology like present-day aeronautics are not likely to think so.

The effect of unsureness in the process of selection, which must likewise always exist, is easier to assess. The vicarious means of trial generated within the nested hierarchy (wind tunnels in aeronautics, for example) typically evolve concurrently with the variations they are being used to select; they are therefore often less than completely sure at a given time.[15] Even direct trials of complicated devices suffer from unsureness; such complication can make it difficult to know how well a component of interest within the device is in fact working. For both vicarious and direct trials, however, the long-term trend is clear. Wind-tunnel methods, for instance, have become increasingly accurate and able to confront a greater variety of problems, providing engineers surer capability for making choices. Theoretical tools in all fields of engineering likewise have increased in ability to handle a greater range of problems with greater precision (again, in recent years, as a result in part of the use of electronic computers). The effectiveness of direct trials similarly increases as instrumentation and the understanding of complex systems improve. There can be little doubt that unsureness in the process of selection tends continually and progressively to decrease.

In the end, decreasing uncertainty in the growth of knowledge in a technology comes, I suggest, mainly from the increase in scope and precision (that is, the decrease in unsureness) in the vicarious means of selection. Just as expanding scope tends, as we saw, to widen the field that can be overtly searched, so also the increase in both scope and precision sharpens the ability to weed out variations that won't work in the real environment. Blindness in the variations may by the same token even increase—engineers have freedom to be increasingly blind in their trial variations as their means of vicarious selection become more reliable. One sees engineers today, for example, using computer models to explore a much wider field of possibilities than they were able to select from just a decade ago.

That, in its essentials, is the variation-selection model. Features of the model pervade our five case studies. I shall point briefly to some of the more evident examples. The reader who wishes can re-examine them in detail and find others as well.

The Davis wing (chapter 2). The variation-selection process in design appears clearly in the work at Consolidated Aircraft. In seeking the wing to use on their long-range airplanes, the company's engineers examined overtly a large number of variations (and doubtless imagined still more). Selection depended on vicarious trial by a combination of analytical study and wind-tunnel test. The variations had a good deal of blindness, and the wind-tunnel methods were far from sure; the combined level of uncertainty was high. Direct trial with the B-24 was completely unsure in its ability to separate out the performance of the Davis airfoil; selection in the environment was in the end inconclusive in this respect.

The variation-selection processes by which Munk, Davis, and Jacobs sought knowledge about how to shape airfoils were nested within the variation-selection process of wing design in the aircraft industry. Details for the last two men were examined in chapter 2. Jacobs's laminar-flow airfoils were selected originally for retention on the basis of vicarious trial in the wind tunnel. Application for their intended purpose was short-lived, however, when direct trial in everyday use proved fruitless. The vicarious process of selection had been unwittingly and completely unsure with regard to manufacturing irregularities.

The "uncertainty" discussed in chapter 2 is basically the same as what I have been talking about here, though use of the word was somewhat different. There I used "uncertainty" in regard *both* to the reasoning and knowledge underlying Consolidated's various overt designs (variation) and to the methods of trial (selection) the company had to employ. Here, for distinction in a general discussion, I have spoken of "blindness" in variation and "unsureness" in selection, with resulting "uncertainty" in the overall process. The concepts to which the words refer are the same in both places.

Flying-quality specifications (chapter 3). This case affords a wealth of examples at various levels of nested hierarchy. The concept that flying-qualities could and should be specified, a novel idea with considerable blindness when Warner proposed it, was selected for retention in airplane design on the basis of direct trials in flight. Vicarious selection was impossible, since the subjective reaction of pilots was crucial. The requirements in Gilruth's report that implemented the concept were selected from a much larger number of variations that Warner, Soulé, Thompson, and Gilruth had devised, some of them considerably blind at first. Sureness in the methods of selection increased over time as the result of improvements in piloting techniques and instrumentation.

Many of the variation-selection processes nested within that for the overall idea were themselves far from simple. Stick force per *g* as the criterion for maneuverability, for example, was the selected outcome from candidate criteria proposed over several years, partly as the result of experience and increasingly refined physical and theoretical reasoning and partly from blind variation of parameters within such analytical framework. The consensus on positive stability was, in different fashion, the result of an informal variation-selection process, with judgment rendered by the aeronautical community on the basis of direct trial, mostly in everyday use; the variations were embodied in the airplanes themselves, which were apparently designed at first rather blindly with regard to stability.

Control-volume analysis (chapter 4). In formulating the control-volume method, Prandtl and the professors at MIT were deliberately expanding the ability to carry out overt trial vicariously by means of theory. They supplied at the same time a cognitive framework that helps engineers imagine which variations of a fluid-flow device might warrant such trial. To achieve the most useful formalism for the method, they had to pursue a variation-selection process of their own. Some of this process is overt in Prandtl's successive formalisms, but more must have gone on in his mind. Since the work was basically a reformulation of existing knowledge, the degree of novelty, and hence blindness, was small. Considerable insight and experience were required, nevertheless, about what was needed and possible.

Propeller data (chapter 5). The parameter-variation studies of Durand and Lesley were clearly a variation-selection process. Like the airfoil research of Jacobs and others, they were nested hierarchically within the larger process of airplane design. They received guidance from experience in that the significant parameters of operation of propellers were already known. The currently important range of these parameters and the desirable geometrical features of the device could also be taken as more or less defined. To the extent that doubt remained in both these matters, however, the variations were still partly blind— arbitrary geometrical choices had to be made and the range of parameters had to be explored uniformly at the outset. Blindness in both respects, however, diminished as the tests progressed. (One of the virtues of parameter variation generally is that it allows fruitful, systematic exploration of a range of variations in the face of prevailing blindness.) Selection took place through vicarious trial at reduced scale in a wind tunnel, with attendant unsureness from the usual scale effects. The investigators sought to reduce this unsureness by making comparative full-scale tests directly in flight. In the end, however, the NACA found it necessary to construct a special wind tunnel for vicarious full-scale trial. Variation-selection evolution of the vicarious testing device (the wind tunnel) thus took place nested within the variation-selection process for the device being tested (the propeller).

Flush riveting (chapter 6). Initial choice of basic geometry and riveting methods at the various aircraft companies (e.g., Douglas, Curtiss-Wright, and Bell) took place with a great deal of blind overt variation. Blindness in a largely prescriptive innovation of this nature was unavoidable despite much prior head scratching and mental winnowing about what might work. Nested variation-selection processes were required for subsidiary problems, such as how to shape dies for pre-dimpling and how to do heated dimpling of increased-strength alloys. Despite considerable vicarious testing, final selection of geometries and methods had to come from direct trial in production and everyday use. Outcomes proved different for different arrangements and relative sizes of rivet and sheet. The Douglas system of dimpling by means of the rivet itself, which originally (and blindly) appeared so sensible, had to be discarded when the resulting joints were found to lose tightness in service. The selection was, in more ways than one, survival of the fittest.

One can thus, without forcing the evidence, see aspects of blind variation and selective retention in all five cases. The historical material plainly supports the theoretical model; indeed, the model grew in part from reflecting on the case studies. The model, I find, helps in return to make sense of the historical process. I think I understand better now what was going on intellectually in the various happenings and how much they share despite their great surface dissimilarities. It strikes me as remarkable, for example, how much alike in structure, though with admitted differences in detail, were the underlying variation-selection processes for the Davis wing, flying-quality specifications, and flush riveting (compare the above summaries). This commonality exists despite marked dissimilarity in the engineering problems and historical circumstances. It tells me, among other things, that the differences, epistemologically speaking, between design, research, and production are not as fundamental as I had thought. In the control-volume and the Davis-wing and propeller-data cases, the model also brings out the similarly vicarious role that theoretical tools and wind-tunnel testing play in the epistemological scheme. These are a few among many insights. Though no theoretical model for the growth of knowledge is likely to satisfy everyone, the one here may be useful for some.

The model outlined here is, I suggest, universal to engineering knowledge: it characterizes all branches of engineering, applies across the categories of knowledge described in chapter 7, and appears, in whole or in part, in the variety of activities that generate such knowledge. Some readers may find the notion of blindness of variation, however essential it may be for more rudimentary kinds of knowledge, difficult to accept for something as seemingly foresighted and self-critical as modern engineering. I find it myself an illuminating and useful—even necessary—idea. The validity of the variation-selection approach, however, does not stand or fall on this concept or the name applied to it. Some other, more sophisticated way may be found to represent the element of unknowing inevitable in any extension of engineering knowledge. The extent to which any such model proves useful must, of course, remain to be seen. Models of cognitive growth too must undergo a variation-selection process (with a considerable element of blindness!).

Some miscellaneous thoughts need to be added. The model and examples suggest that, in the evolution of the variation-selection process in engineering, expanding the processes of selection, especially the vicarious ones, is easier than improving the mechanisms for arriving at promising variations. The former are more deliberate and codifiable and less dependent on individual creativity. This situation is reflected in engineering-school curricula: the tools of analysis and experiment, sophisticated as they may be, are easier to organize and teach than are the mechanisms for thinking up likely variations; modern engineering schools tend accordingly to embrace the former more eagerly.[16] Distinguishing the processes of overt selection from the mechanisms of blind variation may be one of the strengths of the variation-selection model. Differences like the foregoing may then show up more readily.

Comparison of our model for engineering with a corresponding model for science is not practicable at present. Campbell has pointed to the need for "spelling out in detail" how the growth of scientific knowledge reflects a variation-selection process and has made some observations in that direction;[17] as far as I am aware, however, the task remains to be done. Whatever the details, a main difference from engineering must surely be in the criterion for selection. Consistent with earlier discussion, the criterion for retaining a variation in engineering must be, in the end, *Does it help in designing something that works in solution of some practical problem?* (To "designing" we would have to add "producing" and "operating" if we wish to cover all aspects of engineering.) The criterion for scientific knowledge, however we put it, must certainly be different, though its statement raises fundamental and debatable questions in the philosophy of science. Borrowing a phrase used by Alexander Keller in describing the life interest of scientists, I would venture it more or less as follows: *Does it help in understanding "some peculiar features of the universe"?*[18] ("Explaining" could equally well be used in place of "understanding.") Though difficulties are inevitable about details of the two criteria, other writers say much the same thing in different words. Writing about the structure of scientific and technological knowledge, Rachel Laudan observes, for example, that "inconsistent scientific theories leave us unsure what to believe; ill-integrated technologies simply fail to work."[19] The criteria also conform to Herbert Simon's distinction, mentioned in chapter 7, between the sciences as dealing with how things are and design, including engineering design, with how things ought to be. However phrased, the essential difference is one between intellectual understanding and practical utility.[20]

Note an essential asymmetry between the two criteria. In both cases the variation being judged is a means to an end—understanding in science and solution of a practical problem in engineering. In science, however, the means acts directly to the end; in engineering it acts through the intermediary of the "something," usually a material artifact, that is the immediate object of design (or production or operation). This is basically the same asymmetry noted in the statements about the uses of knowledge in science and engineering made in connection with the diagram of knowledge and its generating activities in figure 7-1. As the case studies show, the distinctive nature of the engineering criterion has implications for both the form and content of engineering knowledge.

The two criteria, of course, are not mutually exclusive. The same element of knowledge can both provide understanding for the scientist and help in designing something for the engineer. This commonality underlies much of the complication represented by the diagram in figure 7-1. As we saw with spectroscopic data, for example, some items of knowledge can be both scientifically illuminating and technically useful and appear identically in the two boxes of the diagram. The thermodynamic knowledge used in control-volume analysis also serves both purposes, though it had to be reformulated to work effectively for engineers; it thus appears in both boxes but somewhat differently. Many examples of both kinds can be found.

The criteria can also operate partially or completely independently. Some physically rigorous theories that are, in Polanyi's words, "cultivated in the same way as pure science" exist primarily because they work for engineers; they may or may not find a place in both boxes of the diagram. Phenomenological theories, commonplace for utilitarian purposes in engineering, have little real scientific interest, and some theoretical tools useful in engineering are even known to be wrong— they explain nothing. These items are retained in the engineering box because they help with design; they have no counterparts in the box of knowledge used by scientists. Related or independent, the criteria thus embody a distinction that has existential as well as epistemological meaning.

As indicated by the statement of the engineering criterion, whether or not something "works" has meaning only in relation to some practical problem or goal. The problem may be a purely technical one—to fly in the case of an airplane, to supply lift with the least possible drag in the case of an airfoil, or to hold two pieces of metal together in the case of a rivet. It may also be contextual, that is, originating outside technology. In the past, such considerations have been limited mainly to economic and military concerns. (The latter appeared here in the range requirements of the Consolidated B-24 and the former in the performance specifications of the Douglas DC-4E.) During my career as an engineer I have seen the scope of engineering problems also expand increasingly to include social and environmental matters. Jacques Ellul, Eugene Ferguson, Lewis Mumford, and other analysts of the dilemmas of Western technological culture think it must grow further to embrace cultural norms such as morality and human dignity. In his examination of their thinking, John Staudenmaier sees such writers calling for a broad method for "integrating design characteristics with the non-technological dimensions of the cultural ambience."[21] For that to happen, engineers—and society—will have to define engineering problems, and hence what is meant by "works," differently from how they do today.[22] As these remarks indicate, the notion of what constitutes an engineering problem varies significantly with time and (presumably) place; it is itself a topic for the history of technology. (The content of the engineering criterion also has implications for other aspects of the variation-selection process. As what is meant by an engineering problem changes, so also must the kinds of variations engineers have to consider and the areas of blindness they encounter.)[23]

In its methodological aspects, much of the development of modern engineering can be seen as a vast project to increase the power of the variation-selection process. As we have seen, this increase proceeds by modifying the ability to perceive likely variations in devices and knowledge and by expanding the capability to select from these variations vicariously by means of analysis and experiment. I say "*modern* engineering" because until the past two or three centuries selection had to take place predominantly by direct trial in the environment. This difference speaks to the concern on the part of some scholars: "What is it that is 'modern' about modern technology?"[24] With regard to methodology, it may be the key element. Edward Constant voices much the same idea when he writes, "It may well be that the application of this method of bold total-systems conjecture and rigorous testing (or of variation and selective retention) to large-scale, complex, multilevel systems beginning in the nineteenth century created a fundamentally novel category of knowledge and knowledge processes distinct both from science proper and from craft technology."[25]

The anatomy of knowledge in chapter 7 grew explicitly from considerations of modern engineering design, specifically in the twentieth century. The fundamental variation-selection idea used in the present chapter, if it applies at all, does so in all periods, that is, for craft technology and modern engineering as well as for the transition period between. As elaborated here, the model also says something, albeit somewhat abstract, about how the attributes of the variation and selection processes have changed over time. It thus suggests wherein the transition lay. In this sense, the model is more general than the specifics of the preceding chapters.

In the final paragraph of chapter 5, I also raised the question of whether it might be profitable to look for "engineering method" analogous to but distinguishable from scientific method that has been a fruitful concern for the history of science. Could it be that the variation-selection process outlined here is that method, with its distinctive features lying in the criterion of selection and the vicarious methods used to shortcut direct trial?[26] Whatever the answer, the complex variation-selection process, in both its craft and modern forms, is something human beings had to learn over a period of time, presumably through some higher-order variation-selection process of its own. How this fundamental methodological development took place is an overarching metaproblem for the history of technology.

This and chapter 7, then, have been an attempt to map out in a general way the epistemological domain of engineering. The main impression

I carry away from this effort and from the case studies on which it is based is of the extraordinary and challenging complexity of engineering knowledge and its sources. Obviously a great deal will have to be done—historically, philosophically, and sociologically—to understand this important realm of human knowing.

In the end, of course, engineering knowledge cannot—and should not—be separated from engineering practice. The nature of engineering knowledge, the process of its generation, and the engineering activity it serves form an inseparable whole. What we eventually need to comprehend is the whole of engineering behavior—what it is that "engineers really *do*." We also need to know what part such doing plays in the all-important process of technological evolution and change. All kinds of understanding will be required to achieve these goals. Understanding what engineers know and how engineers think and learn ranks critically among them.

Notes

Chapter 1. Introduction: Engineering as Knowledge

Note: This introduction has benefited from discussion with and comments by James Adams, Rachel Laudan, Robert McGinn, William Rifkin, and an anonymous publisher's reader.

1. For quotations reflecting this view, see editor's introduction to *The Nature of Technological Knowledge. Are Models of Scientific Change Relevant?*, ed. R. Laudan (Dordrecht, 1984), n. 29, and M. Gibbons and C. Johnson, "Science, Technology and the Development of the Transistor," in *Science in Context*, ed. B. Barnes and D. Edge (Cambridge, Mass., 1982), pp. 177–85, esp. p. 177.

2. For Layton's paper, see *Technology and Culture* 15 (January 1974): 31–41, quoted phrase from 32. The adjectives *creative* and *constructive* come from a short, useful note by Barry Barnes: "The Science-Technology Relationship: A Model and a Query," *Social Studies of Science* 12 (February 1982): 166–71. The extended quotation is by J. D. North, "The Case for Metal Construction," *Journal of the Royal Aeronautical Society* 37 (January 1923): 3–25, quotation from 11; it was brought to my attention by Richard Smith.

3. J. Staudenmaier, S.J., *Technology's Storytellers: Reweaving the Human Fabric* (Cambridge, Mass., 1985), chap. 3; G. Wise, "Science and Technology," *Osiris*, 2d ser. 1 (1985): 229–46, quotation from 244. See also Laudan (n. 1 above) and a philosophical discussion leading to the same conclusion by Gary Gutting, "Paradigms, Revolutions, and Technology," in ibid., pp. 47–65.

4. Barnes (n. 2 above).

5. For the science-technology question, in addition to Barnes, Staudenmaier, and Wise cited above, see also A. Keller, "Has Science Created Technology?" *Minerva* 22 (Summer 1984): 160–82, and E. Layton, "Through the Looking Glass, or News from Lake Mirror Image," *Technology and Culture* 28 (July 1987): 594–607. For related philosophical discussion by engineers concerning the autonomy of engineering, see D. Lewin, "The Relevance of Science to Engineering—A Reappraisal," *Radio and Electronic Engineer* 49 (March 1979): 119–24, and G. F. C. Rogers, *The Nature of Engineering: A Philosophy of Technology* (London, 1983), chap. 3.

6. H. Aitken, *The Continuous Wave* (Princeton, 1985), pp. 14–16, quotations from pp. 14 and 15 (Aitken's full discussion of method is well worth reading); Layton (n. 2 above), p. 41; Laudan (n. 1 above), p. 2; N. Rosenberg, *Inside the Black Box: Technology and Economics* (Cambridge, 1982). For the central role of cognitive studies in the study of engineering, see A. Donovan, "Thinking about Engineering," *Technology and Culture* 27 (October 1986): 674–79.

7. Rogers (n. 5 above), p. 51.

8. Layton (n. 2 above), p. 37 and his "American Ideologies of Science and Engineering," *Technology and Culture* 17 (October 1976): 688–701, quotation

from 698. The quotation characteristic of "other scholars" is from comments by the anonymous publisher's reader. For additional discussion of design by Layton, see his "Science and Engineering Design," *Annals of the New York Academy of Sciences* 424 (1984): 173–81.

9. For a detailed account of a specific design, written by engineers who participated in it, see *Case Study in Aircraft Design—The Boeing 727*, in the Professional Study Series, American Institute of Aeronautics and Astronautics, September 14, 1978. For illuminating remarks, references, and quotations about design as a process, see E. W. Constant, *The Origins of the Turbojet Revolution* (Baltimore, 1980), pp. 24–27. Although design is essential to engineering, and examples of its treatment are not hard to find in the historical literature, it is curious that a subject heading by that name rarely appears in either the annual index to *Technology and Culture* or the "Current Bibliography in the History of Technology" published in that journal's April issues. In the United States, historians tend to deal with design in some other context, usually as part of the invention, development, or innovation of some important and/or dramatic device. In Britain, where history of technology is written more by professional engineers, there is more historiographic concern for design per se; see, e.g., *The Works of Isambard Kingdom Brunel: An Engineering Appreciation*, ed. Sir Alfred Pugsley (Cambridge, 1976).

10. Constant (preceding note), quotation from p. 10. See also Layton (n. 2 above), p. 38. Constant's concept of normal technology is analogous to (and derived from) Thomas S. Kuhn's well-known concept of normal science; *The Structure of Scientific Revolutions* (Chicago, 1962). As with *normal design*, Constant does not use the term *radical design* (or *radical technology* or even *revolutionary technology*). Given his and Kuhn's titles, it might have been historiographically consistent for me to contrast normal design with *revolutionary design*. I prefer *radical design*, however, as less extreme and more in line with engineering terminology; it also allows a distinction I wish to make in chapter 7.

11. The quotation is from the anonymous publisher's reader. For a review of historical studies dealing with invention and having bearing on radical design, see Staudenmaier (n. 3 above), pp. 40–61.

12. Mechanical wing design deals with mechanical subcomponents such as those used to extend high-lift devices for takeoff and landing.

13. For hierarchy in design, see also Constant (n. 9 above). The reader may find it suggestive to compare such hierarchical division with the "method of detail" described by Arnold Pacey as one of the key modes of investigation to come from the scientific revolution of the seventeenth century; *The Maze of Ingenuity* (London, 1974), pp. 137–40. For remarks about the scholarly neglect of ordinary or routine engineering (meaning, in present terms, normal and/or lower-level design), see E. W. Constant, "Science *in* Society: Petroleum Engineers and the Oil Fraternity in Texas, 1925–65," *Social Studies of Science* 19 (August 1989): 439–72, esp. 440–41.

14. O. Mayr, "The Science-Technology Relationship as a Historiographic Problem," *Technology and Culture* 17 (October 1976): 663–73, quotation from 664; Rogers (n. 5 above), chaps. 3 and 4, pp. 57–61. Rogers discusses the relation of engineering knowledge to creative art as well as science.

15. W. Stegner, *Angle of Repose* (New York, 1971), p. 24.

16. Staudenmaier (n. 3 above), quoted phrases from pp. 6, 103.

17. For an assortment of readings on how social contexts shape the emergence of particular technologies (though not focused explicitly on design), see *The Social Shaping of Technology*, ed. D. MacKenzie and J. Wajcman (Milton Keynes, 1985).

18. G. Ryle, *The Concept of Mind* (London, 1949), pp. 27–32.

19. Throughout I have resisted the temptation to describe engineering knowledge or some part of it as "know-how," the common catch-all term for a wide (and usually undefined) range of practical knowledge and/or ability; see, e.g., D. Sahal, *Patterns of Technological Innovation* (Reading, Mass., 1981), p. 42, passim. Ingvar Svennilson provides a careful definition of know-how as "the capacity to use technological knowledge," but this definition appears to imply that the capacity of knowing how is not itself part of technological knowledge. I consider such separation inappropriate for my purposes (and Svennilson himself blurs the distinction in the discussion following the definition); "Technical Assistance: The Transfer of Industrial Know-How to Non-Industrialized Countries," in *Economic Development with Special Reference to East Asia*, ed. K. Berrill (New York, 1964), pp. 406–9.

20. The troublesome definition of "technology," which has several meanings besides that of a human activity, is discussed in R. E. McGinn, *Science, Technology, and Society* (forthcoming, 1991), chap. 2.

21. B. Gille, "Essay on Technical Knowledge," in *The History of Techniques*, 2 vols., ed. B. Gille (Montreux, 1986), vol. 2, pp. 1136–85. See also A. R. Hall, "On Knowing, and Knowing how to . . . ," in *History of Technology, Third Annual Volume, 1978*, ed. A. R. Hall and N. Smith (London, 1978), pp. 91–103.

Chapter 2. Design and the Growth of Knowledge

Note: This chapter was published, with minor differences, as "The Davis Wing and the Problem of Airfoil Design: Uncertainty and Growth in Engineering Knowledge," *Technology and Culture* 27 (October 1986): 717–58. Besides the specific assistance acknowledged in the following endnotes, I am grateful for more general help to Edward Constant, James Hansen, Barry Katz, Stephen Kline, Ilan Kroo, Edwin Layton, Laurence Loftin, Mark Mandeles, Russell Robinson, Paul Seaver, and Richard Shevell.

1. K. Popper, *The Logic of Scientific Discovery* (London, 1959), p. 15.

2. Information about Consolidated and the Davis wing, when not noted otherwise, comes mainly from a short unpublished manuscript, "Davis Wing," dated April 3, 1976, generously made available by its author, George S. Schairer, and from personal correspondence with Schairer and Ralph L. Bayless. Both men were members of the Aerodynamics Group at Consolidated in the late 1930s. Schairer moved to Boeing in 1939, where he eventually became vice-president, research; Bayless became chief of aerodynamics at Consolidated Vultee in the 1950s. I have also drawn on W. Wagner, *Reuben Fleet and the Story of Consolidated Aircraft* (Fallbrook, Calif., Aero Publishers, 1976), pp. 203–9, and A. G. Blue, *The B-24 Liberator: A Pictorial History* (New York, n.d.), pp. 11–12. Wagner's journalistic-style study is based primarily on interviews with

Fleet in his late eighties and must be used with caution for present purposes. These and other sources sometimes disagree. Throughout, I have used only information that is confirmed by more than one independent source and/or appears plausible and convincing in light of my experience as an aeronautical engineer. Facts about the various military aircraft come largely from R. Wagner, *American Combat Planes*, 3d ed. (New York, 1982). The situation regarding long-range commercial flying boats is discussed by R. K. Smith, "The Intercontinental Airliner and the Essence of Airplane Performance, 1929–1939," *Technology and Culture* 24 (July 1983): 428–49.

3. This specialized distinction between *airfoil* and *wing* is standard aeronautical-engineering terminology. It departs from usual dictionary usage, which defines airfoil generally as *any* extended surface used to deflect the air and therefore includes wing as a special case.

4. For a detailed history of this development to the mid–1920s, see R. Giacomelli and E. Pistolesi, "Historical Sketch," in *Aerodynamic Theory*, 6 vols., ed. W. F. Durand (Berlin, 1934), vol. 1, pp. 305–94, esp. chap. 3, pp. 336–94.

5. Such a study would not have been carried out as thoroughly and systematically, however, as it would be today. Frank E. Goddard, an engineering colleague of Schairer at Consolidated, writes that "the whole Aerodynamics Group plus the Preliminary Design Group totalled only about 15 people" (personal correspondence).

6. C. B. Millikan, "Report on Wind Tunnel Tests on a 1/24th Scale Model of the Consolidated XPB3Y-1 Flying Boat," *GALCIT Rept. No. 190*, March 11, 1937, and "Report on Wind Tunnel Tests on a Modified Wing for the Consolidated XPB3Y-1 Model," *GALCIT Rept. No. 190-A*, April 13, 1937. (All GALCIT reports for Consolidated are proprietary. Permission to consult them in the GALCIT files was granted by Convair through the help of Walter E. Mooney, an engineering official with that company.)

7. For biographical information on Davis, see obituary article, *Los Angeles Times*, April 18, 1972, part 2, p. 4; T. B. Hoy, "This Wing May Win the War," *Saturday Evening Post*, April 12, 1941, pp. 36, 39, 107–8, 110, 112; and H. Keen, "Mystery Airfoil," *Popular Aviation* (June 1940): 36–37, 64, 66, 68. See also Davis's testimony in a court case cited later (n. 34 below), *Testimony for Plaintiff*, vol. 1, June 18, 1951, pp. 64–77, in National Archives and Records Service (hereinafter NARS), RG 123, Acc. No. 62A176, Case 48775, Box 80. For the Davis-Douglas Company and the Cloudster, see R. J. Francillon, *McDonnell Douglas Aircraft since 1920* (London, 1979), pp. 7–9, 55–59.

8. U.S. Patent 1,942,688, David R. Davis, *Fluid Foil*, filed May 25, 1931, issued January 9, 1934.

9. Ibid., p. 1.

10. Hoy (n. 7 above), pp. 39, 108; Keen (n. 7 above), pp. 37, 64. Although these sources are obviously unsound technically, there is no reason to question their biographical information.

11. These notes are in the possession of Edwin M. Painter, who brought them to my attention. They were given to him by Davis and bear the title "Davis Rotor Airfoils." Davis's daughter Tracy Davis Klahs does not recognize the lettering as that of her father but believes the contents originated with him. She

writes that "his handwriting at best was almost illegible and even when he tried to print, the product did not resemble [these notes]. . . . I imagine that the contents originated with him but that he gave them to someone else to print" (personal correspondence). George Schairer (n. 2 above) thinks the notes are the same ones Davis showed him immediately after signing his eventual contract with Consolidated, though "I only had about ten seconds to look at them before we were interrupted" (personal correspondence). The general style of the notes, which is that of a person without formal mathematical training, supports the conclusion that Davis was the author. That the notes anteceded the patent is indicated by their more complicated and awkward mathematical notation.

12. The curve traced out by a point on a translating-rotating circle is well known in geometry as a *trochoid*. It is described as such in the notes. Under certain relative conditions of translation and rotation, this curve contains a loop that resembles a symmetrical airfoil, that is, an airfoil symmetrical about the line connecting the leading and trailing edges. Davis's scheme first generates from this loop two related symmetrical loops. Using the geometrical construction, it then derives, from one side of these loops, two distinct but related curves that are made to constitute the upper and lower surfaces of a more usual unsymmetrical (or cambered) airfoil. The two surfaces are thus in the end described by two separate equations. The entire procedure could, of course, have begun from observation of the loop of a trochoid, with reference to the Magnus effect added for window dressing, but this seems unlikely. The view that the scheme had nothing to do with basic fluid mechanics is reinforced by the fact that, when Davis applied for a second patent in 1938, he introduced a drastic limitation on his two assignable constants (denoted by A and B in the first patent) by implicitly setting $B = -A$. This reduced the number of constants from two to one in a way that effectively rotated the airstream through 90 degrees relative to the coordinate axes and completely altered the extent of the airfoil described by the two separate equations. Mathematically, the change was a purely algebraic one that has no clear interpretation in terms of the original geometrical construction. At the same time, Davis dropped the statement about the rotor and the Magnus effect. U.S. Patent 2,281,272, David R. Davis, *Fluid Foil*, filed May 9, 1938, issued April 28, 1942.

13. As pointed out to me by Todd Becker, Davis's scheme can be regarded as a kind of "pseudo-science," of which numerous examples exist in the history of medicine, where the right things have often been done for the wrong reasons.

14. Hoy (n. 7 above), pp. 108, 110; Schairer ms. (n. 2 above).

15. Davis's goal would not have been possible, in any event, from pressure measurements alone. Lift can be obtained by integrating the pressure distribution, but profile drag, which depends on fluid friction at the airfoil surface and is crucial for the optimum, requires other measurements.

16. Schairer ms. (n. 2 above). Specifically, Davis took $A = 1$, $B = -1$ in the equations of his first patent. The substitution $B = -A$ that reduced the constants from two to one in the second patent (n. 12 above) was consistent with this choice.

17. C. B. Millikan, "Report on Wind Tunnel Tests on a Davis Tapered Mono-

plane Wing and a Similar Consolidated Corporation Wing," *GALCIT Rept. No. 201-B*, September 13, 1937, quotations from p. 7. Schairer, who was Consolidated's representative at the tests, remembers, contrary to Millikan (and to Millikan's apparent reasoning), that it was the Davis model that was repolished. The difference is immaterial; the significant thing for our purposes is the existence of a concern for surface condition. The designation "21-series" from the GALCIT reports does not appear in NACA publications. It most likely denotes recently developed NACA sections that came to be known as the 210 ("two-ten") series; see E. N. Jacobs and R. N. Pinkerton, "Tests in the Variable-Density Wind Tunnel of Related Airfoils Having the Maximum Camber Unusually Far Forward," *Report No. 537*, NACA (Washington, D.C., 1935).

18. Millikan (n. 17), pp. 6–9, fig. 5. I have also benefited from correspondence with William R. Sears of the University of Arizona, who was head of the wind-tunnel staff at GALCIT, and with his wife, who, as Mabel J. Rhodes, was departmental secretary and responsible for wind-tunnel scheduling, billing, and report typing.

19. Letter, R. H. Fleet to A. B. Cook, September 30, 1937, in NARS, RG 18, Central Decimal File 1917–1938, 400.111; Memo, C. H. Helms to Dr. Lewis, November 22, 1937, in NARS, RG 255, General Correspondence 1915–1942 (Numeric File) 32-3A, Box 174.

20. Memos, G. W. Lewis to Commander Diehl, October 13, 1937, in NARS, RG 255; Memo, A. W. Brock, Jr., to Chief of the Air Corps, December 16, 1937, in NARS, RG 18; Letter, W. S. Diehl to E. B. Koger, December 20, 1938, in NARS, RG 255. By the time of Diehl's letter in 1938, the navy had discussed with Fleet the possibility of requesting such tests. Fleet, however, did not wish to have tests run by the NACA because the results (including presumably the airfoil shape; see next paragraph) "would be public property and available to his competitors" (Diehl letter).

21. Quotation from Schairer ms. (n. 2 above). Memo, Helms to Lewis (n. 19 above). The statements by Fleet presumably explain an otherwise puzzling GALCIT report on tests run on November 12, 1937, which showed no difference in performance between a wing nominally with "Davis sections" and one with NACA sections; C. B. Millikan, "Report on Wind Tunnel Tests of Two Wings for a Consolidated Medium Range Patrol Boat," *GALCIT Rept. No. 205*, April 23, 1938. A second puzzling circumstance here (and in the entire episode) is the unexplained existence in GALCIT files of a report on tests of Davis airfoils carried out a month before start of the comparative measurements for Consolidated; C. B. Millikan, "Report on Wind Tunnel Tests on Davis Biplane and Monoplane Models," *GALCIT Rept. No. 201-A*, July 27, 1937. (In this report, effects of planform were not subtracted out and, perhaps for this reason, no unusual airfoil characteristics were noted.) Unlike the other GALCIT reports, this one does not state for whom the tests were made or list a company observer. On the other hand, the fact that it shares the general number "*201*" with the later report (see n. 17 above) suggests that it was somehow associated with the Consolidated tests. Professor and Mrs. Sears (n. 18 above) do not recall these prior tests but judge that the two test series must have been "considered [by the people at GALCIT] to be two parts of a single program, for

the same customer." They conjecture that the work reported in *201-A* was "financed by Consolidated but . . . done on a pretty personal basis with only Davis and Laddon (Fleet?) involved." This explanation is rendered doubtful, however, by a figure in the report, labeled as drawn at GALCIT, which includes a section through the model and shows that someone there knew how to lay out the Davis profile fairly accurately even before the comparative tests. If the report was indeed financed by Consolidated, why did people at the company not have a better idea of the shape of the profile? Also, why would anyone at Consolidated be interested in something by then as old-fashioned as a biplane? Clearly, something went on that we do not understand. Understanding it, however, would be unlikely to alter the main elements of the story.

22. Letter, I. M. Laddon to Davis-Brookins Aircraft Company, February 9, 1938, in archives at San Diego Aerospace Museum.

23. Contemporary NACA airfoils, whose thickness distributions were not aerodynamically derived, followed the same rule. Later NACA laminar-flow sections, which were designed completely from theory (see below), had no such simple relationship.

24. In choosing an airfoil, designers, for safety reasons, must also pay attention to its *stalling characteristics*, that is, its properties at high angles of attack at which the flow separates from the upper surface and the lift falls off with increasing angle. Schairer (personal correspondence) states that the Davis airfoil also had attractions in this regard. Another aerodynamic consideration, important for an airplane's longitudinal stability, is the chordwise position of the *center of pressure* (the point at which the total force on the airfoil is effectively located). Secondary concern must also be given to the way in which profile shape affects the volume available for fuel tanks within the structure of the wing. I have ignored these matters throughout, since they would have greatly complicated the account without essential change.

25. C. B. Millikan, "Report on Wind Tunnel Tests of 1/14th Scale Models of a Consolidated 120-ft Span Airplane (Comparison of Four Wings)," *GALCIT Rept. No. 213*, June 20, 1938.

26. Evolution of the wing-section thickness through the various GALCIT reports up to the final design was as follows, where the figures in each case are the percent thickness of the profile at the root and tip sections, respectively: *201-B* (n. 17 above), unspecified, 12; *205* (n. 21 above), 18, 12; *213* (n. 25 above), 27.5, 6 and 9; *222*, final design (next note), 22, 9.3. Consolidated had the reputation of doing a great deal of cut-and-try testing, as compared with, say, Douglas, which went in more for analytical examination of a smaller number of measurements. Different countries are sometimes said to exhibit different engineering styles; so too do different companies within a given country.

27. Anon., "'Hush-Hush' Boat," *Aviation* 38 (June 1939): 38–39, and B. W. Sheahan, "A Simplified Method for Development of Prototype Airplanes as Applied to Consolidated Model 31 Long Range Patrol Boat," preprint of paper for National Aircraft Production Meeting, Society of Automotive Engineers, Los Angeles, October 5–7, 1939; W. R. Sears, "Report on Wind Tunnel Tests on a 1/12th Scale Model of the Consolidated Model 31 Flying Boat," *GALCIT Rept. No. 222*, October 13, 1938. The prototype Model 31 was extensively

modified by Consolidated and the design adopted by the navy for wartime antisubmarine use in 1942, with two hundred airplanes scheduled for procurement under the designation P4Y-1. The order was canceled, however, when the submarine threat diminished. C. Hansen, "The Consolidated Model 31/XP4Y-1 'Corregidor' Flying Boat," *Journal of the American Aviation Historical Society* 27 (Spring 1982): 136–47.

28. Blue (n. 2 above), p. 11; Laddon to Arnold, May 11, 1939, in NARS, RG 18, Central Decimal File 1939–1942, 452.1K; C. B. Millikan, "Wind Tunnel Test on a 1/12th Scale Model of the Consolidated Model XB-24 Bomber," *GALCIT Rept. No. 243*, June 26, 1939, and "Report on Comparative Wind Tunnel Tests on Two Alternative Wings for the Consolidated Model 31 Airplane," *GALCIT Rept. No. 254*, April 15, 1940.

29. Blue (n. 2 above), p. 189; R. Wagner (n. 2 above), pp. 213–18, 320–21. For a popularized operational history of the B-24, see S. Birdsall, *Log of the Liberators: An Illustrated History of the B-24* (New York, 1973); the brief discussion of the Davis airfoil, however, is largely fanciful. The crucial antisubmarine role of the B-24 in the Battle of the Atlantic is emphasized by W. Churchill, *Closing the Ring* (Boston, 1951), p. 8. For the relative overall merits of the B-24 and B-17, see Blue, pp. 180–88, and L. K. Loftin, Jr., *Quest for Performance: The Evolution of Modern Aircraft* (Washington, D.C., 1985), pp. 121–23.

30. Bayless correspondence (n. 2 above).

31. Anon., "Davis Low-Drag Wing," *Aviation* (June 1939): 68; C. B. Millikan, "Report on Comparative Wind Tunnel Tests on Five Comparative Wings for the Douglas XTB2D-1 Airplane," *GALCIT Rept. No. 248-B*, November 22, 1939; F. H. Clauser, H. LaMar, and A. B. Croshere, "Theoretical Investigation of a Family of Airfoils," *Report No. 2698*, Douglas Aircraft Company, Santa Monica, Calif., February 29, 1940; E. V. Laitone, "A High-Speed Investigation of a Proposed Hughes Aircraft Company Design," *Memorandum Report* (unnumbered), NACA, March 8, 1940. (The reports on the Douglas work are proprietary. Access to them at GALCIT and Douglas was obtained through help, respectively, of R. E. Pendley and D. L. Tillotson, engineering officials at Douglas.) R. B. Beisel to G. W. Lewis, February 27, 1942, and H. J. E. Reid to NACA, March 13, 1942, in NARS, RG 255, General Records (Decimal File), 618.1, 1942; Schairer ms. (n. 2 above).

32. E. C. Draley, "High-Speed Tests of the XB-32 Bomber Model Including Three Wing Variations," *Memorandum Report for Army Air Corps* (unnumbered), NACA, September 15, 1941. For these problems in relation to later aerodynamic difficulties with the B-32, see Langley Field Memo, E. C. Draley to Chief of Research, April 24, 1944, and I. H. Abbott to Chief of Research, May 20, 1944, in NARS, RG 255 General Records (Decimal File) 613.11 (B-32). Wing-section percent thicknesses for the B-32 and B-29 at root and tip sections, respectively, were B-32: 23, 9 (Draley); B-29: 22, 9 (Schairer, personal correspondence). Consolidated also considered the Davis airfoil in the early 1940s for a four-engine flying boat, which was wind-tunnel tested by the NACA but never built; G. B. McCullough and R. E. Woodworth, "Tests of Three Alternate Wing Designs for the Consolidated XPB3Y-1 Airplane," *Memorandum Report for the Bureau of Aeronautics, Navy Department* (unnumbered), NACA, January

11, 1943. It is not clear whether this was the same basic airplane as tested earlier at GALCIT under the same designation (n. 6 above).

33. The first and most significant of the Washington reports is V. J. Martin, "Wind Tunnel Tests on a Series of Davis Airfoils," *Report 104*, University of Washington Aeronautical Laboratory, Seattle, August 2, 1939. (The reports are proprietary and were made available from the university through the kind permission of Mrs. Klahs [n. 11 above], who is Davis's legal heir.) E. Churchill, "Manta," *Flying* 31 (November 1942): 54, 100, 103.

34. In 1943 Davis signed an agreement with the government limiting the royalty paid to him on Davis-wing aircraft bought by the United States. In the early 1950s he sued the government, claiming that additional payments were owed him for B-24s sold as surplus to private buyers at the end of the war. The government contested the suit, arguing that the claims of Davis's patents were not in fact valid. The Court of Claims ruled both patents invalid, the second on the grounds of double patenting and the first because it required experimentation with the values of the constants to be used in the formulas. Davis's claim was therefore denied. Davis Airfoils, Inc., v. The United States, No. 48775, U.S. Court of Claims, October 5, 1954, in *Federal Supplement* 124 (1955): 350–54.

35. N. 7 above. A brief, highly idealized cartoon biography also appeared anonymously under the title "Here's My Story: The Career of David R. Davis," *Popular Science* 140 (February 1942): 112–13.

36. An excellent contribution in this direction, especially regarding the all-important NACA work, has been made by James R. Hansen in his account of airfoil research at the Langley Laboratory, *Engineer in Charge: A History of the Langley Aeronautical Laboratory, 1917–1958* (Washington, D.C., 1987), pp. 78–84, 97–118. I have drawn from this account in numerous places in this section. J. V. Becker, *The High-Speed Frontier* (Washington, D.C., 1987), chap. 2, provides an illuminating personal account of the development of high-speed airfoils. Mark Levinson is currently engaged in a comparative study of developments in the United States and Germany in the period 1880 to 1922.

37. E. P. Warner, *Airplane Design—Performance* (New York, 1936), p. 158.

38. Ibid., pp. 158–63; C. B. Millikan, *Aerodynamics of the Airplane* (New York, 1941), pp. 66–67.

39. M. M. Munk, "General Theory of Thin Wing Sections," *Report No. 142*, NACA (Washington, D.C., 1922); M. M. Munk and E. W. Miller, "Model Tests with a Systematic Series of 27 Wing Sections at Full Reynolds Number," *Report No. 221*, NACA (Washington, D.C., 1925); H. Davies, "Wing Tunnel Test of Aerofoil R.A.F. 34," *Reports and Memoranda No. 1071*, Aeronautical Research Committee (London, October 1926). I am indebted to Mark Levinson for recognition of the crucial nature of Munk's contribution.

40. E. N. Jacobs, K. E. Ward, and R. M. Pinkerton, "The Characteristics of 78 Related Airfoil Sections from Tests in the Variable-Density Wind Tunnel," *Report No. 460*, NACA (Washington, D.C., 1933), and E. N. Jacobs, R. M. Pinkerton, and H. Greenberg, "Tests of Related Forward-Camber Airfoils in the Variable Density Wind Tunnel," *Report No. 610*, NACA (Washington, D.C., 1937). Both Munk's and Jacobs's work are examples of the general method of *parameter variation* described in chapter 5. For a fuller description of the meth-

ods and importance of the work by Jacobs and his group, see Hansen (n. 36 above), pp. 97–118.

41. Jacobs's idea was arrived at independently in Japan by Itiro Tani and Satosi Mituisi, "Contributions to the Design of Aerofoils Suitable for High Speeds," *Reports of the Aeronautical Research Institute, Tokyo Imperial University* 15 (1940): 399–415.

42. T. Theodorsen, "Theory of Wing Sections of Arbitrary Shape," *Report No. 411*, NACA (Washington, D.C., 1931); E. N. Jacobs, "Preliminary Report on Laminar-Flow Airfoils and New Methods Adopted for Airfoil and Boundary-Layer Investigations," *Wartime Report L-345*, originally *Advanced Confidential Report* (unnumbered), NACA (Washington, D.C., June 1939). For wind tunnels, see Hansen (n. 36 above), pp. 101–5, 109–11, and D. D. Baals and W. R. Corliss, *Wind Tunnels of NASA* (Washington, D.C., 1981), pp. 37–41.

43. The 747A315 was designed to have laminar flow to 40 and 70 percent of the chord on the upper and lower surfaces, respectively, over the range of lift within which significantly low drag can be maintained. It was thus atypical among laminar-flow airfoils, most of which had the same amount of laminar flow on both surfaces. Many also had their maximum thickness farther aft. As far as I know, the 747A315 was never used on an airplane. I employ it here because it brings out most clearly the ways in which the Davis airfoil resembled a laminar-flow design. For an example of a more usual laminar-flow profile, see Hansen (n. 36 above), p. 115.

44. I. H. Abbott, A. E. von Doenhoff, and L. S. Stivers, Jr., "Summary of Airfoil Data," *Report No. 824*, NACA (Washington, D.C., 1945). For secondary sources on the laminar-flow airfoil, see Hansen (n. 36 above), p. 111–18; I. H. Abbott, "Airfoils," in *The Evolution of Aircraft Wing Design* (Air Force Museum, Dayton, March 18–19, 1980), pp. 21–24; and G. W. Gray, *Frontiers of Flight* (New York, 1948), pp. 104–12.

45. For the measured pressure distributions, see I. H. Abbott, "Pressure-Distribution Measurements of a Model of a Davis Wing Section with Fowler Flap Submitted by Consolidated Aircraft Corporation," *Wartime Report L-678*, originally *Memorandum Report* (unnumbered), NACA (Washington, D.C., January 1942). The calculated pressure distributions were made for me by James Baeder using a modern computer version of Theodorsen's method.

46. Abbott, von Doenhoff, and Stivers (n. 44 above), fig. 38.

47. Another possible influence appears in the pressure rise over the rear upper surface of the Davis profile. Both the measured and calculated distributions are of a nature (i.e., rate of pressure rise decreasing progressively toward the rear) that tends to reduce the friction of the turbulent boundary layer in that region. This effect became understood with the work of B. S. Stratford in England in the late 1950s; "An Experimental Flow with Zero Skin Friction throughout Its Region of Pressure Rise," *Journal of Fluid Mechanics* 5 (January 1959): 17–35. The "Stratford effect" also tends to thicken the boundary layer, however, and this could increase the adverse effects near the trailing edge. The net effect on drag is difficult to assess.

48. Abbott, von Doenhoff, and Stivers (n. 44 above), fig. 46.

49. Schairer in his manuscript (n. 2 above) states that Boeing as part of its

B-29 studies (ca. 1942) also supplied Jacobs with a number of Davis-like airfoils (including the RAF 34; see below), which he tested with favorable results. Unfortunately, I am unable to find an account of these tests among NACA records.

50. Sears (n. 18 above), personal correspondence.

51. Schairer ms. (n. 2 above).

52. Abbott, von Doenhoff, and Stivers (n. 44 above), figs. 30–32.

53. For an account of these and later developments, see Becker (n. 36 above). Kármán's remark was related to me in the 1950s by H. Julian Allen.

54. For particulars on airfoil design since 1950, see R. S. Shevell, *Fundamentals of Flight* (Englewood Cliffs, N.J., 1983), pp. 217–28. For examples of modern design for specific requirements, see R. H. Liebeck, "Design of Airfoils for High Lift," and P. B. S. Lissaman, "Wings for Human-Powered Flight," both in *The Evolution of Aircraft Wing Design* (n. 44 above), pp. 25–56.

55. This observation brings to mind a difference between science and engineering that may have epistemological implications: In science what you don't know about is unlikely to hurt you (except possibly in some unfamiliar experimental situations). In engineering, however, bridges fall and airplanes crash, and what you don't know about can hurt you very much.

56. P. Rhinelander, *Campus Report*, Stanford University, January 18, 1984, p. 7.

57. E. Constant, *The Origins of the Turbojet Revolution* (Baltimore, 1980), p. 15.

58. The term is also from Constant (ibid., pp. 12–13). The quoted phrase is from Rachel Laudan's general philosophical discussion of cognitive change in technology as the result of problem-solving activity, which makes a number of points pertinent to the present study; "Cognitive Change in Technology and Science," in *The Nature of Technological Knowledge: Are Models of Scientific Change Relevant?*, ed. R. Laudan (Dordrecht, 1984), pp. 83–104, quotation from p. 85.

59. Campbell's complex and ramified work, of which the above does not pretend to be an adequate summary, came to my attention via Constant (*Origins*, pp. 6–8). Campbell's quotation is from his "Evolutionary Epistemology," in *The Philosophy of Karl Popper*, ed. P. A. Schilpp, vols. 14I and II of *The Library of Living Philosophers* (LaSalle, Ill., 1974), vol. 14I, p. 422. Popper's quotation comes from his "Replies to My Critics," vol. 14II, p. 1062. For an earlier essay by Campbell, see chap. 8, n. 1.

60. The variation-selection aspect of engineering devices also appears unmistakably in the multitude of photographs of American combat aircraft in R. Wagner's book (n. 2 above), which includes an amazing variety of airplane types and configurations that were tried but for one reason or another found wanting.

61. The Davis/Jacobs contrast also exemplifies two other concepts about the manner in which knowledge grows in engineering. (1) Donald A. Schön, in his study of problem solving in various professions (*The Reflective Practitioner* [New York, 1983], pp. 94–95, 131–32, 208–9, passim), has discussed how the reframing of a problem is often critical to its solution. Jacobs's reframing of airfoil design in terms of pressure distribution instead of shape was the critical step in solving the problem of laminar flow. Davis, like those before him, thought

directly in terms of shape. (2) Constant, as a part of a model for technological change (n. 57 above, pp. 10, 24–27), has emphasized the related importance of hierarchy and traditions of practice in engineering design. Airfoil design from the early 1900s to the present appears as a single continuing tradition when viewed from the level of overall airplane design (see chap. 1); from this level, Jacobs's idea was an important but incremental improvement within a continuing tradition. At the level of section design, however, his shift from shape to pressure distribution marked a revolutionary change from one tradition to another. For remarks about technological traditions, see Laudan (n. 58 above), pp. 93–95.

62. For remarks on the historiography of technological failures, see J. M. Staudenmaier, "What SHOT Hath Wrought and What SHOT Hath Not: Reflections on Twenty-five Years of the History of Technology," *Technology and Culture* 25 (October 1984): 707–30, esp. 718–19, and M. Kranzberg, "Let's Not Get Wrought Up about It," ibid., 735–49, esp. 739–41. For recent studies of failures, indicating a growing interest in this topic, see B. E. Seely, "The Scientific Mystique in Engineering: Highway Research at the Bureau of Public Roads, 1918–1940," *Technology and Culture* 25 (October 1984): 798–831; H. Petroski, *To Engineer Is Human: The Role of Failure in Successful Design* (New York, 1985); and I. B. Holley, Jr., "A Detroit Dream of Mass-produced Fighter Aircraft: The XP-75 Fiasco," *Technology and Culture* 28 (July 1987): 578–93.

Chapter 3. Establishment of Design Requirements

Note: Limited excerpts from this chapter were published in "How Did It Become 'Obvious' that an Airplane Should Be Inherently Stable?" *American Heritage of Invention & Technology* 4 (Spring/Summer 1988): 50–56. In addition to specific acknowledgments in the endnotes, I am indebted for more general assistance to Roger Bilstein, Waldemar Breuhaus, Arthur Bryson, Edward Constant, George Cooper, Virginia Dawson, Warren Dickinson, Robert Gilruth, Richard Hallion, James Hansen, Robert Jones, Barry Katz, Stephen Kline, Ilan Kroo, Edwin Layton, William Milliken, Jim Papadopoulos, Courtland Perkins, William Phillips, William Rifkin, Russell Robinson, Paul Seaver, Richard Shevell, Richard Smith, Maurice White, Howard Wolko, and Robert Woodcock. The topic arose out of discusion with my Stanford engineering colleague Holt Ashley.

1. L. Wade, "Performance Test of Morane Saulnier Type A.R. Airplane with Two Sets of Wings Equipped with 80-H.P. Le Rhone Engine," *Air Service Information Circular* 3, no. 285, October 1, 1921, p. 3.

2. The distinction employed here is different from that used in the study of artificial intelligence. There, for a problem to be "well defined," it must have a defined initial state, a specified set of moves allowed for its solution, and a known, unique goal state. All real-life design problems are by this definition "ill defined" (or "ill structured") and are so regarded in artificial intelligence; J. M. Carroll, J. C. Thomas, and A. Malhotra, "Presentation and Representation in Design Problem-Solving," *British Journal of Psychology* 71 (1980): 143–53, esp.

143; see also H. A. Simon, "The Structure of Ill Structured Problems," *Artificial Intelligence* 4 (1973): 181–201. For analysis and examples of how engineers set and solve problems in ill-defined situations (in the sense employed here), see D. A. Schön, *The Reflective Practitioner* (New York, 1983), pp. 168–203.

3. E. W. Constant, *The Origins of the Turbojet Revolution* (Baltimore, 1980), pp. 8–10, passim.

4. *Flying qualities* was the accepted term in the period treated here. In recent times *handling qualities* is sometimes used instead.

5. B. Markham, *West with the Night* (San Francisco, 1983), p. 16. The passage was brought to my attention by Joyce Vincenti.

6. The reader may have noticed an ambiguity in usage. The noun *stability* in standard terminology can denote either a general topic subsuming instability (as in the phrase "stability and control") or a specific condition opposite to *instability*. One quickly learns to live with this ambiguity. The meaning is usually clear from the context.

7. The account in this section, though somewhat extended, is still necessarily oversimplified; I believe, however, it is essentially correct. Details on some matters can be found in the following lectures for engineering audiences, which supplied my initial orientation in the events in this and the subsequent historical sections: 1970 Von Kármán Lecture by C. D. Perkins, "Development of Airplane Stability and Control Technology," *Journal of Aircraft* 7 (July–August 1970): 290–301, and 1984 Wright Brothers Lecture by R. P. Harper, Jr., and G. E. Cooper, "Handling Qualities and Pilot Evaluation," *Journal of Guidance, Control, and Dynamics* 9 (September–October 1986): 515–29. I have also profited from the historical appendix to W. F. Milliken, Jr., "Progress in Dynamic Stability and Control Research," *Journal of the Aeronautical Sciences* 14 (September 1947): 493–519, and from the historical sections of D. McRuer, I. Ashkenas, and D. Graham, *Aircraft Dynamics and Automatic Control* (Princeton, 1973), pp. 21–42.

8. *Fourth Annual Report of the National Advisory Committee for Aeronautics, 1918* (Washington, D.C., 1920), pp. 42–43. For the history of the NACA, see A. Roland, *Model Research: The National Advisory Committee for Aeronautics, 1915–1958*, 2 vols. (Washington, D.C., 1985); for the committee's charge, see vol. 2, p. 394; for the members, see vol. 2, pp. 431–35; see also J. C. Hunsaker, "Forty Years of Aeronautical Research," *Annual Report of the Board of Regents of the Smithsonian Institution*, year ended June 30, 1955 (Washington, D.C., 1956), pp. 241–71.

9. H. D. Fowler, "Stability in General," *Aerial Age Weekly*, December 17, 1917, pp. 606–7, 617, quotation on p. 606; J. B. Rathbun, *Aeroplane Construction and Operation* (New York, 1918), pp. 312–13.

10. C. C. Turner, *Aerial Navigation of To-day* (London, 1910), p. 118. *Automatic* is ordinarily used to describe stabilization by an automatic control mechanism, such as mentioned earlier; from the context, Turner clearly means *inherent* in our terminology.

11. These terms and the ideas and facts of the next three paragraphs come mainly from Gibbs-Smith's books *The Aeroplane: An Historical Survey of Its Origins and Development* (London, 1960), *The Invention of the Aeroplane (1799–1909)*

(New York, 1966), and *Aviation: An Historical Survey from Its Origins to the End of World War II* (London, 1970). For the distinction between chauffeurs and airmen, see in particular *Aviation*, p. 58.

12. Gibbs-Smith, *Aeroplane*, p. 20; *Aviation*, pp. 68–69, 126; *Invention*, chaps. 1–3. The quotation is from Perkins (n. 7 above), p. 290.

13. Gibbs-Smith (*Invention*, p. 35) says the Wrights opted deliberately for instability; Perkins (n. 7 above, p. 291) concludes from study of their letters that "they didn't understand what stability really meant." Neither of these opinions distinguishes between longitudinal and lateral properties. In a recent collection of engineering analyses of the Wright airplane, Frederick E. C. Culick and Henry R. Jex indicate that the inherent lateral instability was indeed largely deliberate. With reference to the more controversial longitudinal situation, however, they conclude after carefully reasoned consideration that "the Wrights did not understand stability in the precise sense that we do now," that without such understanding "they could not formulate precise ideas of stability in contrast to equilibrium," and that "whether their aircraft were stable or unstable was an accidental matter." In connection with his analysis of longitudinal dynamics in the same collection, however, Frederick J. Hooven states that the Wrights were "entirely familiar with static stability inasmuch as they were acquainted with the work of Pénaud and they had made and flown many models" and judges that "they made the deliberate choice of [longitudinal] instability." I find the Culick-Jex analysis the more compelling. F. E. C. Culick and H. R. Jex, "Aerodynamics, Stability, and Control of the 1903 Wright Flyer," and F. J. Hooven, "Longitudinal Dynamics of the Wright Brothers' Early Flyers," both in *The Wright Flyer: An Engineering Perspective*, ed. H. S. Wolko (Washington, D.C., 1987), respectively, pp. 19–43, esp. pp. 22, 31, and 41, and pp. 45–77, quotations from pp. 58 and 51. Tom D. Crouch sees a conceptual link between the deliberate lateral instability and the sideways instability of a bicycle, with which the Wrights were familiar as cyclists and bicycle manufacturers; "How the Bicycle Took Wing," *American Heritage of Invention & Technology* 2 (Summer 1986): 11–16.

14. Gibbs-Smith, *Invention*, chaps. 6, 9–11, 13–14, closing quotation from p. 58; Perkins (n. 7 above), pp. 291–93.

15. Gibbs-Smith, *Invention*, chaps. 16, 17, 43, 44. For the role of Ferber, who was unusual for Europeans in sharing something of the airmen mentality, see F. E. C. Culick, "Aeronautics, 1898–1909: The French-American Connection" (forthcoming).

16. G. H. Bryan, a British aeronautical theoretician whose work we shall come to presently, wrote in 1911 (n. 20 below, p. 165) that "it does not appear to be denied that some [existing, successful airplanes] are unstable." L. Bairstow et al., in an important research report in 1913 (n. 21 below, p. 136) stated, by contrast, that "it is probable that all types of aeroplanes now in common use are longitudinally stable under the usual conditions of normal flight." Historian Gibbs-Smith, looking back, contends that at the first international flying competition, held at Reims in August 1909, the machines of pilot-designers such as the Europeans Louis Blériot, Gabriel and Charles Voisin, and Henri Farman and the American Glenn Curtiss (in fact, of all but the Wrights) had "a consider-

able amount of stability built into them" (*Invention*, chaps. 50–62, quotation from p. 2; see also next note).

17. A. H. Wheeler, *Building Aeroplanes for 'Those Magnificent Men'* (Sun Valley, Calif., n.d.), esp. p. 72, and N. Williams, "On Gossamer Wings," *Flight International* 107 (January 1975): 45–48, esp. 46. Some of the replicas (Antoinette and Bristol Boxkite) closely resemble the machines flown at Reims, thus tending to contradict the statement of Gibbs-Smith quoted in the preceding note. For details about automatic pilots, see McRuer, Ashkenas, and Graham (n. 7 above), p. 29.

18. F. W. Lanchester, *Aerodynamics* (London, 1907), pp. 231–33, 368–70. For a typical modern textbook treatment, see R. S. Shevell, *Fundamentals of Flight* (Englewood Cliffs, N. J., 1983), p. 300.

19. A statically unstable machine, like the previous statically stable machine in which oscillations build up, is also described as dynamically unstable.

20. G. H. Bryan, *Stability in Aviation* (London, 1911). This book was an outgrowth of earlier, more limited work by Bryan and W. E. Williams, "The Longitudinal Stability of Aerial Gliders," *Proceedings of the Royal Society of London* 73 (1904): 100–16. Such stability analyses make the simplifying approximation that the amplitude of oscillation is, in some relative sense, small.

21. The Aeronautics Staff of the Engineering Department of the National Physical Laboratory, "Method of Experimental Determination of the Forces and Moments on a Model of a Complete Airplane: With the Results of Measurements on a Model of a Monoplane of a Blériot Type," *Reports and Memoranda No. 75*, in *Technical Report of the Advisory Committee for Aeronautics, 1912–13* (London, 1914), pp. 128–32; L. Bairstow, B. M. Jones, and A. W. H. Thompson, "Investigations into the Stability of an Aeroplane, with an Examination into the Conditions Necessary in Order that the Symmetric and Asymmetric Oscillations Can Be Considered Independently," *Reports and Memoranda No. 77*, in ibid., pp. 135–71; J. C. Hunsaker, "Experimental Analysis of Inherent Longitudinal Stability for a Typical Biplane," *Report No. 1, Part 1*, NACA (Washington, D.C., 1915) and "Dynamical Stability of Aeroplanes," *Smithsonian Miscellaneous Collections*, vol. 62, no. 5 (Washington, D.C., June 3, 1916). The MIT work was summarized for nontechnical readers by W. H. Ballou, "The Instability of American Airplanes," *Scientific American* 120 (February 1919): 118–19, 128.

22. Supplementary tests of the JN-2, reported in 1918 by the aeronautical research department at McCook Field (and presumably carried out also at MIT, though the record is not explicit), did use a model with a movable elevator; Aeronautical Research Department, "Forces in Diving and Looping," *Bulletin of the Airplane Engineering Department U.S.A.* 1 (June 1918): 89–103, esp. 93–95. As indicated by the title, however, the concern was for the forces on the airplane in maneuvers rather than on maneuverability itself. The results were used later by Warner and Norton in analysis of their pioneering flight data on stability and control (n. 32 below).

23. Bairstow, Jones, and Thompson (n. 21 above), p. 136; Hunsaker, "Dynamical Stability" (n. 21 above), p. 5.

24. Bairstow, Jones, and Thompson (n. 21 above), p. 137; *Fourth Annual Report* (n. 8 above), p. 42.

25. Perkins (n. 7 above), p. 295; for present-day assessment of the JN-4 and S-4C by James N. Nissen, restorer and pilot of antique aircraft and former NACA test pilot, see Harper and Cooper (n. 7 above), p. 517; for the Camel, see B. Lecomber, "Flying the Sopwith Camel and Fokker Triplane," *Flight International* 113 (April 1978): 998–1001, quotation from 998–99; for early evidence on the JN-4 and the quotation by the design engineer, see, respectively, Warner and Norton (n. 32 below), pp. 23–31 and Korvin-Kroukovsky (n. 41 below), p. 267. Some of Gibbs-Smith's statements (e.g., *Aeroplane* [n. 11 above], p. 20) unfortunately suggest that after 1913 the compromises between stability and control were worked out more quickly and readily than was apparently the case.

26. Hunsaker, "Dynamical Stability" (n. 21 above), pp. 37–38, 45.

27. Bairstow, Jones, and Thompson (n. 21 above), p. 137; Captain Schroeder's complete report is reproduced in R. P. Hallion, *Test Pilots: The Frontiersmen of Flight* (Garden City, 1981), pp. 60–61. This reference (pp. 57–64) gives an interesting general account of flight testing at McCook Field. The adjectives "lateral and directional" in Schroeder's report pertain together to what I have called simply "lateral." The reasons behind this alternative terminology are too involved to go into here and are not important to our concerns.

28. "Stability and Balance in Airplanes," *Aviation* 8 (April 1920): 193–94, quotation on 193.

29. Harper and Cooper (n. 7 above), p. 518.

30. Perkins (n. 7 above), p. 296.

31. The British work as of 1920, mostly done at the Royal Aircraft Establishment at Farnborough, was summarized by H. Glauert, "Summary of the Present State of Knowledge with Regard to Stability and Control of Aeroplanes," *Reports and Memoranda No. 710*, in *Technical Report of the Advisory Committee for Aeronautics, 1920–21* (London, 1924), vol. 1, pp. 339–51. For the history of the Langley Laboratory, see J. R. Hansen, *Engineer in Charge: A History of the Langley Aeronautical Laboratory, 1917–1958* (Washington, D.C., 1987).

32. E. P. Warner and F. H. Norton, "Preliminary Report on Free Flight Tests," *Report No. 70*, NACA (Washington, D.C., 1920). For a popular account of this and related work at Langley, see Hallion (n. 27 above), pp. 72–75.

33. Explanation adapted from B. M. Jones, "Dynamics of the Airplane," in *Aerodynamic Theory*, 6 vols., ed. W. F. Durand (Berlin, 1934–35), vol. 5, pp. 1–222, esp. p. 28.

34. Note that static stability, which has to do with the initial response to a *transitory* disturbance, is revealed in both cases by the control action needed to accomplish a change in *steady* speed. Such relationship may at first seem paradoxical. It comes from the fact, implicit in the earlier exposition of moment-curve slope, that the *initial* response to *any small* departure from equilibrium may be treated to first approximation as a succession of steady (or static) states. The previously noted similarity between an airplane's resistance to a controlled change and to a transitory disturbance is a reflection of the same fact.

35. In Britain at about the same time, Hermann Glauert derived the stick-

fixed and stick-free criteria by relating them mathematically to the moment-curve slope (measured in the stick-fixed and stick-free conditions, respectively); "The Longitudinal Stability of an Aeroplane," *Reports and Memoranda No. 638*, in *Technical Report of the Advisory Committee for Aeronautics, 1919–20* (London, 1923), vol. 2, pp. 439–59. Glauert and the Americans, reacting to the same need, appear to have arrived at their results independently.

36. E. P. Warner, "Static Longitudinal Stability of Airplanes," *Report No. 96*, NACA (Washington, D.C., 1920), and "Notes on Longitudinal Stability and Balance," *Technical Note No. 1*, NACA (Washington, D.C., April 1920), quotation from pp. 2–3.

37. All reports from NACA (Washington, D.C.): F. H. Norton and E. T. Allen, "Accelerations in Flight," *Report No. 99* (1921); Norton and T. Carroll, "The Vertical, Longitudinal, and Lateral Accelerations Experienced by an S.E.5A Airplane while Maneuvering," *Report No. 163* (1923); H. J. E. Reid, "A Study of Airplane Maneuvers with Special Reference to Angular Velocities," *Report No. 155* (1922); Norton and Allen, "Control in Circling Flight," *Report No. 112* (1921); Norton and W. G. Brown, "Controllability and Maneuverability of Airplanes," *Report No. 153* (1923); Norton, "Practical Stability and Controllability of Airplanes," *Report No. 120* (1921); Norton and Brown, "Complete Study of the Longitudinal Oscillation of a VE-7 Airplane," *Report No. 162* (1923); Norton, "A Study of Longitudinal Dynamic Stability in Flight," *Report No. 170* (1923). See also Norton, "The Measurement of the Damping in Roll on a JN4h in Flight," *Report No. 167* (1923).

38. All reports from NACA (Washington, D.C.): F. H. Norton and E. P. Warner, "Accelerometer Design," *Report No. 100* (1921); Norton, "N.A.C.A. Recording Air Speed Meter," *Technical Note No. 64* (October 1921) and "N.A.C.A. Control Position Recorder," *Technical Note No. 97* (May 1922); H. J. E. Reid, "The N.A.C.A. Three-Component Accelerometer," *Technical Note No. 112* (October 1922); W. G. Brown, "The Synchronization of N.A.C.A. Flight Records," *Technical Note No. 117* (October 1922); K. M. Ronan, "An Instrument for Recording the Position of Airplane Control Surfaces," *Technical Note No. 154* (August 1923).

39. Warner and Norton (n. 32 above), p. 15; E. T. Allen with C. B. Allen, "Tons Aloft: Test Piloting the Transatlantic Clipper," *Saturday Evening Post*, September 17, 1938, pp. 12–13, 86, 88, 90–91, quotation on p. 90. See also Hallion (n. 27 above), pp. 72–75.

40. See the reports of the Committee on Aerodynamics in *Annual Report of the National Advisory Committee for Aeronautics* (Washington, D.C.) for the years in question.

41. C. N. Monteith, *Simple Aerodynamics and the Airplane* (Dayton, 1924); E. P. Warner, *Airplane Design—Aerodynamics* (New York, 1927); B. V. Korvin-Krukovsky, "Stability and Controllability of Airplanes—Parts I–III," *Aviation*, March 9, 16, 23, 30, April 6, 13, 20, 1925, pp. 266–69, 296–97, 320–22, 347–48, 380–81, 411–12, 436–38 (covering longitudinal problems only; lateral problems appeared in additional articles).

42. W. S. Diehl, *Engineering Aerodynamics* (New York, 1928); V. E. Clark, *Elements of Aviation* (New York, 1928); A. Klemin, *Simplified Aerodynamics*

(Chicago, 1930); E. A. Stalker, *Principles of Flight* (New York, 1931); C. C. Carter, *Simple Aerodynamics and the Airplane* (New York, 1932); K. D. Wood, *Technical Aerodynamics* (Ithaca, 1933).

43. Norton and Allen, *Report No. 112* (n. 37 above), p. 3; Korvin-Krukovsky (n. 41 above), p. 267; Clark (n. 42 above), p. 49; Klemin (n. 42 above), p. 187.

44. C. L. Johnson, "Longitudinal Stability of a Bi-Motor Transport Airplane," *Journal of the Aeronautical Sciences* 3 (September 1935): 1–6, quotation from 1.

45. For the air-racing experiences, see Harper and Cooper (n. 7 above), p. 519. An additional possible influence was the automatic pilot coming into general use in the early 1930s. These autopilots were designed to provide straight and level flight and required an inherently stable airplane lest the coupled system give rise to undesirable hunting oscillations.

46. Stalker (n. 42 above), pp. 255–56; Johnson (n. 44 above), p. 2; W. S. Diehl, *Engineering Aerodynamics*, rev. ed. (New York, 1936), p. 186; M. M. Munk, "Diehl's Static Stability Coefficient," *Aero Digest* 41 (August 1942): 154, 236–37, quotation on 154; A. Klemin and J. G. Beerer, Jr., "Two New Longitudinal Stability Constants," *Journal of the Aeronautical Sciences* 4 (September 1937): 453–59. Otto Koppen in his article of 1940 (n. 85 below, pp. 136–37) demonstrated clearly that no correlation exists between moment-curve slope and perceived flying qualities. Koppen's article also gave a vivid description of the difference between designers' practices and pilots' perceptions.

47. Norton, *Report No. 170* (n. 37 above), pp. 3, 9.

48. Warner (n. 41 above), pp. 388–405 and Stalker (n. 42 above), pp. 336–66; C. H. Zimmerman, "An Analysis of Longitudinal Stability in Power-off Flight with Charts for Use in Design," *Report No. 521*, NACA (Washington, D.C., 1935), and O. C. Koppen, "Trends in Longitudinal Stability," *Journal of the Aeronautical Sciences* 3 (May 1936): 232–33; H. A. Soulé and J. B. Wheatley, "A Comparison between the Theoretical and Measured Longitudinal Stability Characteristics of an Airplane," *Report No. 442*, NACA (Washington, D.C., 1932); A. G. B. Metcalf, "Airplane Longitudinal Stability—A Resume," *Journal of the Aeronautical Sciences* 4 (December 1936): 61–69.

49. J. H. Doolittle, "Accelerations in Flight," *Report No. 203*, NACA (Washington, D.C., 1925). For an account of Doolittle's tests, see Hallion (n. 27 above), pp. 82–84; Barksdale manual quoted in ibid., p. 108; W. F. Gerhardt and L. V. Kerber, *A Manual of Flight-Test Procedure* (Ann Arbor, 1927); Anonymous, "Determination of Stability from Flight Test Stick Force Data," *Air Corps Information Circular*, August 1, 1929.

50. All reports from NACA (Washington, D.C.): C. H. Dearborn and H. W. Kirschbaum, "Maneuverability Investigation of the F6C-3 Airplane with Special Flight Instruments," *Report No. 369* (1930); "Maneuverability Investigation of an F6C-4 Fighting Airplane," *Report No. 386* (1931); F. L. Thompson and Kirschbaum, "Maneuverability Investigation of an '03U-1' Observation Airplane," *Report No. 457* (1933); F. E. Weick, H. A. Soulé, and M. N. Gough, "A Flight Investigation of the Lateral Control Characteristics of Short Wide Ailerons and Various Spoilers with Different Amounts of Wing Dihedral," *Report No. 494* (1934); Soulé and W. H. McAvoy, "Flight Investigation of Lateral

Control Devices for Use with Full-Span Flaps," *Report No. 517* (1935); quoted phrase from Milliken (n. 7 above), p. 515. For brief description of the angle-of-attack recorder, see Soulé and Wheatley (n. 48 above). In the 1930s and 1940s, in contrast to the situation in the early 1920s (see n. 38 above), NACA ceased publication of details of its measuring techniques. A Langley engineer from the period says the laboratory's associate director told him such restriction was desirable "so that industry would have to come to us to get some of the more advanced research done" (William H. Phillips, personal correspondence).

51. S. Flower, M. Gough, R. B. Maloy, and G. S. Trimble, Jr., "Flight Handling Characteristics of a Modern Transport Airplane—A Symposium," *Aeronautical Engineering Review* 7 (October 1948): 18–31, quotation on 18.

52. C. A. Lindbergh, *The Wartime Journals of Charles A. Lindbergh* (New York, 1970), pp. 230–31.

53. Memos, late April 1941, Comdr. J. E. Ostrander, Comdr. W. S. Diehl, and Capt. R. S. Hatcher; in personal files of Gerald G. Kayten.

54. W. F. Milliken, Jr., and D. W. Whitcomb, "General Introduction to a Programme of Dynamic Research," *Proceedings of the Automobile Division, The Institution of Mechanical Engineers* (1956–57): 287–309, quotation from 291.

55. C. B. Millikan, "On the Results of Aerodynamic Research and Their Application to Aircraft Construction," *Journal of the Aeronautical Sciences* 4 (December 1936): 43–53, quotation from 46.

56. Edward Constant has suggested (personal correspondence) that the situation here might better be described in terms of Thomas P. Hughes's metaphor of a "reverse salient in an expanding technological front." ("The Science-Technology Interaction: The Case of High-Voltage Power Transmission Systems," *Technology and Culture* 17 [October 1976]: 646–59, quotation from 646; see also his *Networks of Power* [Baltimore, 1983], pp. 14–15, passim.) In this case, a rapid and continuing expansion in the elements of aeronautical practice affecting performance in the 1930s—streamlining, wing flaps, retractable landing gear, controllable-pitch propellers, etc.—left stability, control, and flying qualities behind as a reverse salient demanding attention in the overall front. In this view, flying qualities became identified as a critical and limiting problem *as a result of* this expansion. Without denying the continuing dynamic nature of the total process (especially performance technology), it seems to me that this interpretation runs an equal risk of misrepresentation. As we have seen, flying qualities had *already* become a recognized problem at the level of performance prevailing at the end of World War I. Whether the problem became significantly aggravated during the 1930s would be difficult to establish from the available evidence, particularly in view of the problem's subjective aspects.

57. H. A. Soulé, "Flight Measurements of the Dynamic Longitudinal Stability of Several Airplanes and a Correlation of the Measurements with Pilots' Observations of Handling Characteristics," *Report No. 578*, NACA (Washington, D.C., 1936), quotation from p. 5.

58. Metcalf (n. 48 above); R. T. Jones, "Letter to the Editor," *Journal of the Aeronautical Sciences* 4 (February 1937): 153; Soulé (n. 57 above), p. 5.

59. For the complicated circumstances surrounding the inception and fate

of the DC-4E, which had a number of innovative features including the first tricycle landing gear on a large airplane, see Anonymous, *Corporate and Legal History of United Air Lines and Its Predecessors and Subsidiaries, 1925–1945* (Chicago, 1953), pp. 360–63, 602–5, and R. J. Francillon, *McDonnell Douglas Aircraft since 1920* (London, 1979), pp. 277–80. For a semipopular account of the findings of the Hunsaker-Mead group, see J. C. Hunsaker and G. J. Mead, "Around the Corner in Aviation," *Technology Review* 39 (December 1936): 1–8. The airlines involved in the project in addition to United were Transcontinental & Western Air, American, Eastern, and Pan American. For a variety of reasons, the DC-4E did not prove to the liking of the airlines, and only the prototype was built. The airplane at the time was designated the DC-4; the E (for Experimental) was added after the fact to avoid confusion with the later (1942), somewhat smaller, and highly successful DC-4 (C-54 in World War II military designation).

60. *Equipment Agreement*, between Douglas Aircraft Co., Inc., and United Air Lines Transport Corp., Transcontinental & Western Air, Inc., American Airlines, Inc., North American Aviation, Inc., and Pan American Aviation Supply Corp., March 23, 1936, pp. 7–17. Although the agreement is anonymous, it is clear from subsequent revision of the section on flying qualities (n. 72 below) that that section was authored by Warner.

61. Flower et al. (n. 57 above), pp. 18–19.

62. *Agreement*, pp. 7–8; Soulé (n. 70 below), p. 1.

63. Millikan (n. 55 above), pp. 48–49.

64. Warner (1894–1958) is an important but historically neglected figure in American aviation. After the period discussed here, he was vice-chairman of the Civil Aeronautics Board and finally, from 1947 to 1957, first president of the International Civil Aviation Organization, the UN agency responsible for the establishment and regulation of international civil air transport. He also wrote several engineering textbooks and served as chairman or member of a number of other NACA committees. A full biography of Warner would constitute almost a history of American aviation over four formative decades. For brief accounts, see T. P. Wright, "Edward Pearson Warner—An Appreciation," *Journal of the Royal Aeronautical Society* 62 (October 1958): 691–703, and R. E. Bilstein, "Edward Pearson Warner," in *Dictionary of American Biography, Supplement Six, 1956–1960*, ed. J. A. Garraty (New York, 1980), pp. 665–67.

65. *Research Authorization No. 509*, approved by Committee on Aerodynamics, December 9, 1935, by Executive Committee, NACA, January 14, 1936. Although requests for research authorizations frequently originated from Langley itself, Hartley Soulé states that in this case "Langley was requested" to do the study "as a result of the pioneering work of Dr. E. P. Warner" on the DC-4E; H. Soulé, *Synopsis of the History of the Langley Research Center, 1915–1939*, unpublished ms. These and other primary documents cited in this chapter are in the archives of the Langley Research Center of the National Aeronautics and Space Administration. For "translator," see H. G. J. Aitken, *The Continuous Wave: Technology and American Radio, 1900–1932* (Princeton, 1985), pp. 16–17, 20–21, quotation from p. 17, and *Syntony and Spark—The Origins of Radio* (New York, 1976), pp. 332–33. The terms *linker* and *gate keeper* are also used by

people who study the management and the diffusion of innovation, respectively.

66. The three steps are set forth in the first report to come from the program (n. 68 below). Remarks in the memorandum of n. 67, however, indicate the plan was outlined initially in a Langley letter of May 4, 1936, which I am unable to locate.

67. Memo, F. L. Thompson to Engineer-in-Charge, July 14, 1936.

68. H. A. Soulé, "Measurements of the Flying Qualities of the Stinson Model SR-8E Airplane," *Confidential Memorandum Report*, NACA (Washington, D.C., September 9, 1937), declassified; see also Memo, H. A. Soulé to Engineer-in-Charge, February 24, 1937. NACA policy after the early 1930s provided for issue of reports (sometimes with government security classification) to a restricted audience, followed frequently by reissue at a later time of a publicly unrestricted (and sometimes variously modified) version of the same material. Reports were also often published initially in unrestricted form. This two-pronged policy, which was conditioned both by military secrecy requirements and by desire to give U.S. industry a head start over its foreign competitors in exploiting NACA research, makes it occasionally difficult to track the chronology of NACA work from the published record. For an outline of the complex, multitiered system of NACA reporting, see Roland (n. 8 above), vol. 2, pp. 551–54.

69. Soulé (n. 68 above), pp. 12, 43. Although their work was the most extensive and influential, the group at Langley was not alone in developing methods for flying-quality evaluation. Eddie Allen, who was at this time a prominent freelance test pilot, mentions (n. 39 above, p. 90) that he developed similar flight techniques when testing a Sikorsky four-engine flying boat built for the navy (not identified but apparently from the description the XPBS-1, first flown on August 13, 1937; see R. Wagner, *American Combat Planes*, 3rd ed. [Garden City, 1982], pp. 314–15). Despite considerable effort, I am unable to locate details of these tests.

70. Letter, C. J. McCarthy to G. W. Lewis, February 24, 1938; Memo, R. R. Gilruth to Chief, Aerodynamics Division, March 23, 1938; H. A. Soulé, "Preliminary Investigation of the Flying Qualities of Airplanes," *Report No. 700*, NACA (Washington, D.C., 1940).

71. Letter, Maj. F. O. Carroll to G. W. Lewis, March 24, 1937; Memo, F. L. Thompson to Engineer-in-Charge, July 14, 1937; R. R. Gilruth, "Measurements of the Flying Qualities of the Martin B-10B Airplane (A.C.R. 34–34)," *Confidential Memorandum Report for Army Air Corps*, NACA (Washington, D.C., January 11, 1938), declassified.

72. Letter, E. P. Warner to G. W. Lewis, July 26, 1938; Warner, "Further Notes on Flying-Quality Requirements for the DC-4," August 1938; S. J. Kline, "Innovation Is not a Linear Process," *Research Management* 28 (July–August 1985): 36–45, esp. 38–39. Warner also attempted to arrange for NACA participation in flight testing the DC-4E, but, for a variety of reasons, nothing came of these efforts; see, e.g., Letter, Warner to Lewis, December 28, 1937; Letter, H. J. E. Reid to Lewis, January 10, 1938; Letter, Warner to Lewis, January 24, 1938; Letter, A. E. Raymond to Lewis, February 16, 1938; Memo, Lewis to

Langley Laboratory, March 11, 1938; Letter, Lewis to Warner, August 3, 1938.

73. M. N. Gough and A. P. Beard, *Technical Note No. 550*, NACA (Washington, D.C., January 1936), quotations from p. 11; W. H. McAvoy, "Maximum Forces Applied by Pilots to Wheel-Type Controls," *Technical Note No. 623*, NACA (Washington, D.C., November 1937).

74. M. N. Gough, talk given to USS *Yorktown* squadrons VB-5 and VB-6, October 7, 1937, unpublished; Letter, L. V. Kerber to G. W. Lewis, March 18, 1938; Memo, H. J. E. Reid to NACA, March 26, 1938; Memos, H. A. Soulé to Engineer-in-Charge, April 2, 1938, and F. L. Thompson to Engineer-in-Charge, April 7, 1938.

75. Following its practice in this period (n. 50 above), NACA did not publish details of these instruments. Sometime in the period 1943 to 1946, a comprehensive set of descriptions was compiled anonymously at Langley Field under the title *Manual of NACA Flight-Test Instruments*. A copy of this manual, which was for use by laboratory staff only, is in the private papers of Isidore Warshawsky, an instrumentation engineer who was one of the contributors. I am indebted to Mr. Warshawsky for this information.

76. "Flight Investigation of Control and Handling Characteristics of Various Airplanes," *Research Authorization No. 608*, approved by Committee on Aerodynamics, May 23, 1938, by Executive Committee, NACA, June 21, 1938. This gave general authorization; specific aircraft were tested under different, individual authorizations.

77. The question in this paragraph was raised for me by Edward Constant.

78. Interviews of White by the author, Palo Alto, Calif., October 16, 1985, May 2, September 15, 1986.

79. Interview of Robert R. Gilruth by James R. Hansen, Kilmarnock, Va., July 10, 1986.

80. M. N. Gough and R. R. Gilruth, "Measurements of the Flying Qualities of the Lockheed 14-H Airplane (Navy Designation XR40-1, Airplane No. 1441)," *Confidential Memorandum Report* (Washington, D.C., April 22, 1939), declassified, quotations from pp. 8, 9, 12; Gough, "Notes on Stability from the Pilot's Standpoint" (paper presented at Air Transport Meeting of Institute of Aeronautical Sciences, Chicago, November 19, 1938), *Journal of the Aeronautical Sciences* 6 (August 1939): 395–98, and "Safety in Flight-Control and Maneuverability," unpublished paper presented to National Safety Council, Aeronautical Section, New York City, March 29, 1939. Testing of the Lockheed airplane took place between presentations of Gough's two papers, which probably explains why he neglected the short-period oscillation in the first but described it as potentially "very annoying" and even "unsafe" (p. 7) in the second. This difference suggests it was the Lockheed that drove home the lesson about the short-period mode. Other contemporary remarks from a pilot are contained in B. O. Howard, "Desirable Qualities of Transport Airplanes from the Pilot's Point of View," *Journal of the Aeronautical Sciences* 6 (November 1938): 15–19.

81. "Requirements for Satisfactory Flying Qualities of Airplanes," *Advance Confidential Report* (unnumbered), NACA (Washington, D.C., April 1941), de-

classified, and *Report No. 755*, NACA (Washington, D.C., 1943); Gilruth interview (n. 79 above).

82. Such longitudinal trim is attained by angular adjustment either of the horizontal stabilizer (the nominally "fixed" portion of the horizontal tail) or of a trim tab (a narrow adjustable surface at the trailing edge of the elevator).

83. R. R. Gilruth and M. D. White, "Analysis and Prediction of Longitudinal Stability of Airplanes," *Report No. 711*, NACA (Washington, D.C., 1941); White interviews (n. 78 above).

84. *Report No. 755* (n. 81 above), p. 2.

85. R. T. Jones and D. Cohen, "An Analysis of the Stability of an Airplane with Free Controls," *Report No. 709*, NACA (Washington, D.C., 1941). The theory was validated by specially designed flight tests in W. H. Phillips, "A Flight Investigation of Short-Period Longitudinal Oscillations of an Airplane with Free Elevator," *Wartime Report L-444*, originally *Advance Restricted Report* (unnumbered), NACA (Washington, D.C., May 1942). At least one other investigator downgraded the long-period mode in the late 1930s. Otto C. Koppen, a recognized authority on stability and control at MIT, in 1936 followed the received tradition by recommending requirements on the period and damping of the long-period oscillation on airplanes for amateur flyers ("Smart Airplanes for Dumb Pilots," paper presented at Annual Meeting, Society of Automotive Engineers, Detroit, January 13–17, 1936). By contrast, four years later, as a result of unspecified "experience" and analysis of the pilot-airplane system on an analogue computer, he wrote in his usual vivid style that "a lack of damping of the [long-period] longitudinal oscillation is usually unnoticed by the pilot, and the result is that any pilot can satisfactorily fly anything that looks like an airplane" ("Airplane Stability and Control from a Designer's Point of View," *Journal of the Aeronautical Sciences* 7 [February 1940]: 135–40, quotation from 138). Koppen, however, made no statement about the short-period oscillation. Information on the subsequent imposition of requirements on damping of the long-period mode comes from Waldemar O. Breuhaus, for many years an authority on flying qualities at the Cornell Aeronautical Laboratory (now Arvin/Calspan) in Buffalo.

86. Perkins (n. 7 above, p. 297) credits Gates with the criterion. Gilruth (personal correspondence) says he is sure they both arrived at it independently. Gates's publication ("Proposal for an Elevator Manoeuvrability Criterion," *Reports and Memoranda No. 2677*, Aeronautical Research Committee [London, June 1942]) cited Gilruth's classified report of a year earlier (which incidentally raises interesting but unresolvable questions about practices concerning classification) but in a regard that gives no indication of where the idea originated. Gates presumably could have arrived at it on his own sometime earlier.

87. Ibid., p. 3.

88. The product of the acceleration of gravity and the mass of a body equals the weight of the body in a gravitational field. A multiple of g is thus a convenient measure of the apparent (or virtual) increase in weight of the aircraft due to acceleration and hence also of the required increase in lift.

89. This was one of only two places in which the requirements differed from

one type of airplane to another. In the other, the aileron roll-rate criterion, statement of the requirement could still be made formally universal, since wing span, the relevant difference between airplanes, could be subsumed into a dimensionless parameter. The validity of this result was based on correlations from twenty different airplanes.

90. *Report No. 755* (n. 81 above), p. 1.

91. Memos (n. 53 above); interview of Robert R. Gilruth by Michael D. Keller at NASA, Manned Spacecraft Center, Houston, Tex., June 26, 1967; I. L. Ashkenas, "Twenty-Five Years of Handling Qualities Research," *Journal of Aircraft* 21 (May 1984): 289–301; quotation from 290. A seeming inconsistency arises here in that the B-25 was designed in 1939–40 (R. Wagner [n. 69 above], p. 219), prior to Gilruth's classified report of 1941. The aileron roll-rate criterion that Ashkenas presumably used (n. 89 above) appeared also, however, in an NACA report published in generally available form in 1941; R. R. Gilruth and W. N. Turner, "Lateral Control Required for Satisfactory Flying Qualities Based on Flight Tests of Numerous Airplanes," *Report No. 715*, NACA (Washington, D.C., 1941); consistent with NACA policy, this report would presumably have been available to Ashkenas earlier in restricted form. The Gilruth-Turner report, incidentally, affords an excellent example of how the flying-quality requirements evolved. Roll rate was a particularly troublesome characteristic, and twenty-eight airplanes were tested to arrive at the published criterion. For Breuhaus, see n. 85 above.

92. "Specification for Stability and Control Characteristics of Airplanes," *SR-119*, Bureau of Aeronautics, Navy Department (Washington, D.C., October 1, 1942); "Stability and Control Requirements for Airplanes," *Specification No. C-1815*, Army Air Forces (Dayton, Ohio, August 31, 1943).

93. Civil Aeronautics Authority, "Amendment 56, Civil Air Regulations," *Federal Register*, June 1, 1940, pp. 2100–3, and Civil Aeronautics Board, "Amendment 04–0, Airplane Airworthiness, Transport Categories," ibid., January 3, 1946, pp. 71–102, esp. 74–76. For another example of airline specifications, see Flower et al. (n. 51 above), pp. 19–21. The difference between military and civil specifications was pointed out to me by Richard S. Shevell, one-time director of advanced design at Douglas.

94. Reported by Richard V. Rhode, Langley flight-research engineer, on witnessing a flight test of the airplane on July 1, 1938; Memo to Engineer-in-Charge, July 7, 1938.

95. The number of NACA tests comes from an important summary report by W. H. Phillips, "Appreciation and Prediction of Flying Qualities," *Report No. 927*, NACA (Washington, D.C., 1949), p. 1. For engineering surveys of developments following the period of the present story, see Harper and Cooper (n. 7 above), pp. 520–23, and W. H. Phillips, "Flying Qualities from Early Airplanes to the Space Shuttle," *Journal of Guidance, Control, and Dynamics* 12 (July–August 1989): 449–59. For the variable-stability airplane, see W. O. Breuhaus, "The Variable Stability Airplane, from a Historical Perspective" (forthcoming). For representative military specifications, see "Specifications for Flying Qualities of Piloted Airplanes," *NAVAER SR-119B*, Bureau of Aeronautics, Navy Department (Washington, D.C., June 1, 1948), and "Military

Specification—Flying Qualities of Piloted Airplanes," *MIL-F-8785C*, Department of Defense (Washington, D.C., November 5, 1980); see also D. J. Moorhouse, "The History and Future of U.S. Military Flying Qualities Specifications," *Paper No. 79-0402*, presented at 17th Aerospace Sciences Meeting of American Institute of Aeronautics and Astronautics, New Orleans, January 15–17, 1979.

96. *Equipment Agreement* (n. 60 above), pp. 4–7. Delineation of Warner's role as mediator comes from personal correspondence with Warren T. Dickinson, flight-test engineer on the DC-4E.

97. Letter, W. A. Patterson to executive heads of the four other airlines, quoted in *Corporate and Legal History of United Air Lines* (n. 59 above), p. 360.

98. This and the next four paragraphs have profited especially from discussion with Ilan Kroo.

99. From a letter written some years ago by W. O. Breuhaus (n. 85 above) to a peer commenting on proposed changes in military specifications for flying qualities (personal correspondence with Mr. Breuhaus).

100. Ibid. This problem was brought to my attention by Mr. Breuhaus. See also Moorhouse (n. 95 above), p. 4.

101. Milliken and Whitcomb (n. 54 above), p. 292.

102. Schön (n. 2 above), p. 172.

103. As far as I am aware, historical study of the automotive case has yet to be made. The engineering situation as of 1956 was reviewed in five articles summarizing a research program at the Cornell Aeronautical Laboratory, Inc., and introduced by Milliken and Whitcomb (n. 54 above), see esp. pp. 299–301. For telemanipulators, see J. Vertut and P. Coiffet, *Teleoperation and Robotics: Evolution and Development* (London, 1986), pp. 44–45. The information about pianos comes from discussion with Edwin M. Good; see also his *Giraffes, Black Dragons, and Other Pianos: A Technological History from Cristofori to the Modern Concert Grand* (Stanford, 1982), pp. 56–58. For human-factors engineering, see G. Salvendy, ed., *Handbook of Human Factors* (New York, 1987).

104. Constant (n. 3 above), quotations from pp. 8 and 9; the quotation on test pilots and engineers is from Perkins (n. 7 above), p. 297.

105. Constant (n. 3 above), p. 10. See also R. Laudan, "Cognitive Change in Technology and Science," in *The Nature of Technological Knowledge: Are Models of Scientific Change Relevant?*, ed. R. Laudan (Dordrecht, 1984), pp. 83–104, esp. pp. 93–95 and J. M. Staudenmaier, S.J., *Technology's Storytellers: Reweaving the Human Fabric* (Cambridge, Mass., 1985), pp. 64–69.

106. Apropos of comparative testing of airplanes, Robin Higham has written, "It is hard to think of another field, except perhaps the auto industry, in which the fate of a new design has been so determined by the skill, observation, and opinions of the testing staff." In review of Hallion (n. 27 above), *Technology and Culture* 24 (January 1983): 146–47, quotation from 147.

107. Constant (n. 3 above), pp. 20–24, 41–51, quotation from pp. 20–21. Constant (p. 22) also mentions the second type in passing.

108. The judgment described here is how things stood at the end of our period and for much of the time since. It has become modified as engineers have tried to determine how much stability constitutes "too much." Current

understanding is that pilots really want substantial stability but without excessive stick motion or forces and that the two requirements are not (as earlier remarks might suggest) necessarily mutually exclusive.

Chapter 4. A Theoretical Tool for Design

Note: This chapter appeared, with slight differences, as "Control-Volume Analysis: A Difference in Thinking between Engineering and Physics," *Technology and Culture* 23 (April 1982): 145–74. I am especially indebted to Robert Dean, Paul Hanle, Joseph Keller, and Ascher Shapiro. The idea for the topic came from discussions with my Stanford engineering colleague Stephen Kline.

1. The same difference holds, with minor qualifications, if the literature of chemistry is examined in lieu of that of physics. For simplicity, however, I limit my discussion to physics. For examples of the differences described, for engineering texts, see R. E. Sonntag and G. J. Van Wylen, *Introduction to Thermodynamics: Classical and Statistical* (New York, 1971); D. B. Spalding and E. H. Cole, *Engineering Thermodynamics*, 3d ed. (London, 1973); W. C. Reynolds and H. C. Perkins, *Engineering Thermodynamics*, 2d ed. (New York, 1977); K. Brenkert, Jr., *Elementary Theoretical Fluid Mechanics* (New York, 1960); J. K. Vennard and R. L. Street, *Elementary Fluid Mechanics*, 5th ed. (New York, 1975); R. W. Fox and A. T. McDonald, *Introduction to Fluid Mechanics*, 2d ed. (New York, 1978); R. K. Pefley and R. I. Murray, *Thermofluid Mechanics* (New York, 1966). For physics texts, see M. W. Zemansky, *Heat and Thermodynamics*, 5th ed. (New York, 1968); C. J. Adkins, *Equilibrium Thermodynamics* (London, 1968); D. Elwell and A. J. Pointon, *Classical Thermodynamics* (Harmondsworth, 1972); L. D. Landau and E. M. Lifshitz, *Fluid Mechanics*, trans. J. B. Sykes and W. H. Reid (London, 1959); D. J. Tritton, *Physical Fluid Dynamics* (New York, 1977); L. C. Woods, *The Thermodynamics of Fluid Systems* (Oxford, 1975).

2. Reynolds and Perkins (n. 1 above), p. 486. This reference also gives an exemplary step-by-step procedure for applying control-volume analysis and a handy tabulation of the control-volume equations. *Control surface*, the accepted term in English, is a literal translation of the original German word *Kontrollfläche*. *Kontroll* in German, however, has not only the usual English meaning of "regulation" but also that of "auditing," as in bookkeeping. For obvious reasons, writers in English often find it natural, as I have above, to describe control-volume analysis as a means "for doing the bookkeeping" in problems of fluid flow. It seems likely that Ludwig Prandtl had this meaning in mind when he chose the German word (see n. 15 below).

3. Use of a control volume purely for derivation of a general fluid-flow equation does not constitute control-volume *analysis* in the engineering sense described here. A prominent example is the control volume of imagined infinitesimal size used as standard means for deriving the differential equations of fluid motion in many textbooks of fluid mechanics for both physicists and engineers. A few texts use a finite-sized control volume as one step in an alternative form of derivation. In either case, once the derivation is accomplished the control volume is forgotten, and the differential equations are used

by themselves for solution of flow problems. In control-volume analysis, by contrast, the control volume (usually of finite size) is set up appropriately and anew as an essential part of each solution, and the control-volume equations (usually integral rather than differential) are applied to this special volume. In control-volume analysis the control volume is thus used as a direct rather than indirect tool for problem solving.

4. Most writers have used the term *system* or *fixed mass*. *Control mass*, which was introduced relatively recently in the engineering text by W. C. Reynolds, *Thermodynamics* (New York, 1965), seems preferable, however, as more consistent and more descriptive. Physicists usually introduce the concept explicitly under the term *system*, though they sometimes leave it implicit; in contrast to engineers, they usually do not delineate it by a diagram.

5. From a broader point of view, the control volume and control mass are both special cases of a general, moving control region. Such a region moves in an arbitrarily specified way, so that it may be stationary and be a control volume, or it may move with a fluid and contain a control mass. The physical equations pertinent to such a region have appeared at least once that I know of in the specialized research literature in physics in a paper written by an applied mathematician (J. B. Keller, "Geometrical Acoustics I. The Theory of Weak Shock Waves," *Journal of Applied Physics* 25 [1954]: 938–47). So far, however, they are not being taught or used as a standard basis for problem solving in either physics or engineering. If they were to be, the situation I am describing might become modified.

6. A. H. Shapiro, *The Dynamics and Thermodynamics of Compressible Fluid Flow*, 2 vols. (New York, 1953), vol. 1, p. 12.

7. L. Prandtl and O. G. Tietjens, *Fundamentals of Hydro- and Aeromechanics*, trans. L. Rosenhead (New York, 1934), p. 233.

8. C. Truesdell, *Essays in the History of Mechanics* (Berlin, 1968), p. 193. Truesdell also says of the cut principle that "the concept it expresses was groped for and glimpsed and fumbled and rightly applied long before its distillation into clear words." This statement aptly describes the history of many other concepts, including control-volume analysis.

9. D. Bernoulli, *Hydrodynamics*, trans. T. Carmody and H. Kobus (New York, 1968), pp. 327–28.

10. L. Euler, "Théorie plus complète des machines qui sont mises en mouvement par la réaction de l'eau," in *Euleri Opera omnia*, ed. J. Ackeret, ser. 2 (Lausanne, 1957), vol. 15, pp. 157–218; see also editor's preface, pp. xlii–xlvi; J. C. de Borda, "Mémoire sur l'écoulement des fluides par les orifices des vases," *Histoire de l'Academie Royale des Sciences 1766* (Paris, 1769), pp. 579–607; R. E. Froude, "On the Part Played in Propulsion by Differences of Fluid Pressure," *Transactions of the Institution of Naval Architects* 30 (1889): 390–405.

11. H. T. Bovey, *A Treatise on Hydraulics* (New York, 1895), pp. 186–208, 283–98; F. C. Lea, *Hydraulics*, 2d ed. (London, 1911), pp. 39–41, 67–69, 72–73, 166–68, 273–75, 277–78. These books were both for practical engineers. It is significant that in the same period the classic text by Horace Lamb (*Hydrodynamics*, 3d ed. [Cambridge, 1960]), which was written primarily for people with a mathematical interest in fluid mechanics, made no use of control-volume

ideas. Something akin to the dichotomy we observe today between engineers and physicists thus appears to have existed within fluid mechanics for some time.

12. Bovey, *Treatise*, pp. 6–8, 32–45.

13. T. von Kármán, *Aerodynamics* (Ithaca, N.Y., 1954), p. 50.

14. T. von Kármán, "Über den Mechanismus des Widerstandes, den ein Bewegter Körper in einer Flüssigkeit Erfährt," *Nachrichten der K. Gesellschaft der Wissenschaften zu Göttingen* (1912), pp. 547–56; Kármán and H. Rubach, "Über den Mechanismus des Flüssigkeits- und Luftwiderstandes," *Physikalische Zeitschrift* 13 (1912): 49–59. Both articles reprinted in *Collected Works of Theodore von Kármán*, 4 vols. (London, 1956), vol. 1, pp. 331–58.

15. L. Prandtl, "Flüssigkeitsbewegung," *Handwörterbuch der Naturwissenschaften*, 10 vols. (Jena, 1913), vol. 4, pp. 112–14; reprinted in *Ludwig Prandtl Gesammelte Abhandlungen*, ed. W. Tolmien, H. Schlichting, and H. Görtler, 3 vols. (Berlin, 1961), vol. 3, pp. 1438–41. The quotation is translated from p. 1439 of the latter. For Prandtl's possible reason for the term *Kontrollfläche*, see n. 2 above. The most notable reproduction of the material, with certain extensions, was in the book *Führer durch die Strömungslehre* (Braunschweig, 1942), pp. 72–80, 201–4. The 3d ed. of this book (1949) was translated as *Essentials of Fluid Dynamics* (New York, 1952). The author's preface to this translation gave a brief history of the various versions to that time.

16. L. Prandtl, "Allgemeiner Nachweis der Grundgleichung des Wirkungsgrades," *Technische Berichte von der Flugzeugmeisterei der Inspektion der Fliegertruppen*, 3 vols. (Charlottenburg, 1917), vol. 2, pp. 78–80; reprinted in *Prandtl Abhandlungen*, vol. 1, pp. 302–4, and reproduced in English in *Translated Abstracts of Technische Berichte, 1917*, 2 vols. (London, 1925), vol. 2, pp. 193–95. "Bemerkung zu dem Aufsatz von D. Thoma," *Zeitschrift für Flugentechnik und Motorluftschiffahrt* 16 (1925): 208–9; reprinted in *Prandtl Abhandlungen*, vol. 2, pp. 617–19. L. Prandtl and O. G. Tietjens, *Hydro- und Aeromechanik*, 2 vols. (Berlin, 1929, 1931); in English as *Fundamentals of Hydro- and Aeromechanics*, trans. L. Rosenhead, and *Applied Hydro- and Aeromechanics*, trans. J. P. Den Hartog (New York, 1934); see esp. *Fundamentals*, pp. 233–50, and *Applied*, pp. 118–36.

17. A. Föppl, *Vorlesungen über technische Mechanik* (Leipzig, 1897–1910), see esp. vols. 1 and 6. For other German works, see, e.g., Ph. Forchheimer, "Hydraulik," *Encyklopädie der mathematischen Wissenschaften* (Leipzig, 1901–8), bd. 4, t. 3, pp. 327–472; H. Lorenz, *Technische Hydromechanik* (Munich, 1910); F. Prásil, *Technische Hydrodynamik*, 2d ed. (Berlin, 1926); W. Kaufmann, *Angewandte Hydromechanik*, 2 vols. (Berlin, 1931, 1934).

18. H. Rouse and S. Ince, *History of Hydraulics* (Ames, Iowa, 1957), p. 231.

19. In my statements about thermodynamics throughout this chapter I have in mind classical (i.e., continuum) thermodynamics. Paul Hanle has pointed out to me that something like control-volume ideas did appear just before World War I in a field of physics closely related to statistical (i.e., particle) thermodynamics. In his important analysis of density fluctuations due to Brownian motion of colloidal particles, the Polish physicist Marian von Smoluchowski introduced a kind of control volume and examined the flow of parti-

cles in and out of the volume and its effect on the density of particles within. Although the particle motion was probabilistic rather than directed, the nature and use of the volume were akin to those described here. Smoluchowski's application, however, was special and isolated even within physics, and I find no evidence that it played a role in our story (M. von Smoluchowski, "Studien über Molekularstatistik von Emulsionen und deren Zusammenhang mit der Brown'schen Bewegung," *Sitzungsberichte der Akademie der Wissenschaften, Wien—Mathematisch-naturwissenschaftlichen Klasse* 123, ser. 2a [1914]: 2381–2405, esp. 2386–99). For a clear exposition of Smoluchowski's theory, see S. Chandrasekhar, "Stochastic Problems in Physics and Astronomy," *Reviews of Modern Physics* 15 (1943): 1–89, esp. 44–46.

20. M. Schröter and L. Prandtl, "Technische Thermodynamik," *Encyklopädie der mathematischen Wissenschaften* (Leipzig, 1903–21), bd. 5, t. 1, pp. 243–319; the third section ("Strömende Bewegung der Gase und Dämpfe," pp. 287–319) is by Prandtl; A. Stodola, *Die Dampfturbinen und die Aussichten der Wärmekraftmaschinen* (Berlin, 1903), and *Dampf- und Gas-Turbinen*, 6th ed. (Berlin, 1924), in English as *Steam and Gas Turbines*, trans. L. C. Loewenstein (New York, 1927); P. J. Kiefer and M. C. Stuart, *Principles of Engineering Thermodynamics* (New York, 1930), pp. 44–57; J. H. Keenan, *Thermodynamics* (New York, 1941), pp. 32–35; L. J. Gillespie and J. R. Coe, Jr., "The Heat of Expansion of a Gas of Varying Mass," *Journal of Chemical Physics* 1 (1933): 103–13.

21. Evidence of the development in Germany appears in a book by Klaus Oswatitsch (*Gasdynamik* [Wien, 1952]; in English as *Gas Dynamics*, trans. G. Kuerti [New York, 1956]). Chapters 4 and 5 contain a careful and complete exposition of control-volume analysis for mass, momentum, and energy, including numerous applications. Oswatitsch is an Austrian who was a lecturer at the Royal Institute of Technology in Stockholm at the time but who had been on the staff at Göttingen during Prandtl's tenure. This extension apparently took place independently of that in the United States, and I have made no attempt to trace it.

22. J. C. Hunsaker and B. G. Rightmire, *Engineering Applications of Fluid Mechanics* (New York, 1947), chaps. 3, 5, 6; Shapiro (n. 6 above), vol. 1, chaps. 1, 2. Shapiro also appears to have been the first (at least in the United States) to include clearly and systematically in the control-volume equations the terms required to analyze unsteady flow (i.e., flow whose pattern changes with time). Hunsaker and Rightmire made casual and unsystematic allusion to such terms in their derivations, but explicitly restricted their final equations to steady flow. Except for a single, ad hoc treatment of energy in an unsteady flow, Prandtl and Tietjens gave attention only to steady flow.

23. D. A. Mooney, *Mechanical Engineering Thermodynamics* (Englewood Cliffs, N.J., 1953), chap. 6.

24. M. P. O'Brien and G. H. Hickox, *Applied Fluid Mechanics* (New York, 1937), chap. 3. (Even such a popular and influential book as R. L. Daugherty, *Hydraulics*, 4th ed. [New York, 1937] made no consistent use of the control volume.) For Mises's lectures, see R. von Mises and K. O. Friedrichs, *Fluid Dynamics* (Providence, R.I., 1942), pp. 23–26. The Mises-Prandtl correspon-

dence is preserved in part in the Mises Papers at Harvard. Shapiro quotations are from personal correspondence with the writer.

25. Shapiro (personal correspondence) states that for his graduate course in compressible flow, in which "I was stressing heavily the value of control-volume analysis," the roll card for the fall term 1948 shows the names of nine people who went on to become teachers (including G. J. Van Wylen; see n. 1 above). About this time he prepared for the course a chart summarizing in systematic form the equations for the control volume and the control mass, and "for many years I was receiving requests for copies of [this chart] together with permission to reproduce it so that it could be used in other schools." Fred Landis, Robert Eustis, and Stephen Kline, all of whom took their doctorate in mechanical engineering at MIT, brought the control-volume approach to Stanford, where it was emphasized in teaching by them and by A. L. London. Writers who obtained graduate degrees at Stanford and have produced texts incorporating control-volume analysis include (n. 1 above) W. C. Reynolds, H. C. Perkins, K. Brenkert, Jr., R. L. Street, R. W. Fox, R. K. Pefley, and R. J. Murray.

26. Control-volume analysis applied to the momentum equation has also been emphasized in at least one text intended primarily for applied mathematicians: G. K. Batchelor, *An Introduction to Fluid Mechanics* (Cambridge, 1967), pp. 138, 372–76, 386–98. Batchelor, professor of applied mathematics at Cambridge University, maintains contact with mathematically minded engineers and states (personal correspondence) that he acquired the control-volume method from the engineering literature. This evidence suggests that groups outside engineering may gradually pick up on the advantages of control-volume thinking (see also n. 5 above).

27. See, e.g., J. P. Johnston and R. C. Dean, Jr., "Losses in Vaneless Diffusers of Centrifugal Compressors and Pumps," *Journal of Engineering for Power* 88 (1966): 49–62, and C. T. Crowe, M. P. Sharma, and D. E. Stock, "The Particle-Source-in-Cell (PSI-CELL) Model for Gas Droplet Flows," *Journal of Fluids Engineering* 99I (1977): 325–32. The second example illustrates the growing utility of control-volume analysis to formulate computational programs for solving flow problems with electronic computers.

28. *Power Test Code PTC 22-1966, Gas Turbine Power Plants* (New York, 1966), p. 33.

29. Spalding and Cole (n. 1 above), p. 167.

30. M. W. Zemansky, *Heat and Thermodynamics*, 2d ed. (New York, 1943), pp. 214–16; M. W. Zemansky and H. C. Van Ness, *Basic Engineering Thermodynamics* (New York, 1966), chap. 13; M. W. Zemansky, *Heat and Thermodynamics*, 5th ed. (New York, 1968).

31. The chapter titles are "Pure Substances"; "Phase Transitions—Liquid and Solid Helium"; "Paramagnetism, Cryogenics, Negative Temperatures, and the Third Law"; "Superfluidity and Superconductivity"; "Chemical Equilibrium."

32. In situations where overall results are obtained despite a lack of understanding of the underlying physics, control-volume analysis, like the method of experimental parameter variation examined in chapter 5, enables the engineer to circumvent a lack of science. It can thus provide as before a means, though

here a theoretical rather than an experimental one, to get on with the technological job in the face of incomplete scientific understanding.

33. As pointed out by Arnold Pacey with reference to the newly founded engineering schools in France in the 1700s, "the need to rationalize the procedures involved in designing machines became more urgent when this aspect of engineering had to be taught to students" (*The Maze of Ingenuity* [London, 1974], p. 223). That people like Bovey and Lea, though also teachers, did not move to control-volume analysis before Prandtl suggests that the time was not yet ripe—the mass of accumulated problems was still too small and the exacting theoretical demands from aeronautics had not appeared.

34. *Statistical Abstract of the United States 1977* (Washington, D.C., 1977), p. 161.

35. In view of the negligible use of control-volume analysis in physics, physics teachers (even if they know of it) probably do not think it worthwhile to spend the time required to teach it as a technique.

36. Robert Dean, in personal correspondence regarding an early draft of this chapter. The ideas in this paragraph have benefited generally from Dr. Dean's comments.

37. S. C. Florman, *The Existential Pleasures of Engineering* (New York, 1976), p. 33.

38. For other examples, see discussion of "propulsive efficiency" in chapter 5 and N. Rosenberg and W. G. Vincenti, *The Britannia Bridge: The Generation and Diffusion of Technological Knowledge* (Cambridge, Mass., 1978), pp. 38–39.

39. As a tool that is not device specific, control-volume analysis resembles the methodology of parameter variation and model testing described in chapter 5.

40. E. T. Layton, Jr., "Mirror-Image Twins: The Communities of Science and Technology in 19th-Century America," *Technology and Culture* 12 (October 1971): 562–80, and "Scientific Technology, 1845–1900: The Hydraulic Turbine and the Origins of American Industrial Research," *Technology and Culture* 20 (January 1979): 64–89, esp. 88–89.

41. F. Rapp, "Technology and Natural Science—a Methodological Investigation," in *Contributions to a Philosophy of Technology*, ed. F. Rapp (Boston, 1974), pp. 93–114, esp. pp. 93–97.

42. Layton, "Mirror-Image Twins" (n. 40 above). The diffusion of knowledge in technology generally is, of course, a much more complex topic even in modern times; see E. T. Layton, Jr., "Technology as Cumulative Knowledge" (paper presented at the Conference on Critical Issues in the History of Technology, Roanoke, Virginia, August 14–18, 1978). (See also C. W. Pursell, Jr., "The Roanoke Conference, I. Summary," *Technology and Culture* 21 [October 1980]: 617–20.)

43. Layton, "Technology as Cumulative Knowledge," contended more broadly that technology considered generally is cumulative (in some unspecified sense) and that this cumulative property can best be explained in terms of technological knowledge.

44. Engineering science thus implies a knowledge-producing activity embedded within a larger problem-solving activity.

45. See also E. T. Layton, Jr., "American Ideologies of Science and Engineer-

ing, *Technology and Culture* 17 (October 1976): 688–701, quoted phrase from 695.

46. H. Skolimowski, "The Structure of Thinking in Technology," *Technology and Culture* 7 (Summer 1966): 371–83, quotation from 376. Skolimowski's meaning for "effectiveness," however, is broader than that used here, encompassing not only the design process but the characteristics of the artifact as well.

47. Layton's paper at the Roanoke conference (see n. 42 above) was followed by animated discussion of the sense, if any, in which the development of technology considered broadly constitutes a cumulative process; for example, in what sense is it valid to say, as Layton did, "that an average engineer today can design a better machine than could da Vinci"? The notion of effectiveness in relation to technological cumulation was brought to my attention by Nathan Sivin in his comments on this discussion; N. Sivin, "The Roanoke Conference, II. Concluding Remarks on Conference," *Technology and Culture* 21 (October 1980): 621–32.

48. E. T. Layton, Jr., review of *Philosophers and Machines*, ed. O. Mayr, *Technology and Culture* 18 (January 1977): 89–91, quotation from 89.

Chapter 5. Data for Design

Note: This chapter appeared originally in essentially the same form under the title "The Air-Propeller Tests of W. F. Durand and E. P. Lesley: A Case Study in Technological Methodology," *Technology and Culture* 20 (October 1979): 712–51. I thank all the many people—too many to list—who helped in one way or another. I am especially indebted, however, to Walter Bonney, Edwin Good, Stephen Kline, Edwin Layton, Otto Mayr, Robert McGinn, Russell Robinson, Howard Rosen, and Nathan Rosenberg.

1. M. M. Munk, "Analysis of W. F. Durand's and E. P. Lesley's Propeller Tests," *Report No. 175*, NACA (Washington, D.C., 1923), p. 3.

2. For examples by philosophers, see F. Rapp, "Technology and Natural Science—A Methodological Investigation," in *Contributions to a Philosophy of Technology* (Boston, 1974), pp. 93–114; and H. Skolimowski, "The Structure of Thinking in Technology," ibid., pp. 72–85, also in *Technology and Culture* 7 (Summer 1966): 371–83.

3. The term *technological convergence* was introduced by Nathan Rosenberg to describe the application of common productive processes in different industrial technologies, but it seems useful also for knowledge and methodology ("Technological Change in the Machine Tool Industry, 1840–1910," *Journal of Economic History* 23 [1963]: 414–46). For the role of incremental change, see A. P. Usher, "Technical Change and Capital Formation," reprinted in *The Economics of Technological Change*, ed. N. Rosenberg (Harmondsworth, 1971), p. 63. See also G. H. Daniels, "The Big Questions in the History of American Technology," *Technology and Culture* 11 (January 1970): 1–21.

4. J. Smeaton, "An Experimental Enquiry Concerning the Natural Powers of Water and Wind to Turn Mills, and Other Machines, Depending on a Circular Motion," *Philosophical Transactions of the Royal Society* 51, pt. 1 (1759): 100–74.

For discussion and assessment of Smeaton's work, see D. S. L. Cardwell, *Turning Points in Western Technology* (New York, 1972), pp. 79–84; A. Pacey, *The Maze of Ingenuity* (London, 1974), pp. 205–15; and N. Smith, *Man and Water* (London, 1975), pp. 153–58. For an illuminating discussion of Smeaton's waterwheel experiments (including matters referred to later in nn. 6, 102, and 104), as well as of the more specialized and less well known tests of his contemporary, Antoine de Parcieux, see T. S. Reynolds, "Scientific Influences on Technology: The Case of the Overshot Waterwheel, 1752–1754," *Technology and Culture* 20 (April 1979): 270–95.

5. This terminology is my own. The method is so much taken for granted by engineers that they rarely call it by name. In the theory of the statistical design of experiments that has matured since World War II, the procedure in the elementary form described here is called "factorial experiment." For general purposes, however, parameter variation seems more descriptive and closer to other engineering terminology.

6. B. C. Hacker, "Greek Catapults and Catapult Technology: Science, Technology, and War in the Ancient World," *Technology and Culture* 9 (January 1968): 34–50, quotation from 49. In science Newton used parameter variation in his experiments in optics (Cardwell [n. 4 above], p. 50). Although they disagree about the importance of these experiments in influencing Smeaton (Pacey [n. 4 above], p. 209), Pacey and Cardwell contend that Smeaton's thinking derived in one way or another from Newton's scientific work. Examination needs to be made, however, of possible methodological influences from technology, as, for example, from experiments on the range of artillery fire by Bernard Forêt de Belidor, *Le Bombardier françois* (Paris, 1731) and Benjamin Robins, *New Principles of Gunnery* (London, 1742), or from prior eighteenth-century work on waterwheels (see n. 8 below).

7. N. Rosenberg and W. G. Vincenti, *The Britannia Bridge: The Generation and Diffusion of Technological Knowledge* (Cambridge, Mass., 1978), pp. 14–23, 29; F. W. Taylor, *On the Art of Cutting Metals* (New York, 1906); see also F. B. Copley, *Frederick W. Taylor*, 2 vols. (New York, 1923), vol. 1, pp. xiv–xv, 245, 434–35, passim; M. W. McFarland, ed., *The Papers of Wilbur and Orville Wright*, 2 vols. (New York, 1953), vol. 1, pp. 547–93.

8. A scale (or geometrically similar) model is a model in which all dimensions of the prototype have been reduced in the same proportion. *Working* scale models are made to function the same as the prototype. They are thus distinct from static scale models used, for example, by architects, simply to visualize spatial relationships. Smeaton is sometimes credited with being the first to use scale models for quantitative engineering experiments (see, e.g., A. F. Burstall, *A History of Mechanical Engineering* [Cambridge, Mass., 1965], p. 244). Norman Smith, however, has called attention to the less well known tests of model waterwheels by Christopher Polhem in Sweden in the early 1700s (N. Smith, "The Origins of the Water Turbine and the Invention of Its Name," in *History of Technology, Second Annual Volume, 1977*, ed. A. R. Hall and N. Smith [London, 1977], pp. 221, 256). For historical discussion of static and working scale models in past centuries, see B. Gille, ed., *The History of Techniques*, 2 vols. (Montreux, 1986), vol. 2, pp. 1156–61.

9. Smeaton (n. 4 above), p. 101.

10. Because of such use, laws of similitude were referred to until around 1920 as laws of "correspondence or comparison."

11. For an account of Froude's work, see H. Rouse and S. Ince, *History of Hydraulics* (Ames, Iowa, 1957), pp. 182–87. Froude's law of similitude derives from consideration of the portion of resistance associated with the formation of surface waves. Rouse and Ince quote Froude's statement of the law as follows: "The diagram which exhibits to scale the resistance of a model at various successive velocities, will express equally the resistance of a ship similar to it, but of (n) times the dimension, at various successive velocities, if in applying the diagram to the case of the ship we interpret all the velocities as ($\sqrt{n}$) times, and the corresponding resistances as (n^3) times as great as on the diagram" (p. 184). Theoretical derivation of the law had been published by Ferdinand Reech in France in 1852 (ibid., pp. 154–55), but it is not clear whether Reech's work was known to Froude.

12. For an elementary technical discussion of dimensionless groups and dimensional analysis, see W. J. Duncan, "Dimensional Analysis," *Encyclopaedic Dictionary of Mathematics for Engineers and Applied Scientists*, ed. I. N. Sneddon (Oxford, 1976), pp. 193–95.

13. It follows that the value of the number is unchanged in any given case by a change in the system of units so long as a particular system (e.g., centimeters-grams-seconds or feet-pounds-seconds) is used consistently throughout.

14. Rouse and Ince (n. 11 above), pp. 206–10. Reynolds also made influential early application of laws of similitude, though without use of dimensionless groups, in his model studies of silting in estuaries (1887–91) (see A. T. Ippen, "Hydraulic Scale Models," in *Osborne Reynolds and Engineering Science Today*, ed. D. M. McDowell and J. D. Jackson [Manchester, 1970], pp. 199–208).

15. For the history of dimensional analysis, see E. O. Macagno, "Historico-critical Review of Dimensional Analysis," *Journal of the Franklin Institute* 292 (December 1971): 391–402.

16. Prior to the development of the variable-pitch propeller in the 1930s, fixed-pitch propellers were usually optimized for either cruise, climb, or take-off, depending on which was most critical for the airplane's mission; operation at other conditions was then not optimum. Alternatively, the propeller might be chosen to obtain a desirable compromise performance over the entire flight range, with no one condition being optimum.

17. The process here is typical of the way an engineering device must often be integrated into a larger system of devices. The requirements for effective performance over a range and for accurate prediction over that range are characteristic of engineering design. They help distinguish the work of the designer from that of the inventor, who is more likely to be concerned with establishing initial feasibility, usually at one condition.

18. This distinction is rarely made explicitly in the technical literature. Engineers tend not to be analytical about their methodologies and take this sort of thing for granted.

19. For the Wright brothers, see McFarland (n. 7 above), vol. 1, pp. 315, 594–640. The Wrights made relatively advanced (for the time) design calculations

for their propellers, as well as performance measurements on their first designs without forward motion and tests of an unspecified nature on later propellers. Almost the only experiments in the United States after the Wrights' were those of D. L. Gallup at Worcester Polytechnic Institute beginning in 1911. Gallup used full-sized propellers mounted at the end of an arm whirling on a central pivot (*Annual Report of the National Advisory Committee for Aeronautics, 1915* [Washington, D.C., 1916], p. 12). In 1912 the apparatus was also used by two MIT students, F. W. Caldwell and H. F. Lehmann, who made tests of five propellers, including a Wright propeller ("Investigation of Air Propellers," M.S. thesis, Massachusetts Institute of Technology, 1912).

20. For reference to Drzewiecki, see F. W. Weick, *Aircraft Propeller Design* (New York, 1930), p. 37. Although the Wrights used ideas similar to (but considerably more approximate than) those of Drzewiecki, they apparently arrived at them independently (McFarland [n. 7 above], p. 315). For the state of European experimental work, see A. F. Zahm, "Report on European Aeronautical Laboratories," *Smithsonian Miscellaneous Collections*, vol. 62, no. 3 (Washington, D.C., 1914). The trip on which this report was based took place in 1913. For confirmation of theory, see F. H. Bramwell and A. Fage, "Experiments on Model Propellers," *Reports and Memoranda No. 82, Technical Report of the Advisory Committee for Aeronautics, 1912–13* (London, 1914), p. 202. These volumes are hereafter cited as *ACA*.

21. G. Eiffel, *Nouvelles recherches sur la résistance de l'air et l'aviation*, text and atlas (Paris, 1914), text p. 353; Bramwell and Fage, "Tests on a Four-Bladed Propeller for Thrust and Efficiency," in *ACA, 1913–14*, pp. 288–89.

22. For Eiffel's tests, see *Nouvelles recherches*, text pp. 335–41, atlas pl. XXXI–XXXIII. For his use of similitude, see G. Eiffel, *The Resistance of the Air and Aviation*, 2d ed., trans. J. C. Hunsaker (Boston, 1913), pp. 191–96. It appears likely that Eiffel here followed the lead of Dimitri Riabouchinsky, who had introduced the law of similitude in Russia in 1909 and France in 1910 (D. Riabouchinsky, "Méthod des variables de dimension zéro et son application en aérodynamique," *L'Aerophile*, September 1, 1911, pp. 407–8).

23. The previously cited report by Zahm in 1914 does not include specific results or references. He states, however, that he and his companion, J. C. Hunsaker, "took copious notes" (p. 1), and these would presumably have included the results of propeller tests. Hunsaker also translated Eiffel's book cited in the preceding note. Durand, on p. 37 of his first report on the Stanford tests in 1917 (n. 43 below), cited Eiffel, *Nouvelles recherches*, but this work was undoubtedly known to him earlier. Lucien Marchis, professor at the University of Paris, in September 1915 presented an extensive review of French aeronautical research, including the work on propellers, at the International Engineering Congress organized by Durand for the Panama-Pacific Exposition in San Francisco. This report was translated by Durand and published: L. Marchis, "Experimental Researches on the Resistance of Air," *Annual Report of the NACA, 1916* (Washington, D.C., 1917), pp. 553–630.

24. H. Glauert, "Airplane Propellers," in *Aerodynamic Theory*, ed. W. F. Durand, 6 vols. (Berlin, 1935), vol. 4, p. 178. Weick (n. 20 above) says that Drzewiecki worked out his ideas independently of Froude.

25. Jerome Hunsaker, who had intimate knowledge of both aeronautical and marine practice, could still write in 1924, "Marine propeller design is based on data obtained from models and from trial trips . . . and the detail mechanism of propeller action is apparently considered too hopelessly complicated for successful analysis" (J. C. Hunsaker, "Aeronautics in Naval Architecture," *Transactions of the Society of Naval Architects and Marine Engineers* 32 [1924]: 1–25).

26. W. F. Durand, "Experimental Researches on the Performance of Screw Propellers," *Transactions of the Society of Naval Architects and Marine Engineers* 13 (1905): 71–85, and "Researches on the Performance of the Screw Propeller," *Publication No. 79*, Carnegie Institution (Washington, D.C., 1907); R. E. Froude, "Results of Further Model Screw Propeller Experiments," *Transactions of the Institution of Naval Architects* 50 (1908): 185–204; D. W. Taylor, *The Speed and Power of Ships* (New York, 1910), vol. 1, pp. 170–75. These three studies, plus a fourth by Karl Schaffran in Germany in the early 1920s, were reviewed by Taylor in "Comparison of Model Propeller Experiments in Three Nations," *Transactions of the Society of Naval Architects and Marine Engineers* 32 (1924): 61–83.

27. For history of the NACA, see chap. 3, n. 8. For Durand's proposal, see his autobiography, *Adventures in the Navy, in Education, Science, Engineering, and in War: A Life Story* (New York, 1953), p. 53. For the committee's view, see *Annual Report of the NACA, 1915* (Washington, D.C., 1916), p. 15.

28. Durand's productive career covered a remarkable span of sixty-five years, from wooden ships to jet aircraft. A biography of Durand would thus afford a fascinating insight into much of American mechanical engineering in the period when it was changing from a largely empirical activity to the theoretically based profession of today. For the facts of Durand's life, see his interesting but personally unrevealing autobiography cited above. See also F. E. Terman, "William Frederick Durand," *Biographical Memoirs* (Washington, D.C., 1976), vol. 48, pp. 153–93, which contains a chronological bibliography of his many publications, and H. L. Dryden, "Contributions of William Frederick Durand to Aeronautics," *Aeronautics and Astronautics, Proceedings of the Durand Centennial Conference*, ed. N. J. Hoff and W. G. Vincenti (Oxford, 1960), pp. 9–17.

29. The remarks in these sentences are based on conversations with the late Lydik Jacobsen, who was a younger colleague of Durand at Stanford. In commenting (on the whole favorably) on a proposal from Felix Pawlowski of the University of Michigan to do research on aircraft propellers, Durand wrote, "I understand that the proposed investigation is solely theoretical and *will not therefore add anything by way of fundamental data*" (emphasis added) (letter to S. W. Stratton, December 2, 1916, Record Group 255, National Archives, Washington, D.C. All letters cited are from this source).

30. *Annual Report of the NACA, 1916* (Washington, D.C., 1917), p. 14.

31. "Memorial Resolution, Everett Parker Lesley" (1945), Stanford University Archives, Stanford, Calif.

32. Durand to Richardson, November 2, 1916. Prior preliminary correspondence is also in the record group cited in n. 29 above.

33. Durand had earlier begun to exploit this similarity in a theoretical paper,

"The Screw Propeller: With Special Reference to Aeroplane Propulsion," *Journal of the Franklin Institute* 178 (September 1914): 259–86. For a general discussion of the borrowings of aeronautics from naval architecture and vice versa, see Hunsaker (n. 25 above).

34. For the choice of tunnel, see Durand to Richardson, August 21, 1916; Richardson to Durand, September 1, 1916; Durand to Richardson, September 5, 1916. This correspondence is particularly interesting for illustrating the kinds of considerations that go into choice of experimental apparatus. For detailed considerations behind the design, see Durand to Richardson, November 2, 1916. Allusions in the various letters and reports make it clear that Eiffel's experience had initial influence on other aspects of the Stanford tests as well. This influence appears to have come (see Durand to Richardson, August 21, 1916) via correspondence between Durand and Marchis (see n. 23 above).

35. This is the usual terminology, which derives from flight in still air. In the wind tunnel, V is in fact the speed of the moving test stream. While retaining the term *forward speed* throughout, I shall where convenient in my subsequent descriptions adopt the wind-tunnel point of view, in which the location of the propeller is fixed and the air moves past it.

36. I have omitted air density from the primary quantities as unnecessary for the present discussion.

37. Varying V and n corresponds to testing a single propeller; additionally varying D and $r_1, r_2, \ldots$ corresponds to testing families of related propellers.

38. Durand to Lesley, April 19, 1917; Lesley to Durand, April 24, 29, 1917. The results of these tests were included in the initial Stanford report (see n. 43 below). The free-jet test stream (as against a stream completely bounded by solid walls, then favored by the British) turned out to be a fortunate choice. Later theoretical work in Britain showed the free jet to have much the smaller limited-stream effect (see H. Glauert and C. N. H. Lock, "On the Advantages of an Open Jet Type of Wind Tunnel for Airscrew Tests," *Reports and Memoranda No. 1033, Aeronautical Research Committee, 1926–27* [London, 1928], pp. 318–27). This report quoted the Stanford experiments to substantiate their theoretical findings. Note: The Advisory Committee for Aeronautics (cited as ACA) changed its name in 1920 to Aeronautical Research Committee (hereafter cited as ARC).

39. Durand to Richardson, November 2, 1916. The *pitch* of an individual blade section, more precisely, is the distance that section would advance in one propeller revolution if it were moving along a helix having an angle the same as the angular orientation of the section. The pitch *ratio* of a blade section is then its pitch divided by the propeller diameter. The *mean* pitch ratio is the pitch ratio of the section at the standard representative radius (13/18 of the total propeller radius in the Durand-Lesley tests). In having five shape parameters, the propellers were more diverse than in Durand's work on marine propellers, where only the mean pitch ratio and blade width were varied. This greater generality may have stemmed from the influence of Eiffel (*Nouvelles recherches* [n. 21 above], pp. 335–41).

40. Today sophisticated statistical methods are used increasingly for choosing the values of the parameters and analyzing the measured data (see, e.g.,

E. M. Bartee, *Engineering Experimental Design Fundamentals* [Englewood Cliffs, N.J., 1968], chap. 5). Most engineers, however, still employ the simple, "commonsense" procedures described here.

41. Thus all parameters in equation (1) were varied except *D*. Variation of *D* was unnecessary for reasons that will appear later. Approximately half the models mentioned in this paper are on permanent exhibit in the Terman Engineering Library, Stanford University.

42. Lesley to Durand, April 27, June 9, 20, 23, 1917. Lesley reported homely items such as his directions to the Durands' Stanford housekeeper "about the iris and other plants" and his election to the city council of Palo Alto, "beating my opponent by a majority of one vote."

43. "Experimental Research on Air Propellers," *Report No. 14*, NACA (Washington, D.C., 1917).

44. Specifically, when written in the form of $(V/n)/D$ it can be recognized as giving the distance the propeller moves forward in one revolution, measured in number of propeller diameters instead of ordinary units.

45. This property of dimensionless representation, which is not peculiar to propellers, is of great importance for parametric testing, at whatever scale, since it greatly reduces the number of tests that need be run to cover a required range of the individual parameters. It often makes parametric testing economic where it would otherwise be prohibitively costly.

46. W. F. Durand, *The Resistance and Propulsion of Ships* (New York, 1903), pp. 128–45, 300–3.

47. V/nD later replaced slip in marine work as physically more meaningful (see Hunsaker [n. 25 above], pp. 10–11).

48. Durand to Richardson, November 2, 1916; Durand to Lesley, May 1, 1917.

49. See, e.g., J. C. Hunsaker et al., "Reports on Wind Tunnel Experiments in Aerodynamics," *Smithsonian Miscellaneous Collections* 62, no. 4 (1916): 15–26, 29, 86–88. Buckingham contributed a theoretical section to this work. For publication of his theorem, see his "On Physically Similar Systems: Illustrations of the Use of Dimensional Equations," *Physical Review* 4 (October 1914): 345–76. The theorem derived by Buckingham was first established by Aimé Vaschy in France in 1892 and rediscovered independently by Dimitri Riabouchinsky, who gave an account of it in a French aeronautical journal in 1911 (n. 22 above). Buckingham later (1921) credited an abstract of Riabouchinsky's work published by the British Advisory Committee for Aeronautics with providing a "hint" leading to his own derivation (see Macagno [n. 15 above], pp. 397–400).

50. Inclusion of viscosity and compressibility leads to two additional dimensionless groups, known, respectively, as Reynolds number and Mach number.

51. The ability to do this explains why it was unnecessary to vary *D* in the tests (see n. 41 above). Moreover, since both *V* and *n* were varied, the same values of V/nD were attained in more than one way. This afforded an experimental check of the validity of the law of similitude.

52. Some new values of the parameters were incorporated in only a single propeller.

53. Durand and Lesley, "Experimental Research on Air Propellers, IV," *Report No. 109*, NACA (Washington, D.C., 1921), p. 3.

54. Durand and Lesley, "Experimental Research on Air Propellers, II, III, IV," *Report[s] No. 30, 64, 109*, NACA (Washington, D.C., 1919, 1920, 1921). Besides the normal performance measurements of the parameter-variation tests, the first of these reports also included tests of the following: (1) an adjustable-pitch propeller; (2) three pairs of oppositely rotating propellers, each having one propeller behind the other on the same axis; (3) twelve propellers, selected from the total number, to measure the braking effect (or negative thrust) at high V/nD; (4) sixty-seven selected propellers to measure thrust and power at zero forward speed. In 1921, as a sideline, Durand and Lesley reported tests of propellers with their axis of rotation at nearly 90 degrees to the airstream, as for a helicopter rotor ("Tests on Air Propellers in Yaw," *Report No. 113*, NACA [Washington, D.C., 1921]).

55. Lesley to Durand, January 17, 1918.

56. Durand and Lesley, "Experimental Research on Air Propellers, V," *Report No. 141*, NACA (Washington, D.C., 1922).

57. Efficiency is calculated from the measured thrust T and torque Q by the formula $\eta = TV/2\pi nQ$.

58. The original hub with a second set of blades is in the unexhibited collection at the National Air and Space Museum, Washington, D.C. (accession no. NAM 705, catalog no. 1951-51). Tests of this model are sometimes said to have been the first of their kind, apparently in reiteration of Durand's own statement (*Adventures* [n. 27 above], p. 140). In fact, a somewhat similar model had been tested in Britain in 1912–13, though with confirmation of propeller theory rather than optimum performance as the apparent goal (Bramwell and Fage [n. 20 above], pp. 199–217). Flight tests of a variable-pitch propeller at the Royal Aircraft Factory were also reported in 1918 (Anonymous, "The Variable Pitch Propeller—Experiments Conducted at the Royal Aircraft Factory," *ACA, 1917–18*, 2:515–17).

59. The text of the lecture was reprinted by the NACA, *Annual Report of the NACA, 1918* (Washington, D.C., 1920), pp. 33–44.

60. Ibid., p. 41.

61. In the airplane, because of the critical importance of weight, the margin between success and failure was narrower than in any previous device. The aeronautical engineer was accordingly forced to more thorough and refined methods of design. Some of these methods have been adopted subsequently in other branches of engineering.

62. Durand and Lesley, "Comparison of Model Propeller Tests with Airfoil Theory," *Report No. 196*, NACA (Washington, D.C., 1924).

63. For detailed explanation of the theory, see Glauert (n. 24 above), pp. 178–81. For discussion of the more sophisticated vortex theory (and instructive description of both airfoil and propeller action), see E. E. Larrabee, "The Screw Propeller," *Scientific American* 243 (July 1980): 134–48.

64. The necessary airfoil data came from wind-tunnel tests specially run at the California Institute of Technology.

65. For a résumé of British work to 1919, including reference to prior reports, see R. McK. Wood, "Summary of Present State of Knowledge with Regard to Airscrews," *ACA, 1918–19*, 2:549–68.

66. One might also interpret the difference as confirming sometimes supposed differences in engineering style between the two nations, the British being thought to be more theoretical and the Americans more empirical. Such interpretation would be dubious in the present case. For whatever reasons, it *is* true that the British approach to aeronautics at this time relied more on theory than the American. On the other hand, the British marine tradition was as much oriented toward experimental parameter variation as that in America. Durand's text on ship propulsion, in fact, drew heavily on the British experimental work of William and Robert Froude.

67. C. N. H. Lock and H. Bateman, "Experiments with a Family of Airscrews. Part III—Analysis of the Family of Airscrews by Means of the Vortex Theory and Measurements of Total Head," *ARC, 1923–24*, 1:377–408; see also Weick (n. 20 above), pp. 73–74.

68. A. Fage, C. N. H. Lock, R. G. Howard, and H. Bateman, "Experiments with a Family of Airscrews. Part I—Experiments with the Family of Airscrews Mounted in Front of a Small Body," *ARC, 1922–23*, 1:174–239. This report mentions the parameter-variation studies of both Durand and Eiffel.

69. W. C. Nelson, *Airplane Propeller Principles* (New York, 1944), pp. 97–121; see also Weick (n. 20 above), pp. 257–83.

70. *Annual Report of the NACA, 1923* (Washington, D.C., 1924), p. 17.

71. Durand and Lesley, "Comparisons of Tests on Air Propellers in Flight with Wind Tunnel Model Tests on Similar Forms," *Report No. 220*, NACA (Washington, D.C., 1926).

72. In addition to ibid., two other Stanford reports appeared dealing with the obstruction problem: E. P. Lesley and B. M. Woods, "The Effect of Slipstream Obstructions on Air Propellers," *Report No. 177*, NACA (Washington, D.C., 1924) and W. F. Durand, "Interaction between Air Propellers and Airplane Structures," *Report No. 235*, NACA (Washington, D.C., 1926).

73. Marine engineers, including Durand, had struggled with the similar but more difficult problem of a propeller operating in the wake of a ship's hull (*Resistance and Propulsion of Ships* [n. 46 above], pp. 230–38). Marine thinking had proceeded along different lines, however, and the concept of propulsive efficiency had not been arrived at.

74. See, in particular, "Experimental Research on Air Propellers, V" (n. 56 above), pp. 15–19.

75. For a discussion of the ideas of Alexandre Koyré regarding technology as a system of thought, see E. T. Layton, Jr., "Technology as Knowledge," *Technology and Culture* 15 (January 1974): 31–41, esp. 35–37. Layton also adds some views of his own concerning the relationships between technological thought and design.

76. W. F. Durand, "Tests on Thirteen Navy Type Model Propellers," *Report No. 237*, NACA (Washington, D.C., 1927).

77. See, e.g., the propeller collection at the Science Museum, London.

78. Metal propellers of steel or aluminum alloy, which have practical advan-

tages over wood, came into use in the latter half of the 1920s.

79. While the parametric tests were in progress at Stanford, Eiffel was doing similar work in Paris, published circa 1920 as *Etudes sur l'hélice aérienne* (Paris, n.d.). This included propellers of three, four, and six as well as two blades and employed adjustable- instead of fixed-pitch models to minimize the number needed to cover the desired pitch range. The equivalent of about ninety fixed-pitch models was tested, with results essentially similar to those at Stanford. Eiffel also tested a single model constructed to reproduce one of the Durand-Lesley designs, which he described without amplification as "une de nos meilleures hélices" (p. 121). Though no reference is given, it seems likely that Eiffel obtained the necessary design information from Durand when the latter was on war duty in Paris in 1918. By the same token Durand must also have known of this work by Eiffel. Neither of them, however, attempted a correlation of each other's results. For whatever reason, this neglect was characteristic of the international situation in propeller research (and other areas of aerodynamics) at the time—occasional isolated mention of foreign work but little of the careful referencing or cross comparison taken for granted today.

80. Shortly before his death in 1976, Walter S. Diehl, a key engineering officer with the navy's Bureau of Aeronautics from 1917 to 1951, wrote in personal correspondence, "The Durand-Lesley data were of tremendous value to aeronautical engineers in the period 1920–40. . . . We in BuAer made much use of these data. I think that every aeronautical engineer did." Fred Weick, who joined the bureau in 1924, writes, "One of the first things I did after I got my feet on the ground was to work out a simple system of blade-element analysis using only a single element . . . but obtaining the airfoil lift and drag characteristics by working the analysis backwards from the model propeller data [i.e., Durand's results on thirteen navy-type propellers, which had been made available prior to publication]. . . . The method was reported in some detail later in NACA *Technical Notes 235* and *236* [1926]. . . . The propellers for many Navy airplanes were designed by this method but I have no records of the details. . . . A little later I worked out a system for selecting propellers for given airplanes directly from the Navy model propeller tests by Dr. Durand, and this was described in NACA *Technical Note 237* [1926]. . . . In 1929 when I was chief engineer of the Hamilton Aero Manufacturing Company I used this same system of selection for both wood and metal propellers used on many different commercial airplanes from small ones like the Monocoupe to large ones like the Boeing 80A. . . . All this is only my experience, but it is apparent that the propeller tests of Durand and Lesley had a great influence on aircraft propeller design and selection."

81. Weick, n. 20 above; Munk, n. 1 above; W. S. Diehl, *Engineering Aerodynamics* (New York, 1928) and "The General Efficiency Curve for Air Propellers," *Report No. 168*, NACA (Washington, D.C., 1923); D. W. Taylor, "Some Aspects of the Comparison of Model and Full-Scale Tests," *Report No. 219*, NACA (Washington, D.C., 1926). The last report was the text of Taylor's Wilbur Wright Memorial Lecture to the Royal Aeronautical Society in 1925 and included Durand and Lesley's recent comparison between wind-tunnel and flight results.

82. F. E. Weick and D. H. Wood, "The Twenty-Foot Propeller Research Tunnel of the National Advisory Committee for Aeronautics," *Report No. 300*, NACA (Washington, D.C., 1928). For the relation to the Stanford work, see, e.g., F. E. Weick, "Full Scale Tests of Wood Propellers on a VE-7 Airplane in the Propeller Research Tunnel," *Report No. 301*, NACA (Washington, D.C., 1929). These tests, the first complete series in the PRT, were run specifically to provide comparison with Durand and Lesley's model and flight results made with the same airplane. For later data from the tunnel, see, e.g., E. P. Hartman and D. Biermann, "The Aerodynamic Characteristics of Full-Scale Propellers Having 2, 3, and 4 Blades of Clark Y and R.A.F. 6 Airfoil Sections," *Report No. 640*, NACA (Washington, D.C., 1938). The tunnel was also used for the investigation of other airplane components and for initial development of the famous NACA cowling for radial air-cooled engines.

83. See, e.g., E. N. Jacobs, K. E. Ward, and R. M. Pinkerton, "The Characteristics of 78 Related Airfoil Sections from Tests in the Variable-Density Wind Tunnel," *Report No. 460*, NACA (Washington, D.C., 1933). Fred Weick, who had moved from the Bureau of Aeronautics to the Propeller Research Tunnel by 1926, writes in personal correspondence to the author, "In regard to the use of a systematic series of independent variables in research, I am sure that I was greatly influenced by my propeller indoctrination using Durand and Lesley's test results. I used the method repeatedly in laying out test programs, and tried to include extremes that were beyond the cases of immediate interest. An example of this is the cowling investigation that we made in the PRT in 1926–8."

84. For a discussion of purpose, knowledge, and method as aspects of technology viewed as a form of human activity, see R. E. McGinn, "What Is Technology?" in *Research in Philosophy and Technology* (Greenwich, Conn., 1978), vol. 1, pp. 79–97.

85. Although propeller theory never became sufficiently accurate to supply design data, other theories sometimes are. An example is the theory of heat conduction in solids, which is used to provide parametric design charts for heat transfer in variously shaped bodies (see, e.g., the much-used textbook by W. H. McAdams, *Heat Transmission*, 3d ed. [New York, 1954], pp. 33–43).

86. H. Geiger and E. Marsden, "The Laws of Deflextion of α Particles through Large Angles," *Philosophical Magazine* 25 (1913): 604–23. For a textbook account, see P. A. Tipler, *Foundations of Modern Physics* (New York, 1969), pp. 147–60.

87. Geiger and Marsden (n. 86 above), p. 623.

88. In such cases the worker may have to be regarded as both scientist and engineer. There is perhaps a kind of "complementarity principle" here, as in physics, where light is regarded alternatively and simultaneously as a wave and a particle, depending on the purpose at hand.

89. Layton (n. 75 above), pp. 40–41. The statements and evidence here are also consistent with the general views expressed by Rapp (n. 2 above), pp. 102–7. I find helpful also the following statement by M. Fores in "Price, Technology, and the Paper Model," *Technology and Culture* 12 (October 1971): 621–27 (emphasis added): "Science is primarily an analytical activity which seeks ultimately

to describe natural phenomena in a series of general relationships, normally as the result of controlled experiment. In contrast, technology is mainly a synthetic process which, making use of the knowledge and general relationships of science (*among other things*), builds useful objects which are judged for their utility and efficiency." The "other things" I take to include engineering knowledge (design data) of the kind discussed here. The terminology used by Layton (and others), however, raises a semantic problem. If indeed there is such a thing as peculiarly engineering (or technological) knowledge sought after by engineers (or technologists), as seems agreed, then it may be confusing to characterize the scientist alone as valuing knowing. It might be more consistent to grant that both communities value knowing, though for different reasons, and speak of the technologist as valuing "doing" whereas the scientist values "understanding." See also A. R. Hall, "On Knowing, and Knowing how to . . . ," in *History of Technology, Third Annual Volume*, ed. A. R. Hall and N. Smith (London, 1978), pp. 91–103.

90. It also implies an important difference between the scientist and the engineer: The scientist can (and sometimes must) wait when a theory is not forthcoming; the engineer frequently does not have this possibility.

91. A particularly clear historical example appears in the account by Rosenberg and myself (n. 7 above) of the design of the Britannia Bridge, where experimental parameter variation was used to circumvent the lack of a theory for the buckling of thin-walled tubes.

92. R. W. Fox and S. J. Kline, "Flow Regimes in Curved Subsonic Diffusers," *Journal of Basic Engineering* 84 (September 1962): 303–16. The theory has since been much improved based, in part, on the systematic data in the report. This example thus also illustrates an advantage of experimental parameter variation in providing systematic data for possible later analysis.

93. Such difficulties have been considerably diminished in recent years by the availability of high-speed electronic computers.

94. L. Bryant, "The Problem of Knock in Gasoline Engines," unpublished ms., 1972.

95. Quoted in M. Josephson, *Edison* (New York, 1959), pp. 233–36. Josephson relates, however, how the range of Edison's experiments was deliberately extended for publicity purposes.

96. It is also useful to distinguish "trial-and-error" methods in which the parameters are known and which are therefore properly included under parameter variation. In contrast to the comprehensive, a priori procedure employed by Durand and Lesley, however, the parameters are varied here in a step-by-step process, using the results of one test to guide the selection of values for the next. The object is usually to design a device for a single situation rather than provide a collection of design data applicable over a range. In terms of an equation such as (1), the experimenter establishes a discrete path of the performance function F through the parameter space rather than a general mapping.

97. This item has been substantially revised from earlier publication as the result of questions by Edward Constant.

98. For discussion and review of uses in earth sciences to the mid–1930s, see M. K. Hubbert, "Theory of Scale Models as Applied to the Study of Geologic

Structures," *Bulletin of the Geological Society of America*, October 1, 1937, pp. 1459–1519, esp. 1460–64, 1496–1519. For a review of model studies of tornados, see R. P. Davies-Jones, "Tornado Dynamics," in *Thunderstorm Morphology and Dynamics*, 2d ed., ed. E. Kessler (Norman, 1986), pp. 197–236, esp. pp. 231–35. For recent uses in meteorology, see, e.g., G.-Q. Li, R. Kung, and R. L. Pfeffer, "An Experimental Study of Baroclinic Flows with and without Two-Wave Bottom Topography," *Journal of the Atmospheric Sciences*, November 15, 1986, pp. 2585–99, and L. P. Rothfusz, "A Mesocyclone and Tornado-like Vortex Generated by the Tilting of Horizontal Vorticity: Preliminary Results of a Laboratory Simulation," ibid., pp. 2677–82. For a summary of model studies of atmospheric circulation, see J. R. Holton, *An Introduction to Dynamic Meteorology*, 2d ed. (New York, 1979), pp. 274–80.

99. The planetarium, which models the cosmos, might be thought of as a possible additional example. It is for demonstration and education, however, rather than knowledge generation. Models of chemical molecules might also be cited. These, however, are static aids to visualization and thinking rather than working models of reality.

100. H. E. Huntley, *Dimensional Analysis* (New York, 1951), p. 43.

101. One could argue, of course, that in so doing they are really doing the engineering necessary to their science, but such distinction seems unjustifiably arbitrary in this context.

102. Some writers—e.g., A. E. Musson and E. Robinson, *Science and Technology in the Industrial Revolution* (Manchester, 1969), p. 73, passim—seem at times to equate "systematic" with "scientific" and thus to regard parameter variation and associated experimental methods as by definition methods of science. Surely this is simplistic and forecloses some useful distinctions.

103. An especially important example, evident since ancient times but still unsolved in a basic scientific sense, is that of flow in a pipe in most conditions of practical operation (see Rouse and Ince [n. 11 above], pp. 151–61, 170, 207–9, 232–35, and J. K. Vennard and R. L. Street, *Elementary Fluid Mechanics*, 5th ed. [New York, 1975], chap. 9).

104. Pacey (n. 4 above), p. 208, states that Smeaton's work on waterwheels "was outstanding as an example of experimental method in science, and how it could be used to shed light on engineering problems." Edwin Layton ("Mirror-Image Twins: The Communities of Science and Technology in 19th-Century America," *Technology and Culture* 12 [October 1971]: 562–80) says that Smeaton "used the experimental methods of science" and that "following Smeaton, technologists might borrow the methods of science to found new sciences built on existing craft practices" (p. 566). The present discussion suggests it would be more precise and complete to say that Smeaton, taking over from experimental science the techniques of controlled experiment and careful measurement, established a method whose function was to serve the design needs of engineering, including when necessary the bypassing of a lack of scientific theory. Smeaton did in fact so use the method (see Cardwell [n. 4 above], p. 80).

105. E. T. Layton, Jr., review of *Philosophers and Machines*, ed. O. Mayr, *Technology and Culture* 18 (January 1977): 89–90.

106. For papers dealing with development in both senses, see T. P. Hughes,

ed., "The Development Phase of Technological Change," *Technology and Culture* 17 (July 1976): 423–81. For development of the technological base in the electric-power industry, see Hughes, "The Science-Technology Interaction: The Case of High-Voltage Power Transmission Systems," ibid., pp. 646–62.

107. Commenting on Smeaton's work, Cardwell (n. 4 above) says, "It is an exemplary method of getting the very best out of established machines but it cannot, or at least generally does not, lead to radically new or revolutionary improvement" (p. 84).

108. See n. 7 above. Taylor's invention is a particularly momentous illustration of Abbott Payson Usher's statement that "acts of insight frequently emerge in the course of performing acts of skill" (n. 3 above, p. 44; see also p. 47). In terms of Usher's four-step theory of the process of invention, parameter variation may serve to "set the stage" for an act of inventive insight (*A History of Mechanical Inventions*, rev. ed. [Cambridge, Mass., 1954], chap. 4).

109. I am indebted to Edwin Layton for the observation, in personal correspondence, that there is no necessary connection in general between methodology and the size, nature, or importance of a technological development. It can be argued that inputs to technology from science, though they sometimes lead to revolutionary developments, also lead typically to incremental changes. Big breakthroughs are rare whatever the methods.

110. Usher (n. 3 above), pp. 43–44, 46, 49–50.

111. N. 7 above.

112. See remarks by Hughes (n. 106 above), p. 423.

Chapter 6. Design and Production

Note: This chapter is only slightly altered from earlier publication as "Technological Knowledge without Science: The Innovation of Flush Riveting in American Airplanes, ca. 1930–ca. 1950," *Technology and Culture* 25 (July 1984): 540–76. I am especially indebted for their help to Todd Becker, James Hansen, Stephen Kline, Edwin Layton, Robert McGinn, George Rechton, David Richardson, Nathan Rosenberg, Eugene Speakman, Howard Wolko, and Charles Wordsworth.

1. See, e.g., P. A. Hanle, *Bringing Aerodynamics to America* (Cambridge, Mass., 1982).

2. Another problem requiring such activity in the aircraft industry is the sealing of pressurized airplane cabins to prevent leakage at high altitude. Possible examples from other industries are the production of round passages with very tight tolerances, as occur in hydraulic cylinders and carburetors for automobiles, and the early changeover from wiring to (nonminiaturized) printed circuits in television receivers. For discussion of the importance and difficulties of the "lower" forms of technological knowledge involved in such problems, see N. Rosenberg, "Problems in the Economist's Conceptualization of Technological Innovation," in his *Perspectives on Technology* (Cambridge, 1976), pp. 61–84, esp. pp. 77–79.

3. S. G. Winter, "An Essay on the Theory of Production," in *Economics and the*

World around It, ed. S. H. Hynan (Ann Arbor, Mich., 1982), pp. 55–93.

4. Flush riveting was a tangible innovation, even if a secondary one, in a product, the airplane; at the same time, it depended critically on innovation in a process, that of airframe production. It thus does not fit neatly into either of the economist's standard classifications of product or process innovation. Classification of an innovation according to where it is embodied may not always be possible. Of course, classification according to where the crucial innovative activity is centered may be expected to have its problems too.

5. S. Hollander, *The Sources of Increased Efficiency: A Study of du Pont Rayon Plants* (Cambridge, Mass., 1965); A. D. Little, Inc., *Patterns and Problems of Technical Innovation in American Industry*, report to National Science Foundation (September 1963). Considerable related scholarship exists concerning production in nineteenth-century industry, notably with regard to the American system of manufactures; see, e.g., M. R. Smith, *Harpers Ferry Armory and the New Technology* (Ithaca, N.Y., 1977), chaps. 4, 8; P. Uselding, "Measuring Techniques and Manufacturing Practice," and D. A. Hounshell, "The *System*: Theory and Practice," both in *Yankee Enterprise: The Rise of the American System of Manufactures*, ed. O. Mayr and R. C. Post (Washington, D.C., 1981), pp. 103–52. In its concern for details of production, the present paper also relates to the vast literature on technique, most of it descriptive, amassed by archaeologists, industrial archaeologists, and historians of technology: e.g., *Dictionnaire archéologique des techniques* (Paris, 1963), and classification 10 in any annual "Current Bibliography in the History of Technology," *Technology and Culture*, April (now July) issues.

6. Riveting, of whatever sort, is used in aluminum airplanes, in preference to the welding that has replaced it in industries that employ other metals, because the aluminum alloys needed in aircraft tend to be affected adversely in their material properties and structural behavior by the high temperatures of the welding process.

7. R. Miller and D. Sawers, *The Technical Development of Modern Aviation* (New York, 1970), pp. 18–20, 47–50, 63–65; C. G. Grey and L. Bridgman, eds., *Jane's All the World's Aircraft 1931* (London, 1931), pp. 303c–4c, and *Jane's . . . 1936*, pp. 276c–77c. For recent valuable discussion of the role in this development of long-range commercial flying boats, largely neglected by Miller and Sawers and others, see R. K. Smith, "The Intercontinental Airliner and the Essence of Airplane Performance, 1929–1939," *Technology and Culture* 24 (July 1983): 428–49.

8. In Europe flush riveting appeared at least as early as 1922 on the Bernard C.1, a racing plane that was the hit of the Paris Air Show of that year; T. G. Foxworth, *The Speed Seekers* (New York, 1976), pp. 83–84. I have made no attempt to trace the details of the application in European commercial and military aircraft, but a cursory examination suggests it was more or less simultaneous with that in the United States. The innovation appears to have taken place essentially independently on the two sides of the Atlantic, until flush riveting had become well established in the 1940s. This fact further emphasizes the simultaneous, widespread nature of the development as mentioned in the introduction and elaborated below.

9. U.S. Patent 1,609,468, C. W. Hall, *Metallic Construction for Aircraft and the Like*, December 7, 1926, pp. 3, 7. For Hall's equipment, see Aluminum Company of America, *The Riveting of Aluminum* (Pittsburgh, 1929), pp. 21–22, and Hall, "Weight Saving by Structural Efficiency," *SAE Journal* 28 (1931): 77–83, esp. 79, 80. A photograph in the files at the National Air and Space Museum (information from Howard Wolko) shows that the XFH-1, a navy single-engine fighter and Hall's first airplane (1929), had protruding rivets. Photographs of the PH-1 are not conclusive but strongly suggest flush riveting. Boeing information is from personal correspondence with the late Edward C. Wells, an engineer and leading engineering executive with that company after 1931.

10. Airplanes I have been able to identify as using flush rivets extensively in these years (though there were undoubtedly more) are the mammoth Hall XP2H-1 four-engine military flying boat of 1934 ("Navy's Giant," *Aviation* 33 [October 1934]: 317–19), the Sikorsky multiengine commercial flying boats and amphibians of 1934 and 1935 (*Jane's . . . 1934*, pp. 311c–12c, *Jane's . . . 1935*, p. 335c; the S-42 of 1934 is an exception to the earlier statement about the lack of extensive flush riveting in the revolutionary aircraft of the first half of the 1930s), the Hughes H-1 single-engine racer of 1935, in which Howard Hughes set a land-plane speed record of 352 MPH (*Jane's . . . 1935*, p. 311c), and the Boeing 299 of 1935, prototype of the four-engine B-17 Flying Fortress (Wells correspondence). The news item is "Flush Riveting," *U.S. Air Services* 19 (February 1934): 32. An instance of partial use is the Boeing 247D commercial transport of 1933, a key airplane in the mainstream of development, which had flush rivets limited to the engine cowling (see example in the National Air and Space Museum).

11. C. Dearborn, "The Effect of Rivet Heads on the Characteristics of a 6 by 36 Foot Clark Y Metal Airfoil," *Technical Note No. 461*, NACA (Washington, D.C., May 1933), p. 5.

12. T. Hughes, "The Science-Technology Interaction: The Case of High-Voltage Power Transmission Systems," *Technology and Culture* 17 (October 1976): 646–59, quotation from 646; see also his *Networks of Power* (Baltimore, 1983), pp. 14–15, passim.

13. H. Hodges, *Technology in the Ancient World* (New York, 1970), pp. 143–44, 187.

14. The first comprehensive experiments on the strength of riveted joints, carried out by Eaton Hodgkinson in 1838 under the direction of William Fairbairn, included flush rivets; W. Fairbairn, "An Experimental Inquiry into the Strength of Wrought-Iron Plates and Their Riveted Joints as Applied to Shipbuilding and Vessels Exposed to Severe Strains," *Philosophical Transactions of the Royal Society of London* 140 (1850): 677–725, esp. 698. For use in ships and boilers, see, e.g., E. J. Reed, *Shipbuilding in Iron and Steel* (London, 1869), pp. 196–97, 329–31, 340, and C. H. Peabody and E. F. Miller, *Steam-Boilers* (New York, 1908), p. 200. For a complete history and bibliography of the (mainly nonaeronautical) literature up to 1944 on the strength of riveted joints generally, see A. E. R. de Jonge, *Riveted Joints: A Critical Review of the Literature Covering Their Development* (New York, 1945). The bibliography contains 1,209 items,

each with an abstract carefully prepared by de Jonge—an impressive monument to engineering knowledge with little or no science.

15. The question of materials was central in both production and design. Considerations in this regard, however, were mostly no different for flush than for protruding rivets. I shall mention them only when something arises peculiar to flush riveting.

16. *Aircraft Fabrication. Aircraft Riveting*, Part 1 (Scranton, Pa., 1943), p. 1; E. S. Jenkins, "Rational Design of Fastenings," *SAE Transactions* 52 (September 1944): 421–29.

17. For some of the considerations involved prior to the shift to flush riveting, see "Riveting of Aluminum and Its Alloys," *Aero Digest* 30 (April 1937): 44–46.

18. The number 15 comes from experiments on the strength of flush-riveted joints made by the National Bureau of Standards from 1941 to 1943 using specimens supplied by the manufacturers; see, e.g., W. C. Brueggeman, "Progress Summary No. I. Mechanical Properties of Flush-riveted Joints Submitted by Five Airplane Manufacturers," *Wartime Report W-79*, originally *Restricted Bulletin* (unnumbered), NACA (Washington, D.C., February 1942). I find no record of the overall summary this report promised, but file cards, of individual progress reports, in the library of the Langley Research Center, NASA, reveal the head angles used by twelve of the manufacturers as follows: 78 degrees, Curtiss-Wright, Grumman; 100 degrees, Lockheed, Douglas, Vultee, Vought-Sikorsky; 110 degrees, Brewster; 115 degrees, Martin, Republic, Consolidated; 120 degrees, Bell; 130 degrees, Boeing. For differences in dimensions for the same head angle, see E. B. Lear and J. E. Dillon, *Aircraft Riveting* (New York, 1942), p. 72.

19. The specified diameter of a rivet is always the shank diameter.

20. *Aircraft Fabrication. Aircraft Riveting*, p. 10; part 2, pp. 7–12; A. H. Nisita, *Aircraft Riveting* (New York, 1942), pp. 83–89; interview (November 4, 1980) with Edward Harpoothian, from 1939 to 1962 chief of Structures Section, Douglas Aircraft Company, Santa Monica, Calif.

21. A. A. Schwartz, "Flush Rivets for Speedy Airplanes," *American Machinist*, May 14, 1941, pp. 443–48, quotation from p. 446; G. C. Close, "Flow-Form Dimpling at Northrop," *Machinery* (N.Y.) 60 (February 1954): 191–93, quotation from 191.

22. Miller and Sawers (n. 7 above), pp. 131–32; *Jane's . . . 1939*, p. 245c; Harpoothian interview; interview (February 5, 1981) with John Buckwalter, longtime engineer with Douglas and project engineer on the DC-4E. P. W. Brooks, *The Modern Airliner: Its Origins and Development* (London, 1961), p. 94, states that the DC-4E was "the first American airliner with flush-riveted skin." This statement is hardly correct in view of the use of flush rivets in the Sikorsky commercial aircraft of 1934 to 1935 (n. 10 above).

23. V. H. Pavlecka, "Final Report on Flush Riveting," *Report No. 1645*, Douglas Aircraft Company (Santa Monica, Calif., July 12, 1937). For evidence of European work, see M. Langley, *Metal Aircraft Construction*, 2d ed. (London, 1934), pp. 306–7. In contrast to the Douglas procedure of pushing the rivet from the outside (figure 6-3a, b), Boeing used a special tool that clamped onto

the shank of the rivet and pulled it into the die from the inside (Wells correspondence, n. 9 above). This presumably was done to avoid possible scarring of the skin from the flat dimpling tool required by the Douglas system. I have seen no record of the Boeing scheme being used elsewhere.

24. G. B. Haven and G. W. Swett, *The Design of Steam Boilers and Pressure Vessels* (New York, 1915), p. 87. I am unable to discover why the seemingly idiosyncratic value of 78 degrees was originally adopted. Perhaps a prominent early manufacturer of boilers with flush rivets happened to have a conical cutter with that angle available for his countersinking (which, of course, only displaces the mystery), and everyone afterward simply followed suit. Things sometimes happen in technology for no better reason.

25. Douglas patented its riveting system in 1941; U.S. Patent 2,233,820, V. H. Pavlecka, assignor to Douglas Aircraft Company, *Method of Riveting*, March 4, 1941. Shortly thereafter Douglas released the patent to the aircraft industry in connection with the national defense program; Anonymous, "Flush Riveting in Aircraft," *Automotive Industries*, May 15, 1941, p. 518. I find no evidence, however, that the Douglas system was widely imitated except for the rivet head, as we shall see later.

26. D. R. Berlin and P. F. Rossman, "Flush Riveting Considerations for Quantity Production," *SAE Transactions* 34 (1939): 325–34.

27. Schwartz (n. 21 above), quotation from p. 443.

28. R. L. Templin and J. W. Fogwell, "Design of Tools for Press-Countersinking and Dimpling of 0.040-Inch-Thick 24S-T Sheet," *Technical Note No. 854*, NACA (Washington, D.C., August 1942); interview (September 2, 1981) with George D. Rechton, process engineer at Douglas, Santa Monica, Calif., during the period in question. Although the Templin-Fogwell work was done in the early 1940s, it may be taken as representative of earlier activities elsewhere.

29. *Aircraft Fabrication. Aircraft Riveting* (n. 16 above), pp. 18–46.

30. *Aircraft Fabrication. Aircraft Riveting*, part 2, pp. 1–15; Pavlecka (n. 23 above), pp. 10–12; U.S. Patent 2,257,267, A. T. Lundgren, A. B. Rogers, and V. H. Pavlecka, assignors to Douglas Aircraft Company, *Percussive Tool*, September 30, 1941; U.S. Patent 2,274,091, V. H. Pavlecka, G. D. Rechton, and C. C. Misfeldt, assignors to Douglas Aircraft Company, *Bucking Tool*, February 24, 1942.

31. M. J. Hood, "The Effects of Some Common Surface Irregularities on Wing Drag," *Technical Note No. 695*, NACA (Washington, D.C., March 1939), and "Surface Roughness and Wing Drag," *Aircraft Engineering* 11 (September 1939): 342–44; C. P. Autry, "Drag of Riveted Wings," *Aviation*, May 11, 1941, pp. 53–54.

32. E. E. Lundquist and R. Gottlieb, "A Study of the Tightness and Flushness of Machine Countersunk Rivets for Aircraft," *Wartime Report L-294*, originally *Restricted Bulletin* (unnumbered), NACA (Washington, D.C., June 1942); R. Gottlieb, "A Flush Rivet Milling Tool," *Restricted Bulletin* (unnumbered), NACA (Washington, D.C., June 1942); R. Gottlieb and M. W. Mandell, "An Improved Flush-Rivet Milling Tool," *Restricted Bulletin No. 3E18*, NACA (Washington, D.C., May 1943); U.S. Patent 2,393,463, R. Gottlieb, assignor of one-half to A.

Sherman, *Milling Tool*, January 22, 1946. For a popular account, see G. W. Gray, *Frontiers of Flight* (New York, 1948), pp. 197–202. The idea of having the upset head be the flush head had also appeared in the European aircraft industry before 1934 in the "De Bergue rivet" used in fuel tanks (Langley [n. 23 above], p. 267) and in a U.S. patent on a riveting procedure granted to a German in 1939 (U.S. Patent 2,147,763, E. Becker, *Flush or Countersunk Riveting*, February 21, 1939).

33. U.S. Patent 2,395,348, A. Sherman and R. Gottlieb, *Flush-Riveting Procedure*, February 19, 1946; interviews with Eugene Speakman (August 31, 1981), employed at Douglas Aircraft Company, Long Beach, Calif., since 1947 and currently senior structural engineer specializing in fasteners, and with Neil Williams (September 2, 1981), materials and process engineer with the same company.

34. Alcoa 24S-T, common in the early 1940s, had 4.5 percent copper, together with 1.5 percent magnesium and 0.6 percent manganese. The higher-strength 75S-T introduced in the mid-1940s had 5.6 percent zinc, 2.5 percent magnesium, 1.6 percent copper, and 0.3 percent chromium. Aluminum Company of America, *Alcoa Aluminum Handbook* (Pittsburgh, 1956), p. 34.

35. K. F. Thornton, "Standard Dimpling Methods Adapted for 75S-T Aluminum Sheet," *American Machinist*, August 2, 1945, pp. 106–8; "Hot Dimpling Widens Metal Use," *Aviation Week*, April 2, 1951, pp. 21–22; T. A. Dickinson, "How to Hot-Form a Dimple," *Steel Processing* 38 (April 1952): 172–74; Close (n. 21 above), pp. 191–93.

36. Thornton, (n. 35 above), pp. 106–7; "Spin-Dimpling," *Aircraft Production* 13 (January 1951): 3–7; interview (August 12, 1982) with David Richardson, for over twenty years an engineering specialist on fasteners with the Lockheed-California Company, Burbank, Calif.

37. J. E. Cooper, "Production Riveting by Machine," *Western Machinery and Steel World* 36 (November 1945): 494–97; E. O. Baumgarten, "Riveting as Applied to Aircraft Production," *Automotive Industries*, March 15, 1950, pp. 176, 265–66, 268, 270, 272; "Automatic Riveter Speeds Plane Assembly," *Aviation Week*, December 24, 1951, pp. 34–35; "Subassembly Riveting," *Aircraft Production* 13 (January 1951): 12–17. For remarks and references on the tendency toward mechanization generally, see R. R. Nelson and S. G. Winter, *An Evolutionary Theory of Economic Change* (Cambridge, Mass., 1982), p. 260.

38. "Spin-Dimpling" (n. 36 above), p. 3; S. F. Hoerner, *Fluid-Dynamic Drag* (Brick Town, N.J., 1965), pp. 5–8; "Coin-Dimpling. Part I. Basic Principles and Governing Factors," *Aircraft Production* 12 (June 1950): 181–83. For remarks on the evolutionary nature of innovative activity and references to other cases, see D. Sahal, *Patterns of Technological Innovation* (Reading, Mass., 1981), pp. 36–38.

39. Interview (September 3, 1981) with William Barker, engineer with North American in the 1940s and 1950s and for some years leader of the SAE panel; Buckwalter interview (n. 22 above); R. L. Hand, "Notes on the Use of Rivets and Other Fasteners," *SAE Journal* (June 1953): 41–43; E. Baumgarten and C. A. Sieber, "Fasteners—Conventional and Special," *Report SP-317, SAE Production Forum*, Society of Automotive Engineers (New York, 1956), pp. 62–

66. The importance of the fastener community is indicated by the fact that it has its own technical magazine, *Assembly Engineering*, formerly *Assembly and Fastener Engineering*, begun in 1958. For a review essay on the rather unsatisfactory state of research concerning the relationship between generation and diffusion in technological innovation, see C. Ganz, "Linkages between Knowledge, Diffusion, and Utilization," *Knowledge—Creation, Diffusion, Utilization* 1 (June 1980): 591–612.

40. Lear and Dillon (n. 18 above); Nisita (n. 20 above); A. M. Robson, *Airplane Metal Work*, vol. 4, *Airplane Pneumatic Riveting* (New York, 1942); Anonymous, *Riveting and Drilling*, Consolidated Vultee Aircraft Corporation (San Diego, 1943); Anonymous, *Aircraft Fabrication. Aircraft Riveting*, parts 1 and 2 (Scranton, Pa., 1943).

41. G. Rechton, *Aircraft Riveting Manual* and *Addendum I. Riveting Methods*, Douglas Aircraft Company, (Santa Monica, Calif., 1941, 1942). For related nonverbal elements in technology, see E. S. Ferguson, "The Mind's Eye: Nonverbal Thought in Technology," *Science*, August 26, 1977, pp. 827–36, and B. Hindle, *Emulation and Invention* (New York, 1981), pp. 133–38.

42. Buckwalter interview (n. 22 above).

43. The production and design engineers interviewed for this chapter show a distinctly different temperament and mind-set from academic and research engineers who function closer to the interface between engineering and science, as explored in the preceding chapters. For the difficulties in building an airplane "from the drawings," see also I. B. Holley, Jr., "A Detroit Dream of Mass-produced Fighter Aircraft: The XP-75 Fiasco," *Technology and Culture* 28 (July 1987): 578–93, esp. 583.

44. The assumption is not in fact true. Attempts have been made to allow for unequal loading (see, e.g., Jenkins [n. 16 above], pp. 421–29), but they are complicated and do not appear to have caught on in routine design.

45. S. R. Carpenter, "Study to Determine the Suitability of Countersunk Rivets in Duralumin Airplane Construction," *Engineering Section Memorandum Report No. AD-51-114*, Air Corps Matériel Division (Dayton, Ohio, April 18, 1933); D. M. Warner, "Investigation of Efficiency of Countersunk Rivets with 110 Degree Heads," *Engineering Section Memorandum Report No. M-56-2601*, Air Corps Matériel Division (Dayton, Ohio, January 4, 1934). A typical company report is H. B. Crockett, "LS 1105 Flush Rivet Tests," *Report No. 2377*, Lockheed Aircraft Corp. (Burbank, Calif., August 4, 1941). The NBS report is W. C. Brueggeman and F. C. Roop, "Mechanical Properties of Flush-Riveted Joints," *Report No. 701*, NACA (Washington, D.C., 1940).

46. R. S. Hatcher, "Strength of Flush Riveted Joints," *Structures Bulletin No. 68*, Bureau of Aeronautics, Navy Department (Washington, D.C., January 30, 1942); Army-Navy-Civil Committee on Aircraft Design Criteria, *ANC-5: Strength of Aircraft Elements*, rev. ed. (Washington, D.C., December 1942), pp. 5-27–5-29.

47. ARC Rivet and Screw Allowables Subcommittee (Airworthiness Project 12), *Report on Flush Riveted Joint Strength*, Aircraft Requirements Committee, Aircraft Industries Association of America (Washington, D.C., original dated April 12, 1946, revisions June 24, 1946, September 20, 1946, December 3,

1946, and February 26, 1947, Appendix F added May 25, 1948) (the Aircraft Industries Association took the activity over from the Aircraft War Production Council at the end of World War II); Subcommittee on Air Force–Navy–Civil Aircraft Design Criteria, *ANC-5a: Strength of Metal Aircraft Elements*, rev. ed. (Washington, D.C., May 1949), pp. 76–78; interview (September 3, 1981) with Milton Miner. The number of specimens could have been greatly reduced and the analysis facilitated if the committee had examined their results in terms of a system of generalized parameters proposed in the mid-1940s at the Martin Company; G. E. Holbeck, "Structural Fastenings in Aircraft," *Detail Aircraft Structural Analysis. Supplemental Pamphlet No. 8.*, Glenn L. Martin Company (Baltimore, n.d., but probably 1945–46 on basis of listed references). This system was not generally adopted, however, until sometime in the 1950s; Richardson interview (n. 36 above).

48. ARC Subcommittee, p. 01.18.

49. D. G. Lingle and G. A. Seitz, "Army-Navy Aeronautical Standardization: The Record So Far," *SAE Journal* 50 (November 1942): 32–33, 66–80; H. Mingos, ed., *The Aircraft Year Book for 1942* (New York, 1942), pp. 253–56; Templin and Fogwell (n. 28 above), p. 2; National Aircraft Standards Committee, "Rivet—Countersunk Head 100°," *NAS 1*, Aircraft Industries Association of America (Washington, D.C., 1945) (the Aircraft Industries Association succeeded the Aeronautical Chamber of Commerce under which the Standards Committee was set up); personal letter to author from Capt. John Ross, Directorate of Equipment Engineering, U.S. Air Force, October 19, 1981.

50. This reasoning is my own, though I have had valuable assistance from Howard Wolko and David Richardson.

51. Speakman, Richardson, and Barker interviews. For the fatigue problem at different stages, see, e.g., L. R. Jackson, H. J. Grover, and R. C. McMaster, "Advisory Report on Fatigue Properties of Aircraft Materials and Structures," *OSRD No. 6600, Serial No. M-653*, Office of Scientific Research and Development (Washington, D.C., March 1, 1946); J. Schijve, "The Fatigue Strength of Riveted Joints and Lugs," *Technical Memorandum No. 1395*, NACA (Washington, D.C., 1956); R. B. Heywood, *Designing against Fatigue of Metals* (New York, 1962), pp. 230–42.

52. Buckwalter interview (n. 22 above); quotation from Baumgarten and Sieber (n. 39 above), p. 62.

53. To borrow expressions current in social historiography, the studies close to science constitute examination of *high* technological culture; the present study examines *popular* technological culture.

54. The divorce from science, though a historical fact here, is not essential in principle for widespread innovative activity. If a high degree of scientific aptitude and training were present throughout an industry, widespread science-based innovation could occur when needed provided the essential ideas about the way to proceed were well known or obvious. Perhaps this may already have happened in the microelectronics industry.

55. Mario Bunge ("Technology as Applied Science," *Technology and Culture* 7 [Summer 1966]: 329–47, esp. 338–41) contrasts the narrower categories of "scientific law" and "technological rule" in somewhat similar fashion. This,

however, oversimplifies the distinction into one between science and technology and overlooks the possibility, evident below, of having to deal with statements of fact that would be hard to describe as "scientific."

56. These statements require minor qualification. In the case of failing loads (though not of yielding loads), the published allowable strengths were obtained by reducing the original experimental values through division by a *judgmental* factor of 1.15 to allow for dispersion caused by experimental inaccuracies and other imponderables. This had the effect of introducing a secondary prescriptive element into the published figures. In this case, the original experimental values constitute the descriptive knowledge.

57. Not all prescriptive knowledge will differ from company to company, however (see under "quantitative data" in chapter 7). Engineering standards agreed upon across an industry (e.g., the proportions of flush-rivet heads in the present study) are a case in point.

58. The term *tacit knowledge* is taken from the writing of Michael Polanyi, *Personal Knowledge* (Chicago, 1962), pp. 69–245. Polanyi, however, uses "tacit" more broadly to describe the unspoken element indispensable to all knowing. For discussion of the relation of skills and tacit knowledge, see Nelson and Winter (n. 37 above), pp. 76–82.

59. The ideas and terminology of the last four paragraphs have benefited crucially from discussion with Todd Becker.

60. Several historians have called attention (though in different words) to the tacit element in mechanical design; see, e.g., A. F. C. Wallace, *Rockdale* (New York, 1978), pp. 237–39, and B. Hindle (n. 41 above).

61. E. Layton, "Technology as Knowledge," *Technology and Culture* 15 (January 1974): 31–41, quotation from 37–38.

Chapter 7. The Anatomy of Engineering Design Knowledge

Note: In this and the next chapter, I am especially indebted to the comments and criticisms of Lynwood Bryant, Edward Constant, Bruce Hevly, Barry Katz, Rachel Laudan, Edwin Layton, and William Rifkin. This chapter has also profited from suggestions by Mark Cutkosky, Henry Fuchs, and Ilan Kroo.

1. R. Laudan, "Cognitive Change in Technology and Science," in *The Nature of Technological Knowledge. Are Models of Scientific Change Relevant?*, ed. R. Laudan (Dordrecht, 1984), pp. 83–104, quotation from p. 84. Laudan reserves "system" to denote the collective technology of historical epochs, such as those sometimes identified by "wood, wind, and water" and "iron, coal, and steam"; she gives this class (third in her categorization) only passing discussion. For illuminating general remarks about engineering as problem solving, see also S. G. Winter, "An Essay on the Theory of Production," in *Economics and the World around It*, ed. S. H. Hynan (Ann Arbor, Mich., 1982), pp. 59–93, esp. pp. 68–71.

2. Laudan (n. 1 above), pp. 84–89. As she points out (with pertinent references), variants of the imbalances included under (3) have been described

under the terms reverse salient, backward linkage, compulsive sequence, and technological co-evolution.

3. I retain here the view, explained in chapter 3 and prevalent at the time of the story, of the airplane as an open-loop system with the pilot as an external (i.e., contextual) agent. In the modern closed-loop view, the pilot and airplane together constitute a single system, and flying qualities constitute a design problem purely within the system.

4. In Staudenmaier's concept of design-ambient tension mentioned in chapter 1 (as I understand the concept), contextual factors provide the primary source of engineering problems and Laudan's "sources" are to be considered part of a broadly defined "design." The issues she discusses are then aspects of the tension between design and its cultural ambience. Either view can be useful, depending on the primary focus of analysis (engineering problem solving in Laudan's work and here, and the cultural function of technology in Staudenmaier's treatise).

5. E. Constant, *The Origins of the Turbojet Revolution* (Baltimore, 1980), p. 15.

6. T. P. Hughes, *Networks of Power* (Baltimore, 1983), chap. 2.

7. C. Ellam, "Developments in Aircraft Landing Gear, 1900–1939," *Transactions of the Newcomen Society* 55 (1983–84): 48–51. The point here has relevance for the study of what recently has come to be called the "social construction of technology." (See, e.g., W. E. Bijker, T. P. Hughes, and T. J. Pinch, *The Social Construction of Technological Systems—New Directions in the Sociology and History of Technology* [Cambridge, Mass., 1987], particularly the articles by Pinch and Bijker.) Though such study clearly has merit, it is hard to see much beyond internal logic in the problem posed to the landing-gear design team—or in any such similar problem. By the same token, little scope remains for social construction to influence the problem's (sometimes difficult) solution. Both specifying the problem and arriving at its solution may, of course, require interaction among numerous people. This social process, however, has little effect on the problem or solution. The negotiations and compromises that take place are predominantly technical ones, and the outcomes are shaped overwhelmingly by technical considerations.

8. The foregoing discussion of the origins and selection of engineering problems has said little about how such problems become formulated. Such formulation may be immediate and explicit, as in the case of airfoil selection at Consolidated or of flush-riveted joints in the aircraft industry generally; or it may start out vaguely and spread over decades, as happened with flying qualities. This important aspect of the problem-posing process undoubtedly deserves more analytical attention than I am giving it.

9. M. Polanyi, *Personal Knowledge* (Chicago, 1962), pp. 174–84, 328–32, quotation from p. 328.

10. C. H. Gibbs-Smith, *Sir George Cayley's Aeronautics, 1796–1855* (London, 1962), p. 48.

11. Polanyi (n. 9 above), pp. 328–32, quotation from p. 330.

12. Constant (n. 5 above), pp. 10–12, quotation from p. 10.

13. Technical specifications as used here (i.e., of design goals) are not to be

confused with the specifications of materials and processes that engineers put on detailed drawings for the direction of shop workers.

14. This difference was pointed out to me by Rachel Laudan.

15. For engineering sciences, see also G. F. C. Rogers, *The Nature of Engineering: A Philosophy of Technology* (London, 1983), pp. 52–54.

16. W. G. Vincenti and C. H. Kruger, Jr., *Introduction to Physical Gas Dynamics* (New York, 1965).

17. Polanyi (n. 9 above), p. 179; Polanyi's term *systematic technology* is perhaps unfortunate, since the theories in our next group are also systematic. Henry Petroski also observes that "engineers find it necessary to study beams and other elements of construction as if they were the natural stuff of science"; *To Engineer Is Human: The Role of Failure in Successful Design* (New York, 1985), p. 49.

18. For a more complicated case than those cited here, see R. Kline, "Science and Engineering Theory in the Invention and Development of the Induction Motor, 1880–1900," *Technology and Culture* 28 (April 1987): 283–313.

19. Empirically obtained data are sometimes represented by "empirical equations," which are mathematical expressions fitted in some way to the data. Such expressions can then be used as analytical tools in the same way as the theoretically obtained expressions in the preceding category. Whether they are then called quantitative data or analytical tools is a matter of choice; I prefer the former.

20. Process specifications are distinct from the technical specifications discussed earlier; see n. 13 above.

21. For an example of prescriptive data from electric power transmission (diameter and spacing of conductors to avoid loss in transmission lines with specified voltages), see T. P. Hughes, "The Science-Technology Interaction: The Case of High-Voltage Power Transmission Systems," *Technology and Culture* 17 (October 1976): 646–62, esp. 653–54.

22. For a widely used example, in compilation of which the author had a hand, see Ames Research Staff, "Equations, Tables, and Charts for Compressible Flow," *Report No. 1135*, NACA (Washington, D.C., 1953). With the availability of personal computers, such data can now be calculated rapidly as required, eliminating much of the need for prior compilation. This tends to blur the distinction made here between theoretical tools and quantitative data.

23. The first example was suggested to me by Richard Shevell, the second by Ilan Kroo. For other examples of rules of thumb and related design knowledge, see B. V. Koen, *Definition of the Engineering Method* (Washington, D.C., 1987), pp. 46–48; this reference was brought to my attention by Jonathan Coopersmith.

24. For representative engineering treatises, see E. V. Krick, *An Introduction to Engineering and Engineering Design*, 2nd ed. (New York, 1969); P. Gasson, *Theory of Design* (New York, 1973); T. F. Roylance, ed., *Engineering Design* (Oxford, 1966). For an illuminating philosophical discussion and analysis of design complementary to present concerns, see D. Pye, *The Nature and Aesthetics of Design* (London, 1978). For research from the standpoint of artificial intelli-

gence, see J. M. Carroll, J. C. Thomas, and A. Malhotra, "Presentation and Representation in Design Problem-Solving," *British Journal of Psychology* 71 (1980): 143–53; and J. M. Carroll, J. C. Thomas, L. A. Miller, and H. P. Friedman, "Aspects of Solution Structure in Design Problem Solving," *American Journal of Psychology* 93 (June 1980): 269–84.

25. H. A. Simon, *The Sciences of the Artificial*, 2nd ed. (Cambridge, Mass., 1981), pp. 138–40.

26. The procedure for laying down useful technical specifications, described earlier, is yet another example. While specifications for individual projects would hardly be counted as knowledge, knowing how to arrive at them certainly should.

27. For a historical example involving synthetic thinking in structural design, see N. Rosenberg and W. G. Vincenti, *The Britannia Bridge: The Generation and Diffusion of Technological Knowledge* (Cambridge, Mass., 1978), pp. 38–39.

28. The weathercock analogy was pointed out to me by Ilan Kroo. For a simple analogy that played a specific historical role, see ibid., pp. 24–25.

29. E. S. Ferguson, "The Mind's Eye: Nonverbal Thought in Technology," *Science*, August 26, 1977, pp. 827–36, quotations from p. 835; see also B. Hindle, *Emulation and Invention* (New York, 1981), pp. 133–38. For an engineering text, see R. H. McKim, *Experiences in Visual Thinking*, 2nd ed. (Monterey, Calif., 1980).

30. In discussing skills needed for design, I have left aside the obvious manual skills required to make the sketches and drawings mentioned above. Computer-aided design, in any event, is making these skills less important than they used to be. The quoted words are from personal correspondence from John Staudenmaier.

31. G. Gutting, "Paradigms, Revolutions, and Technology," in R. Laudan, ed. (n. 1 above), pp. 47–65, quotation from p. 63.

32. The final three categories have much in common with the "four characteristics of technological knowledge" that John Staudenmaier sees reflected in the historiographic literature on technology. Theoretical tools, quantitative data, and design instrumentalities here contain (though obviously not in one-to-one correspondence) most of what he calls scientific concepts, problematic data, engineering theory, and technical skill. Staudenmaier's historiographic concerns, however, differ from my epistemological ones. He looks at the whole of technological knowledge as it has been viewed by historians; I have examined the portion of that knowledge used in normal engineering design from the standpoint of the needs of the designer. The two viewpoints lead to somewhat differing anatomies. J. M. Staudenmaier, S.J., *Technology's Storytellers: Reweaving the Human Fabric* (Cambridge, Mass., 1985), pp. 103–20.

33. F. K. Mason, *Harrier* (Cambridge, 1981); T. J. Pinch and W. E. Bijker, "The Social Construction of Facts and Artifacts: Or How the Sociology of Science and the Sociology of Technology Might Benefit Each Other," in Bijker, Hughes, and Pinch, eds. (n. 7 above), pp. 17–50.

34. E.g., M. Kranzberg, "The Disunity of Science-Technology," *American Scientist* 56 (Spring 1968): 21–34, esp. 28–29, and D. de S. Price, *Science since Babylon*, enlarged ed. (New Haven, 1975), pp. 129–30.

35. For a review and discussion of attempts to deal with the science-technology relationship, see O. Mayr, "The Science-Technology Relationship as a Historiographic Problem," *Technology and Culture* 17 (October 1976): 663–73. Mayr concludes (p. 668) that the number of variables in the relationship is so large that "a dynamic model that would do justice to all would be prohibitively complex." Such conclusion may be true, but to give up modeling entirely would make epistemological discussion of the kind attempted here even more difficult. Barry Barnes and David Edge, in reviewing the problem, note the general validity of Mayr's argument but also observe that "a model may, nonetheless, retain some utility for specific purposes"; "Introduction to Part Three, The Interaction of Science and Technology," in *Science in Context: Readings in the Sociology of Science*, ed. B. Barnes and D. Edge (Cambridge, Mass., 1982), pp. 147–54, esp. pp. 147–48. For an assortment of metaphorical models for the science-technology relationship, see also A. Keller, "Has Science Created Technology?" *Minerva* 22 (Summer 1984): 160–82, esp. 175–77.

36. H. Aitken, *Syntony and Spark—The Origins of Radio* (New York, 1976), p. 314.

37. Mayr (n. 35 above, p. 667) denies that such operational distinction is possible. He speaks of engineers and scientists as both working "in [research?] laboratories of like appearance," however, and takes no account of the vast number of engineers occupied in design offices, which have no equivalent in science.

38. L. S. Reich, "Irving Langmuir and the Pursuit of Science and Technology in the Corporate Environment," *Technology and Culture* 24 (April 1983): 199–221, quotation from 201. For a still more complex example of the interpenetration of engineering and science, see J. L. Bromberg, "Engineering Knowledge in the Laser Field," *Technology and Culture* 27 (October 1986): 798–818.

39. The situation is analogous to that with the spectrum of visible light—the fact that one cannot say precisely where yellow becomes green does not mean there is no real and definable difference between red and violet.

40. "The Characteristics of Tungsten Filaments as Functions of Temperature," *Physical Review* 7 (1916): 302–30; "Tungsten Lamps of High Efficiency," *Transactions of the AIEE*, October 10, 1913, pp. 1913–54.

41. One can imagine elaborating the diagram to represent more of the features mentioned. The use of knowledge to generate more knowledge could be indicated by a feedback loop from each use box to the knowledge-generating activity vertically above (a lighter, less prominent loop for engineering than for science). Consistent with this, a dashed box would then be added below and connected to the right-hand use box to represent engineering's primary, end activity of design (as distinguished from its secondary, though essential, knowledge-generating activity). The asymmetry in the uses of knowledge in engineering and science would then be graphically apparent. Since the diagram as it stands is adequate for the present discussion, I have not included these features.

42. E. F. Kranakis, "The French Connection: Giffard's Injector and the Nature of Heat," *Technology and Culture* 23 (January 1982): 3–38; D. F. Chan-

nell, "The Harmony of Theory and Practice: The Engineering Science of W. J. M. Rankine," *Technology and Culture* 23 (January 1982): 39–52; R. Kline (n. 18 above).

43. See chap. 5, n. 49. Buckingham himself encouraged the transfer to engineering by contributing a section on "The Dimensional Theory of Wind Tunnel Experiments" to Hunsaker's 1916 Smithsonian publication.

44. For discussion of a representative sample, see Staudenmaier (n. 32 above), pp. 40–45.

45. For instructive general and specific discussion of how engineers learn from failures in structural design, see H. Petroski (n. 17 above), passim, and Rogers (n. 15 above), p. 55. The case of the Comet is covered on pp. 176–84 of the former. For key portions of the commission report on the Challenger explosion, see the *New York Times*, June 10, 1986, pp. 20–21.

46. Rosenberg and Vincenti (n. 27 above), pp. 9–43.

47. The reader may find it illuminating to think about the table (and the material it summarizes) in relation to the insightful "chain-linked model" of industrial innovation proposed by Stephen J. Kline; "Innovation Is not a Linear Process," *Research Management* 28 (July–August 1985): 36–45. The work here can be seen, in effect, as an attempt to detail a large part of the content of the layer labeled "knowledge" in Kline's diagram.

48. Simon (n. 25 above), pp. 132–33. See also Staudenmaier (n. 32 above), pp. 169–70.

49. For a collection of representative recent efforts, see Bijker, Hughes, and Pinch, eds. (n. 7 above).

50. E. W. Constant, "Communities and Hierarchies: Structure in the Practice of Science and Technology," in Laudan, ed., *Nature of Technological Knowledge* (n. 1 above), pp. 27–46, quotation from p. 29; see also his *Turbojet Revolution* (n. 5 above), pp. 8–10.

51. The quotation is from valuable comments by William Rifkin.

52. The situation may be different in the broader spheres of production and distribution of artifacts and services. Edward Constant argues that in these spheres of technological "function," one kind of institution—the industrial firm—plays the central social role. (See, in particular, his article on "The Social Locus of Technological Practice: Community, System, or Organization?" in Bijker, Hughes, and Pinch, eds. [n. 7 above], pp. 223–42.) Constant also points out that groups of industrial firms dealing with a given product or service (e.g., aircraft, air transportation, automobiles, electric power) can form communities in the same sense as individuals. Such behavior was visible here, for example, in the cooperation of nine aircraft companies to measure the allowable strengths of flush-riveted joints and in the joining of five airlines to order the DC-4E. Including this consideration would complicate the points I have made, but I doubt it would affect their essential validity.

Chapter 8. A Variation-Selection Model

1. For two fundamental essays from Campbell's extensive writings on the topic, see his "Blind Variation and Selective Retention in Creative Thought as

in Other Knowledge Processes," *Psychological Review* 67 (November 1960): 380–400, and "Evolutionary Epistemology," in *The Philosophy of Karl Popper*, ed. P. A. Schilpp, vols. 14I and 14II of *The Library of Living Philosophers* (La Salle, Ill., 1974), vol. 14I, pp. 413–62. Campbell's work is part of a large tradition of trial-and-error and natural-selection models for knowledge growth; a pertinent bibliography is appended to the latter essay. Both essays have been reprinted in G. Radnitzky and W. W. Bartley, III, eds., *Evolutionary Epistemology, Rationality, and the Sociology of Knowledge* (La Salle, Ill., 1987), pp. 47–114.

2. Campbell, "Blind Variation" (n. 1 above), p. 381.

3. T. J. Gamble, "The Natural Selection Model of Knowledge Generation: Campbell's Dictum and Its Critics," *Cognition and Brain Theory* 6 (Summer 1983): 353–63, quoted phrase from 353.

4. Campbell, "Blind Variation" (n. 1 above), pp. 380–81, quotation from p. 381, and "Evolutionary Epistemology," pp. 421–22; Gamble (n. 3 above), pp. 357–58.

5. Popper's discussion of "blindness," from which his quotation is taken, is well worth reading; see his "Replies to My Critics," in Schilpp, ed., *Philosophy of Karl Popper* (n. 1 above), vol. 14II, pp. 1061–62, also reprinted in Radnitzky and Bartley, eds., *Evolutionary Epistemology* (n. 1 above), pp. 117–18.

6. F. E. C. Culick, "Aeronautics, 1898–1909: The French-American Connection" (forthcoming). The complex details of aircraft and events are covered in C. H. Gibbs-Smith, *The Rebirth of European Aviation, 1902–1908* (London, 1974).

7. Campbell, "Evolutionary Epistemology" (n. 1 above), p. 421.

8. For further brief description of the design process, see G. F. C. Rogers, *The Nature of Engineering: A Philosophy of Technology* (London, 1983), p. 65. For drawings showing successive versions of a design, in particular for the Boeing B-52 bomber, see J. E. Steiner, "Jet Aviation Development: A Company Perspective," in *The Jet Age: Forty Years of Jet Aviation*, ed. W. J. Boyne and D. S. Lopez (Washington, D.C., 1979), pp. 140–83, esp. p. 148.

9. Campbell, "Evolutionary Epistemology" (n. 1 above), pp. 419–21. The hierarchy described here is essentially the same as that for knowledge generation from among the various kinds of hierarchy described in the first section of chapter 7.

10. Ibid., pp. 423–25, 435–36.

11. Gamble (n. 3 above), p. 359. The following statement by Campbell is also pertinent: "But while introducing consideration of intentional problem-solving will explain *constraints* in the randomness of search (introduced by the combination of purpose and assumptions about 'already established' 'knowledge' or tentatively trusted beliefs) it will not obviate the wasteful fumbling among alternatives that characterizes *all* discovery processes." "Selection Theory and the Sociology of Scientific Validity," in *Evolutionary Epistemology: A Multiparadigm Program*, ed. W. Callebaut and R. Pinxten (Dordrecht, 1987), pp. 139–58, quotation from p. 147.

12. For classic discussions of the role of the unconscious in creative work, specifically mathematics, see H. Poincaré, *The Foundations of Science* (New York, 1929), pp. 383–94, and J. Hadamard, *An Essay on the Psychology of Invention in the Mathematical Field* (Princeton, 1945). For review of contemporary research on

the impact of nonconscious processes on conscious thought and action, see J. F. Kihlstrom, "The Cognitive Unconscious," *Science*, September 18, 1987, pp. 1445–52.

13. The distinction between periods of enhancement and exhaustion applies only within the context of the normal, conventional design that is our concern here. Growing experience in a technology is likely to inhibit progressively the radical novelty needed for *un*conventional change, hence the eventual need for such input from outside.

14. H. Petroski, *To Engineer Is Human* (New York, 1985), p. 44. The paragraph in which the quotation appears gives an excellent description of iterative design that illustrates clearly (though not explicitly) how it is a variation-selection process. Petroski's ideas have a good deal in common with the present work, though couched in very different words.

15. Campbell points to a similar situation in vicarious selection in science. "Evolutionary Epistemology" (n. 1 above), p. 436.

16. For a discussion of the logical and substantive difficulties of organizing a curriculum of design, see H. A. Simon, *The Sciences of the Artificial*, 2d ed. (Cambridge, Mass., 1981), chap. 5.

17. Campbell, "Evolutionary Epistemology" (n. 1 above), pp. 434–37.

18. A. Keller, "Has Science Created Technology?" *Minerva* 22 (Summer 1984): 160–82, quoted phrase from 169.

19. "Cognitive Change in Technology and Science," in *The Nature of Technological Knowledge. Are Models of Scientific Change Relevant?*, ed. R. Laudan (Dordrecht, 1984), pp. 83–104, quotation from p. 88.

20. For discussion of basically the same distinction from a different frame of reference, see E. W. Constant, "Communities and Hierarchies: Structure in the Practice of Science and Technology," in Laudan, ed., *Nature of Technological Knowledge* (n. 19 above), pp. 27–46, esp. pp. 35–38.

21. J. Staudenmaier, *Technology's Storytellers: Reweaving the Human Fabric* (Cambridge, Mass., 1985), pp. 136–39, quotation from p. 139.

22. Some experts in artificial intelligence already regard technological imitation of some aspects of human understanding (reasonably defined) as a valid goal. Such effective definition of understanding as a practical engineering problem might appear to present difficulty in the distinction between the two selection criteria. What might help in *designing* an artifact to imitate understanding, however, might very well be different from what would help a human being to understand *directly*. And even if the same knowledge should serve both purposes, the criteria are not required to be exclusive in any event. There is thus no real difficulty.

23. This and the preceding four paragraphs have benefited from discussion with Robert McGinn.

24. See, e.g., the summary by Robert E. McGinn of a conference presentation by Howard S. Rosen, from whom the question is taken; "Organizational Notes," *Technology and Culture* 16 (July 1975): 431–35.

25. E. Constant, *The Origins of the Turbojet Revolution* (Baltimore, 1980), p. 21.

26. For a very different approach to engineering method by another engineer, see B. V. Koen, *Definition of the Engineering Method* (Washington, D.C., 1987).

Index